Frequency-Domain Characterization of Power Distribution Networks

For a listing of recent titles in the *Artech House Microwave Library*,
turn to the back of this book.

Frequency-Domain Characterization of Power Distribution Networks

Istvan Novak
Jason R. Miller

ARTECH HOUSE
BOSTON | LONDON
artechhouse.com

Library of Congress Cataloging-in-Publication Data
A catalog record for this book is available from the U.S. Library of Congress.

British Library Cataloguing in Publication Data
A catalogue record for this book is available from the British Library.

ISBN-13: 978-1-59693-200-5

Cover design by Yekaterina Ratner

685 Canton Street
Norwood, MA 02062

10 9 8 7 6 5 4 3 2 1

To our families

Contents

Preface

Overview of the Book

The book provides practical, how-to guidance on making frequency-domain measurements and modeling and simulating power distribution networks (PDN) and components. Accurate frequency-domain measurements are fraught with difficulties. This book aims to break the measurement process down step by step, highlighting the many pitfalls that can be encountered along the way. Selection of instruments, probes, cables, and calibration processes is explained, and choices are listed and evaluated. The book focuses equal attention on modeling and simulation. In Chapter 2, which is on simulation, a few selected tools are considered, and through practical examples, their strengths and weaknesses are evaluated and parameters and options settings are discussed. Finally, several chapters are devoted to approaches to modeling every aspect of a PDN, including planes, vias, bypass capacitors, inductors, and dc-dc converters. The book focuses on frequency domain characterization exclusively. The reasons behind the book's focus on frequency domain are numerous and discussed in detail in Chapter 1. Throughout the book you will see an emphasis placed on closing the loop between these three elements of characterization: measurement, modeling, and simulation. Components are characterized using multiple techniques, and the results are compared and the discrepancies discussed. It is through the process of correlation that we show how to develop trust in measurement, modeling, and simulation techniques.

Scope of the Book

This book focuses on the frequency domain characterization of power distribution networks. Design approaches and design methodologies of power distribution networks and semiconductor (silicon) power distribution, as well as time-domain characterization are not covered in this book. Discussing the design methodologies and silicon applications would be relevant and important, but it would go well beyond the size and scope of this book. Though the principles and rules discussed in the book are applicable to a wide range of digital and analog electronic systems, the illustrations are primarily taken from high-power digital boards. Time-domain characterization is intentionally left out from the topic list, as explained in Chapter 1.

It should be noted that although the book heavily relies on examples taken from board design, much of the underlying simulation, modeling, and measurement techniques are applicable to silicon packaging technology. Notwithstanding the difference in scale, packages share many similarities to boards in terms of their building blocks, for example, vias, planes, and decoupling capacitors. Where appropriate, references will be made to, and examples taken from, package characterization to demonstrate the similarities that exist between packages and printed circuit boards.

Ultimately, the three basic PDN requirements (discussed further in Chapter 1) are to: (1) deliver sufficiently clean supply voltage to the ICs; (2) provide low-noise reference path for signals (optional); and (3) *not* radiate excessively. Insuring these requirements are met involves characterizing specific aspects of the design. The scope of the book is thus focused on developing the techniques and methodologies necessary to examine these specific aspects of the PDN. For example, to insure that the first PDN requirement is met (deliver sufficiently clean supply voltage to the ICs), we must characterize the impedance of the PDN, insuring that it is flat and sufficiently low. Thus the scope of this book encompasses techniques to accurately model, simulate, and measure the impedance profile of the PDN.

Target Audience

This book is for practicing engineers. Power distribution network characterization in electronic systems encompasses many disciplines from control-loop theory to material science, from assembly technology to metrology, and so forth. Devoted specialists of any of these disciplines may find this book overly simplistic or lacking the necessary details. On the other hand, practicing engineers, facing the task of PDN design and characterization may find the task too complex and overwhelming. This book is an attempt to bring together the important aspects of frequency-domain PDN characterization for the practicing engineer.

Throughout the book we assume that the reader has generic familiarity with the basic concepts of electronic circuits; network matrices, notably the impedance matrix and the scattering matrix, as well as with fundamentals of high-frequency measurements and simulations.

Organization of the Book

The book starts with the explanation, why frequency-domain characterization is the focus. The pros and cons of frequency and time-domain characterizations are summarized. Exceptions are pointed out, when time-domain characterization is needed. The modeling approach used in the subsequent chapters is outlined.

Chapter 2 is devoted to frequency-domain power-distribution simulations using various approaches, including MATLAB, SPICE, and field-solvers. Simulations are correlated to measurement results to help the reader understand how the simulation parameters and settings impact accuracy.

Chapters 3 and 4 focus on characterizing and modeling vias, via arrays, planes, and laminates.

Chapters 5, 6, and 7 describe in great detail proper measurement practices. From selecting the right probe or cable to calibrating an instrument—these chapters covers all aspects of how to make accurate PDN measurements.

Chapter 8 is devoted entirely to the characterization and modeling of bypass capacitors, starting with the simplest model and working through various complexity levels to frequency dependent causal models.

Chapter 9 focuses on the characterization and modeling of inductors, dc-dc converters, and systems.

Disclaimers

Throughout this book, for purposes of guiding the reader with as much practical information as possible, manufacturers, vendors, and models are often mentioned. This, however, does not imply that the cited instruments, software tools, or components are the only ones available for the particular purpose or that they would be the best for that purpose. It merely reflects the experience of the authors. To help the reader to make the proper choice, wherever applicable, the considerations, pros and cons, and potential pitfalls to avoid are elaborated upon.

Like every practical guide, this book has a lot of dated details: software tools are constantly improved upon, new instruments become available each year, and components evolve at a rapid pace. For this reason, some of the details, in particular the illustrations with specific instruments, tools, and components, may become obsolete in a few years. What we hope will stay and serve as a helpful guide in a more timeless manner is the thought process and the methodology that goes into the characterization work.

Although every effort has been made to ensure that this text is free from errors, unfortunately, some errors will not have been caught during the review process. If you do find an error in the text, please let us know. Furthermore, we invite your questions and feedback; all will be carefully considered if a new edition is warranted. The authors can be reached by e-mail at istvan.novak@ieee.org.

Tools Used and Trademarks

The book relies on many software packages to analyze and simulate the PDN. For electromagnetic analysis, the bulk of the simulations were performed using Ansoft HFSS. Additional simulations were performed using CST Microwave Studio and Ansoft SIwave. Synopsis Hspice was used for circuit simulations. Microsoft Excel was used for all of the spreadsheet macro simulations. We used MathWorks MATLAB to perform several numeric computations described in the book.

Interra and Pyralux, ForadFlex, ZBC2000 and ZBC1000, C-Ply, PicoProbe, Q3D Extractor, HSPICE, CST Microwave Studio, and MATLAB are trademarks of DuPont, Oak-Mitsui, Sanmina-SCT, 3M, GGB Industries, Ansoft, Synopsis, CST of America, and Mathworks, respectively.

Acknowledgments

My curiosity and attention to detail date back to my early childhood, when my late father, who was a physician and pharmacist, introduced my elder brother and me to simple electrical circuits. His wide knowledge of chemistry and physics, together with my late mother's encouragement to always do things right, and my elder brother's advanced knowledge in the field of radio-frequency circuits, created a lifelong passion for electronics. During my first job in the Space Research Group of the Technical University of Budapest, Budapest, Hungary, which lasted almost 30 years with a few intermissions, I had the honor to work with many talented colleagues and coworkers who shaped my knowledge and interest. Richard Redl, the late Bela Molnar, Gabor Rajkai, and Andras Gschwindth are among the many whom I have to thank the foundation of my knowledge in power distribution. In later years this basic knowledge got a significant boost when I was visiting Design Autiomation, Inc., and had the privilege to work with and learn from Nathan Sokal and Roy Sallen.

The bulk of this book is based on work done during the past 10 years at Sun Microsystems. The interesting and challenging discussions with the various engineering teams, most notably Larry Smith, Raymond Anderson, and Barry K. Williams, shaped my knowledge more towards power distributions in server applications. To do our daily job, we have to interface with component, instrument, and tool vendors. I want to acknowledge the support and help from a large number of companies and individuals who support our efforts in creating better power distribution for our products. The complete list would fill pages, but I want to mention the key contributors providing measurement and simulation support, as well as samples: Vahe Adamian and Brad Cole from Ultimetrix, Richard Hoft of Agilent Technologies, Hemant Shah and Roger Cleghorn from Cadence, Carl Eggerding and John Galvagni of AVX Corporation, Hideki Isida from Sanyo, John Prymak and Mike Randall from Kemet, Steve Ruscak from Linear Technologies, Masayuki Shimizu from Taiyo-Yuden, Akira Uchida from Maruwa, Takashi Chiba and Chris Burket of TDK, Mark Pieper and John Salzillo from Texas Instruments, Bruce Lee and Mihee Lee from Samsung, Tom Lantzer and David McGregor from DuPont, Nick Biunno, Frank Alberto, Gary Young, and George Dudnikov from Sanmina-SCI, Joel Pfeifer of 3M, Bob Greenlee from Merix, and Robert Galli of Panasonic.

Our coworkers at Sun have been instrumental in widening our knowledge: Valerie St. Cyr, Frank Alberto (then at Sun), Karl Sauter, and Mike Freda provided invaluable insight into the intricacies of printed circuit board fabrication. Gustavo Blando contributed his talent in MATLAB scripting and probe-station measurements. Special thanks goes to David Greenhill for his reviews and comments and to the management team of Sun Microsystems for supporting this project.

I am deeply indebted to Jason R. Miller; his excellence in simulations and his tireless efforts made the simulation part of this book possible.

Last but not least, I am very grateful for the support and encouragement of my loving wife, Agnes.

Istvan Novak
Maynard, Massachusetts
July 2007

First and foremost, I wish to express my gratitude to Istvan Novak for permitting me to collaborate with him on this book. It has been a great pleasure for me to work with him on this project and many other projects over the past years.

I have also had the privilege to work with a number of highly talented people both in industry and academia, whom I would like to recognize for their indirect or direct contributions to this book: Gustavo Blando, Aykut Dengi, and Professor Q.Y. Ma.

We would like to acknowledge the people at Ansoft and CST who generously provided support and resources for some of the simulation runs included in Chapter 2, in particular, Jim DeLap, Chris Herrick, and Mike Cole at Ansoft and Antonio Ciccomancini and Roy Arsenault at CST.

We extend our thanks to Dan Swanson for his valuable comments on simulation tools and to Bruce Archamabeault, who provided us with a copy of EZ-PowerPlane for evaluation. We are grateful for the helpful discussions with Yuriy Shlepnev on broadband analysis of plane pairs.

This book would not be possible without the people at Artech House; Mark Walsh, Barbara Lovenvirth, Judi Stone, and Rebecca Allendorf were a pleasure to work with throughout the project. I would also like to recognize Sun Microsystems for supporting my involvement in this project. Many thanks to Dave Greenhill, who spent countless hours reviewing the text and providing excellent technical critique.

Personally, I wish to thank my friends and family for all of their support during this time-consuming endeavor. In particular, I am very grateful to Joanna for her understanding and constant encouragement.

My curiosity about electronics began when my father bought me my first personal computer and only smiled after seeing it disassembled some weeks later. I will forever be grateful for his support and encouragement in these endeavors and in all my pursuits.

Jason R. Miller
Somerville, Massachusetts
July 2007

CHAPTER 1

Introduction

The design and validation of the power distribution network (PDN) in today's electronic circuits, notably in high-speed digital equipment, is becoming an increasing challenge. The challenge can be best shown by looking at the evolution and functions of PDN.

1.1 Evolution of Power Distribution Networks

During the second half of the twentieth century, the PDN has grown from an almost nonexistent afterthought to an important subsystem. To illustrate this on computers, Figure 1.1 shows two photos side-by-side: an old and a more recent computer board. Figure 1.1(a) shows a board from the 1960s implementing a simple gate function. The discrete components in the diode-transistor-logic (DTL) circuitry were laid out on an 11 × 18 cm (4.3 × 7 inch), two-sided board. The bill of material is simple: 40 glass-encapsulated germanium diodes, two rows of 10 germanium transistors each, in metal housing, and 50 carbon resistors. The power connections are at the lower right edge of the board, but apart from the traces carrying ground and power around the devices, there is no bypass capacitor at all, and there are no planes in the stackup to serve as a signal reference. The operating frequency was probably a few hundred kilohertz. Power distribution, signal integrity (SI) of logic signals, and electromagnetic radiation was not a concern.

Figure 1.1(b) shows a central processing unit (CPU) module from a recent V890 server of Sun Microsystems [1]. The 20 × 50 cm (8 × 20 inch) module has 2 quad-core UltraSparcIV CPUs and 16 memory slots. In the 22-layer printed-circuit board, there are four dedicated thin-laminate plane pairs to distribute power for the eight major supply rails. There are seven dc-dc converters and 1,907 bypass capacitors on the board. The parallel buses run with subnanosecond, sometimes 100–200-ps edges. The CPU cores operate at a clock frequency above 1 GHz. The PDN had to be designed to ensure not only clean power to the electronic circuit, but also to serve as a good reference for high-speed signals and to ensure that legal limits of electromagnetic compatibility (EMC) radiation are not violated. We will look at these functions of PDN in Section 1.2.

(a)

(b)

Figure 1.1 Illustration of digital boards: (a) old and (b) new.

1.2 The Importance of Frequency Domain

In general, there are three basic requirements for PDNs that designers need to consider and meet, either by simulations or by measurements [2]. Namely, the PDN should:

1. Deliver sufficiently clean supply voltage to the ICs;
2. Provide a low-noise reference path for signals (optional);
3. *Not* radiate excessively.

The PDN of electronic circuits has to feed the chips with clean power. The switching circuitry in digital logic and the input-level and activity-dependent power draw in analog circuits create current transients, which across the PDN impedance generate voltage fluctuations. The voltage fluctuations (i.e., transient noise) must be kept below a predefined limit so that it does not interfere with the analog or digital signaling. For each signal line, there is a time window within which any supply-rail noise will reduce the noise margin. Outside the sampling window, the noise will not harm the signal with the exception of extra large noise levels. These levels may drive the attached devices into nonlinear regions or even cause breakdown in low-voltage devices. However, noise propagation from multiple noise sources may be too complex to analyze in detail. Therefore a conservative approach assumes that the specified supply-rail noise limit should be kept at all times.

The reference path for signals is not an inherent function of the PDN, but in many designs the PDN acts also as a reference for one or more signals. The PDN may be of significant size in terms of wavelength of the highest frequency of interest. Therefore, a full-wave solution may be necessary to obtain the noise. Here, the user may want to simulate the impact of reference-layer transitions, reference-plane changes over split planes, simultaneous switching noise (SSN) due to shared vias (due to finite plane resistance and inductance, including plane perforations).

As for radiation, again the user may be interested in the impedance profile of the PDN to avoid resonances that may get excited by the signals or noise sources. There are some differences between PDN functions (1) and (3): common-mode currents that may not create SI problems may create excessive radiation. Also, point-of-load PDN structures tend to have a progressively band-limited filtering as we move away from the active device through the package and onto the board. Therefore, high-frequency noise appearing on the board may not find its way back to the silicon, but it can create too much radiation from the board.

Since digital signaling is defined by its voltage or current levels in the time domain, it may seem obvious to characterize the behavior of the PDN also in the time domain. While the characterization itself can be done in the time domain, validation of a real system in the time domain by comparing measured and simulated behavior is much harder.

One practical difficulty with time-domain PDN measurements comes from data acquisition. Usually we do not have a hardware trigger signal to collect samples only in well-defined time windows. Oscilloscopes triggered from the measured noise signal are prone to pick up noise erroneously from the environment, especially when we need to measure noise levels in the millivolt range. To illustrate this potential problem, Figure 1.2 shows two time-domain waveforms captured with a 2 GS/s digital real-time oscilloscope. Figure 1.2(a) was taken with a 182-cm- (72-inch-) long RG174 coaxial cable with SMA (subminiature version A) connectors at both

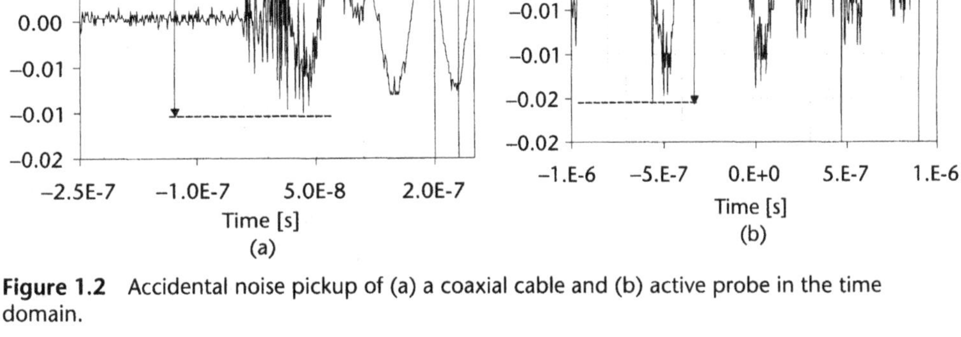

Figure 1.2 Accidental noise pickup of (a) a coaxial cable and (b) active probe in the time domain.

ends. One end was terminated with a 50Ω SMA termination; the other end was connected to the oscilloscope input with an SMA-BNC adaptor. The noise source was a fluorescent light 1m (40 inches) away from the cable. The noise bursts occurred when the light was switched on or off. The noise pickup happened through the finite surface-transfer impedance of the coaxial cable. Figure 1.2(b) was measured with the same oscilloscope with a 1.7-GHz differential active probe. The pins from the probe ends were removed and replaced with a wire short. The probe head was placed 2 cm (0.8 inch) close to a five-phase 100-A dc-dc converter, but otherwise it was not making any connection. The noise pickup happened through the front-end electronics in the probe head. In both cases, the oscilloscope reading without the external noise source was below 1 mVpp.

Having multiple active devices connected to the same PDN rail, the noise voltage at each location becomes the sum of the products of the appropriate self- and transfer impedances and noise currents. As long as the PDN components are linear and time-invariant, we can use Ohm's law and superposition. We know from basic circuit theory that, under these assumptions, any of the network matrices could be used to fully characterize the PDN. In traditional radio-frequency characterization the impedance matrix is not preferred because by definition it requires open termination to obtain the matrix elements. Fringing fields and the finite impedances of probes and instruments at the ports create errors at high frequencies. The solution is the scattering parameter matrix (S-matrix), which uses matched terminated ports to obtain the matrix elements. This suggests that we may want to use the S-matrix instead of the Z-matrix for wideband PDN characterization. This is certainly true for measurements: at high frequencies we can measure S-parameters more accurately than impedances on a multiport PDN.

1.3 The Impedance Matrix

Still, to characterize the PDN over a wide band of frequencies, we may want to use the impedance matrix. The first reason for that is that a well-designed PDN should approximate a voltage source: the noise currents should create only small voltage fluctuations on top of the dc voltage. The electronics injecting the noise at the ports can then be approximated as current sources or open terminations. The second reason is practical: to obtain the S-matrix, we have to obtain the appropriate S-parameters by matched termination on all ports. This is not only inconvenient in case the PDN has a large number of ports, but also creates the problem that the characterization has to be repeated each time when we change the port assignment (i.e., if we add ports or change their locations). This is not the case when we use the Z-matrix; since the Z-matrix requires open terminations (or "nothing") at the ports, reassigning or adding ports does not require a recalculation other than obtaining the matrix elements for the newly added ports.

If we had the S-parameters for the PDN, using the generalized definition of voltage reflection coefficient, we could translate the results into Z-parameters [3]. As long as we find a way to obtain the S-parameters with infinite precision, this provides the correct result for any possible values of the impedance matrix elements. However, as we will see later, the inevitable finite accuracy of measured values will severely limit the applicability of the conversion. Also, if we have more than two ports, sometimes many ports, in the PDN to characterize, the correct procedure would be to obtain the scattering matrix elements by connecting a matched (usually 50Ω) termination to all of the ports. If we have the proper connection points to terminate all ports, this is a tedious but practicable requirement. Still, usually we do not want to terminate the unconnected ports based on the assumption that the overall PDN impedance is significantly less than 50Ω and therefore the terminations would change the result only in a negligible way. This would not be true, however, for low-power and/or high-impedance PDNs, or any PDN with high-impedance peaks. The generic solution will be shown in Section 5.4.

Figure 1.3 shows a simplified sketch of a system, where the PDN consists of a printed board circuit (PCB), one power-ground plane pair, three bypass capacitors, and two active devices. For sake of simplicity, let us assume that we are interested in the noise voltages at the active devices only, and the size of the devices and the highest frequency of interest allow us to use only one test point for each active device. Furthermore, we can assume from simulations and measurements that we know the impedance matrix of the PDN for the two test points. With a few exceptions, the PDN interconnects and components are electrically reciprocal, therefore $Z_{12} = Z_{21} = Z_{transfer}$. However, Z_{11} in general is not equal to Z_{22}.

Most of the time our ultimate (but seldom reached) goal is for the elements of the Z-matrix to be frequency independent (resistive). If they are, the noise voltages at test points 1 and 2 can be calculated directly as the sum of the products of the $i_1(t)$ and $i_2(t)$ noise currents and the respective resistive terms:

$$\begin{aligned} v_1(t) &= R_{11} i_1(t) + R_{transfer} i_2(t) \\ v_2(t) &= R_{transfer} i_1(t) + R_{22} i_2(t) \end{aligned} \tag{1.1}$$

For the generic case, when the elements of the Z-matrix are frequency dependent, we have to take the $i_i(f)$ frequency spectrum of each of the current excitation, do the multiplication in the frequency domain, and get the time-domain transient noise by transforming the result into time domain (for instance, using the inverse fast Fourier transform):

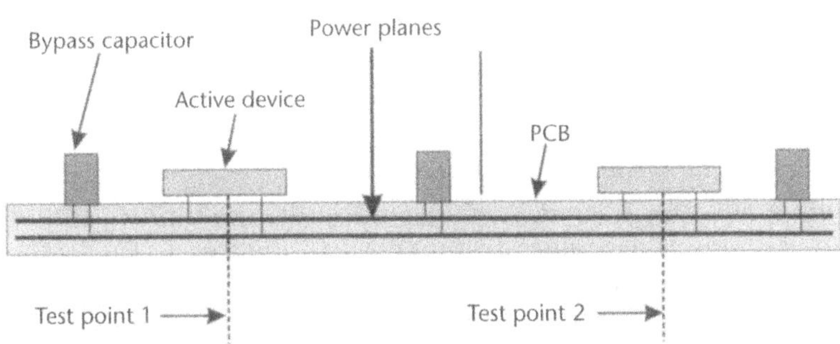

Figure 1.3 Simple sketch of PDN: only two active devices, three capacitors, and one pair of power planes are shown.

$$\begin{aligned} v_1(t) &= invFFT\left[Z_{11}(f)I_1(f)\right] + invFFT\left[Z_{transfer}(f)I_2(f)\right] \\ v_1(t) &= invFFT\left[Z_{transfer}(f)I_1(f)\right] + invFFT\left[Z_{22}(f)I_2(f)\right] \end{aligned} \tag{1.2}$$

For M ports on the PDN, (1.2) can be extended and the noise voltage at the kth test point becomes:

$$v_k(t) = \sum_{i=1}^{M} invFFT\left[Z_{ik}(f)I_i(f)\right] \tag{1.3}$$

1.3.1 Size of the Matrix

Even though in this simple illustration we assumed only two test points, note that the impedance matrix elements still must be obtained for these two ports *together* with the three additional locations of bypass capacitors, because the locations and electrical parameters of the bypass capacitors all have an influence over the elements of the Z-matrix. This means that even if we may not need the impedance matrix elements at the locations of bypass capacitors, initially we need to calculate and/or measure the network matrix on the PCB for a potentially large number of ports, which equals the sum of bypass capacitors, active devices (noise sources), and test points. Once we combine the matrix of PCB with the PDN components, then we can reduce the number of ports to the number of test points.

If we want to close the PDN design cycle with validation in the time domain, we have to know both the impedance matrix and the transient noise vector. Except for the active voltage sources, the PDN is built of passive components. Traditionally, semiconductor devices show more change of their electrical parameters due to process, voltage, and temperature (PVT) than passive RLC components. Thus, usually the impedance of PDN is more stable than the semiconductor network it has to feed. Certainly, with well-behaving components, it does not vary greatly with time. In contrast, the transient noise in a complex system comes from many packages of different active devices. These sources have their own timing and activity schedule. When we have only a few key components in the PDN, we have a good chance of working through this exercise successfully [4].

On the other hand, on a large board, there may be hundreds of packages with thousands of active cells switching. Predicting or simulating the worst-case transient noise in the entire system is a daunting task, because transfer impedances and propagation delays among the noise sources and test points are usually not negligible. Therefore, the worst-case maximum transient noise will not necessarily occur when (or if) all the sources switch simultaneously.

Instead, if we wanted to find the transient current by measurements, the difficulty is very practical: measuring current in a small-pitch printed-circuit board is not easy. It would require a shunt element in series to each current path or a current-measuring loop around each current-carrying conductor to be measured. Even if we could measure the current, it does not remove the problem stemming from the highly statistical nature of the system activity that creates the excitation current.

Several publications have described ways to measure the impedance profiles and/or transient currents indirectly (see, e.g., [5, 6]). These methods assume that the active devices can be exercised in a controlled manner.

Though eventually we may want to measure and validate the noise in the time domain, it is more straightforward to segment the task by doing the design and validation for the impedance matrix first, followed by time-domain measurements and validation (if necessary and if transient-current data is available).

A methodology suggested in [6] starts with the step transient responses corresponding to the test points and then calculates the absolute worst-case transient noise magnitude. In this case, obtaining the impedance matrix by measurements or simulations is sufficient because the step responses can be obtained from the impedance matrix by linear network calculations. This way we can get the worst-case estimated noise without doing extensive time-domain measurements.

Finally, another reason for using frequency-domain measurements in the characterization process instead of time domain is that external random noise from the environment can be suppressed more readily in the frequency domain. Vector network analyzers (VNA) and impedance analyzers (IA) operate with narrow bandwidth in a synchronous mode. The narrowband receiver circuit helps to filter out noise further away in frequency from the test signal. As we will show, with narrowband synchronous measurements, we can characterize the PDN even when the active circuitry is running and generating a large amount of random noise.

When we try to measure the worst-case transient noise in the time domain (i.e., with an oscilloscope accumulating samples in infinite persistence mode) any external noise getting into the measurement setup will corrupt our data. We can use averaging on time-domain instruments; but it will suppress all noise components that are asynchronous to the averaging window sequence, both the noise that we want to measure and the noise coming from external sources. This is particularly important when we measure large running systems in a noisy lab environment; the finite surface transfer impedance of cables and probes, and the unavoidable ground loops of cables, make it very difficult to measure time-domain PDN noise accurately.

Moreover, wideband time-domain digital instruments necessarily will have fewer effective bits in signal quantization; this increases the noise floor and limits the achievable dynamic range in comparison to narrowband frequency-domain instruments. At the time of this writing, real-time high-speed oscilloscopes achieve a 60 GS/s interleaved sampling rate with 8-bit A/D conversion and an equivalent analog bandwidth up to 18 GHz. Undersampling oscilloscopes (also called equivalent-time oscilloscopes) and time-domain reflectometry (TDR) instruments may have 12-bit quantization and up to 100-GHz front-end bandwidth, but the undersampling process requires a repetitive signal to be measured. Since each quantization bit corresponds to about a 6-dB dynamic range, we may get up to a 48- to 72-dB dynamic range from time-domain instruments without averaging [7]. With averaging we can get another 10- or 20-dB improvement, but this is still less than what we get from a well-built narrowband frequency-domain instrument.

1.4 When Time-Domain Characterization Is Useful

So far we have shown the benefits of frequency-domain characterization: for passive, linear, and time-invariant networks, the frequency-domain characterization is sufficient, interchangeable with time-domain data, and usually easier to perform. However, there are cases when one or more of the initial assumptions are not valid. An obvious item is linearity. In later chapters, we will show that on a micro level many of our PDN components exhibit nonlinearity (more or less). Of the possible components in a PDN, linear voltage regulators and switching power sources (dc-dc converters) are the prime examples. Active sources are also nonpassive and nonreciprocal. This is where time-domain characterization has its important niche role. As we will show in Section 7.1.2.6, both the periodical switching noise and the optional aperiodical ringing will be highly suppressed if the frequency-domain characterization is done by vector network analyzers or narrowband spectrum analyzers. Time-domain characterization is better suited to describe the switching ripple, high-frequency ringing, and large-signal transient behavior of power sources.

1.5 If and When Time-Domain Response Is Needed

If we need the time-domain response of a passive, linear, and time-invariant PDN, we can use (1.3) and (1.4) to calculate the time-domain noise from known noise sources. Also, we can calculate the worst-case response to an arbitrary sequence of rising and falling step excitations with equal step magnitude and set rise time with the *reverse pulse technique* [8]. The reverse pulse technique can also be used to compare the time-domain performances of various PDN designs. Figure 1.4 shows a comparison from [9] among three possible PDN impedance profiles and their resulting step responses and worst-case transient noise numbers. The three PDN design

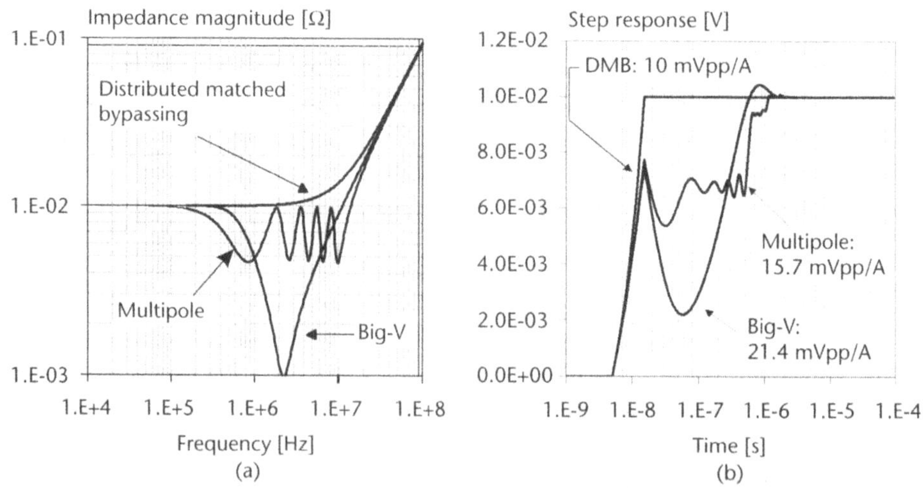

Figure 1.4 Comparison of frequency-domain and time-domain responses of three PDN approaches: distributed matched bypassing, multipole, and Big-V: (a) frequency-domain responses and (b) time-domain responses. The mVpp/A numbers are the worst-case transient-noise responses for an arbitrary sequence of 1-A step noise excitation.

methodologies included in the comparison are: distributed matched bypassing, multipole, and Big-V. The distributed matched bypassing uses low-Q bypass capacitors to synthesize smooth impedance in a manner similar to what we would get from a series R-L circuit. The multipole design approach approximates the target impedance by aligning the series resonance frequencies of bypass capacitors with moderate Q. The Big-V design uses bulk capacitors with moderate ESR to create a flat low-frequency impedance plateau, whereas the high-frequency bypassing is accomplished with a number of low-inductance capacitors of the same value and body size. The identical series resonance frequencies drive a deep notch in the impedance profile, hence the name Big-V. The mVpp/A labels on the time-domain response graph on the right indicate the worst-case transient noise that can occur in the particular PDN with an arbitrary sequence of filtered 10-ns rise-time 1-A current steps. Since we assume linear PDNs, the graphs and the worst-case numbers scale linearly with the excitation magnitude. Contrary to popular beliefs, note that the lowest worst-case transient noise comes from the flattest and not from the lowest impedance profile. The distributed matched bypassing produces the lowest, worst-case transient noise, even though its impedance profile represents the upper bound for the other two impedance responses. Also note that both horizontal scales are logarithmic.

1.6 The Characterization Process

Characterization of a PDN and its components may be done for several different reasons. We may want to characterize materials and components to evaluate and investigate material properties and the effect of various constructions of components. Material-science measurements will require the highest accuracy and most expensive characterization process. In the United States, the National Institute of Standards and Technology (NIST) develops characterization methods and performs accurate measurements. In engineering design work, characterization is necessary to collect data of unknown, undocumented, or new components and to verify and validate designs. Also, this may have different levels: from quick-and-dirty to detailed-and-accurate modeling and characterizations.

Figure 1.5 shows the triangle of the proper engineering execution, code-named here as *SUM* (simulations, understanding, and measurement). This is generally accepted as a best practice, where these contributors should all go hand-in-hand and should result in the same data, or close enough that the differences are acceptable and the reasons for the differences are understood. The three elements of *SUM* are used to cross-correlate the results to ensure the highest quality.

In the optimum scenario, all three components of *SUM* are equally strong. This ideal scenario, however, cannot always be reached. If one of the three elements is missing or weak, we compensate by strengthening the remaining two elements.

For instance, if *M* (measurement) is missing from the *SUM*, we can still do a proper job with trusted and tested simulators and a good knowledge of the underlying physics. However, in this case, we may not be able to do a complete and direct validation (which would require measurements). On the other hand, we may verify

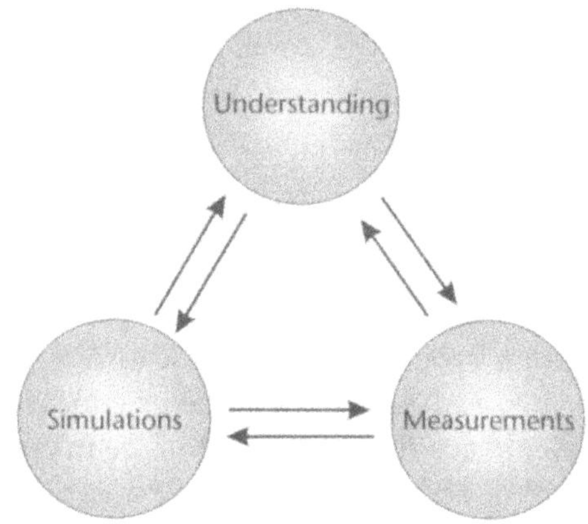

Figure 1.5 The SUM triangle. Simulations, understanding, and measurements all should eventually provide the same answer.

the result indirectly; for instance, instead of measuring the PDN performance, we stress and exercise the entire system extensively during the validation phase.

When we cannot do a robust cross-correlation among simulation, understanding, and measurement, we can enhance one or more of these contributors by cross-correlating within its own boundaries. For instance, we can increase confidence in the simulation space by using simulators based on different principles and then comparing closely their outputs until we get agreement on the PDN parameters we need to characterize. Obviously this requires additional knowledge of the tools to ensure that the ones involved in the exercise are all suitable for the particular task. If we need to increase our confidence in the measured data, we can apply different instruments, calibration methods, or measuring principles to the same problem and compare the results. In later chapters, we will see several illustrations of cross-correlating different simulations and different measurement techniques.

1.7 The Modeling Process

Though we can never achieve error-free measurements and simulations or perfect understanding, we can constantly improve them. When we describe a real-life component, circuit, or system with equations or equivalent circuits, we apply models. Models are simplified replicas of the real things. We use the models to describe the selected phenomena or behavior in a selected range of input parameters. The complexity of the model depends on the accuracy and range of parameters we want to cover. The wide possible range will be very visible in the later chapters as we introduce various models for bypass capacitors. Depending upon the information we want to capture with the model, we can use a wide range of equivalent-circuit models, from a single ideal capacitance to a full 2D or 3D bedspring model with hundreds or thousands of nodes.

References

[1] Novak, I., "Comparison of Power Distribution Network Design Methods: Bypass Capacitor Selection Based on Time Domain and Frequency Domain Performances," TecForum MP3, *Proceedings of DesignCon 2006*, Santa Clara, CA, February 6–9, 2006.

[2] Kaw, R., et al., "Towards Developing a Standard for Data Input/Output Format for PDN Modeling & Simulation Tools," *Proceedings of IEEE EMC 2005 Symposium*, Chicago, IL, August 8–12, 2005.

[3] Pozar, D., *Microwave Engineering*, New York: John Wiley & Sons, 2005, Chapter 4.

[4] Matoglu, E., et al., "Voltage Regulator Module Noise Analysis for High-Volume Server Applications," *Proceedings of the 14th Topical Meeting on Electrical Performance of Electronic Packages*, Austin, TX, October 24–26, 2006.

[5] Taylor, G., et al., "An Approach to Measuring Power Supply Impedance of Microprocessors," *Proceedings of the 10th Topical Meeting on Electrical Performance of Electronic Packaging*, Cambridge, MA, October 29–31, 2001, pp. 211–214.

[6] Kantorovich, I., et al., "Measurement of Milliohms of Impedance at Hundred MHz on Chip Power Supply Loop," *Proceedings of the 11th Topical Meeting on Electrical Performance of Electronic Packaging*, Monterey, CA, October 27–29, 2002.

[7] Smolyansky, D., "Advances in Time and Frequency Domain Measurements, Modeling and Signal Integrity Analysis of Gigabit Interconnects," TecForum TF5, *Proceedings of DesignCon 2004*, Santa Clara, CA, February 2–5, 2004.

[8] Drabkin, V., et al., "Aperiodic Resonant Excitation of Microprocessor Power Distribution Systems and the Reverse Pulse Technique," *Proceedings of the 11th Topical Meeting on Electrical Performance of Electronic Packaging*, Monterey, CA, October 27–29, 2002, p. 175.

[9] Weir, S., "Bypass Filter Design Considerations for Modern Digital Systems, A Comparative Evaluation of the Big 'V,' Multi-Pole, and Many Pole Bypass Strategies," *Proceedings of DesignCon East 2005*, 2005.

CHAPTER 2

Simulation Methods and Tools

There are three major types of simulation tools used to predict the response of power distribution components and networks: (1) analytic expressions, (2) circuit simulators, and (3) electromagnetic field simulators. Closed-form analytic solutions exist for a very small subset of structures, for example, waveguides or coaxial cables. Consequently, their utility in solving real-world problems is limited. Semianalytic solutions can be obtained for a number of useful structures, including, for example, the self-and transfer-impedance profiles of parallel planes. Often, these types of solutions involve infinite series approximations or complicated integrals. Circuit simulators translate conductors and dielectrics into equivalent circuit elements, such as resistances, capacitances, inductances, and their coupling. Then, frequency-domain simulations can be performed on the entire circuit network to capture the device behavior. Although circuit simulators permit the engineer to construct a very sophisticated model of PDN components, realistic 3D structures are difficult to simulate accurately for many reasons. One reason is that creating an equivalent circuit requires some understanding of the current flow and coupling in order to capture its response. Also, circuit simulators require an estimate of the electrical parameters of the model (which is often unknown). On the other hand, electromagnetic field solvers do not share the limitations described above for these two types of simulation tools: they are not limited to solving a class of problems that have closed-form or semianalytic expressions; they can solve a range of problems where no such expression exists. Unlike circuit simulators, they do not require estimates of the equivalent circuit elements or parasitic coupling between the elements. Potentially, field solvers can include all electromagnetic effects from first principles leading to a deeper understanding of how the devices or networks operate. However, these tools can be difficult to operate due to their level of sophistication. For accurate results, the user must be careful to ensure that the geometry is correctly represented, the source in the model is realistic, the correct material properties are included, and that the internal solver options are properly selected.

This chapter will discuss these three types of simulation tools, starting with analytic expressions. These expressions can be evaluated using many techniques, including spreadsheets (Section 2.1) and numerical analysis software, such as MATLAB (Section 2.3).

2.1 Spreadsheet Calculations

A spreadsheet is a convenient simulation tool for performing frequency domain calculations. For example, simulating the frequency-domain response of single and multiple bypass capacitors, simulating the self-impedance of bare rectangular parallel plates at arbitrary locations or simulating the plane's dc plane losses, are all achievable using spreadsheets. As an example of the use of a spreadsheet to calculate the frequency-domain response, the structure shown in Figure 2.1 (consisting of a simple plane pair) was measured and simulated using a Microsoft Excel spreadsheet. A double infinite summation of modal harmonics can be computed using a spreadsheet to obtain the impedance for a parallel plate structure. Several formulations for the impedance of plane pairs are discussed in Chapter 4. Here, the lossless plane impedance expression was evaluated and coded into Excel. The plane pair was measured using a pair of vias located in the center of the plane pair. Figure 2.2 shows the test board measurement arrangement. *S*-parameters were obtained in the frequency range from 100 MHz to 10 GHz. Further details about this type of measurement can be found in Chapter 5.

Figure 2.3 plots the measured and simulated results for the parallel plate structure. Figure 2.3(b) is an enlarged view of Figure 2.3(a), showing the correlation at higher frequencies above the series resonance. The simulation captures the location of the series and parallel resonances up to 10 GHz with reasonable accuracy. However, since the expression for the plane impedance does not include losses, sharper resonances are observed in simulation. The simulation also does not include the frequency dependence of the dielectric constant or the impact of antipad cutouts on the

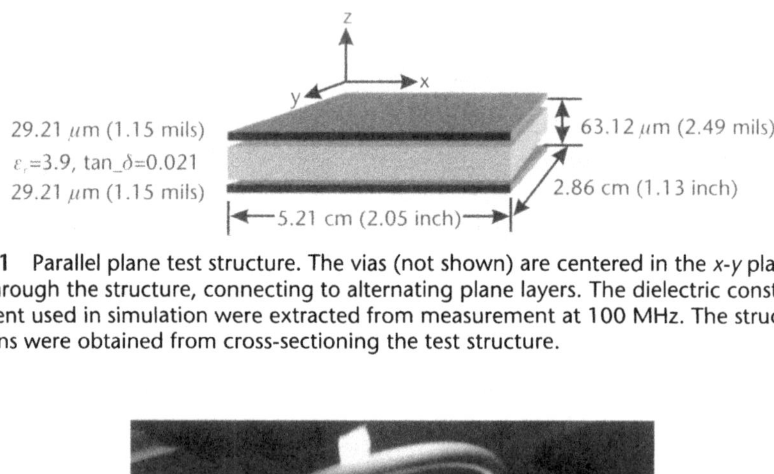

Figure 2.1 Parallel plane test structure. The vias (not shown) are centered in the *x-y* plane and extend through the structure, connecting to alternating plane layers. The dielectric constant and loss tangent used in simulation were extracted from measurement at 100 MHz. The structure dimensions were obtained from cross-sectioning the test structure.

Figure 2.2 Vertical probe arms were used to contact the via pair, which connects to the plane pairs.

impedance profile. The impact of these last two items on the impedance profile is investigated further in Section 2.4.

2.2 SPICE AC

SPICE, an acronym for simulation program with integrated circuit emphasis, is a general-purpose circuit simulation program. SPICE is an eminently useful tool for both analog and digital design. It can be used for nonlinear dc, nonlinear transient, and linear ac analyses. Circuits may contain resistors, capacitors, inductors, mutual inductors, independent voltage and current sources, four types of dependent sources, lossless and lossy transmission lines (two separate implementations), switches, uniform distributed RC lines, and the five most common semiconductor devices (diodes, BJTs, JFETs, MESFETs, and MOSFETs). In this section, we focus on one particular application of SPICE: simulating the ac frequency response of PDN components. With the ac frequency response command (.AC), first SPICE computes the dc node voltages and currents. Next, using the nodal matrix with the real and imaginary parts of the impedance, nodal analysis calculates the results over the desired frequency range. Linearized, small-signal models are obtained for all of the nonlinear devices in the circuit. Points in between can be spaced linearly (LIN) or logarithmically either by decade (DEC) or octave (OCT). Run times in SPICE ac will depend on the frequency span, the number of points requested, the number of nodes, and the accuracy setting.

As an example of the use of SPICE to simulate the ac response of PDN components, an equivalent circuit for the structure shown in Figure 2.1 was constructed. For rectangular plane shapes, a SPICE grid consisting of an array of unit cells can be used to capture the frequency domain response. Each unit cell can consist of RLC elements or transmission line elements; details about calculating the unit cell param-

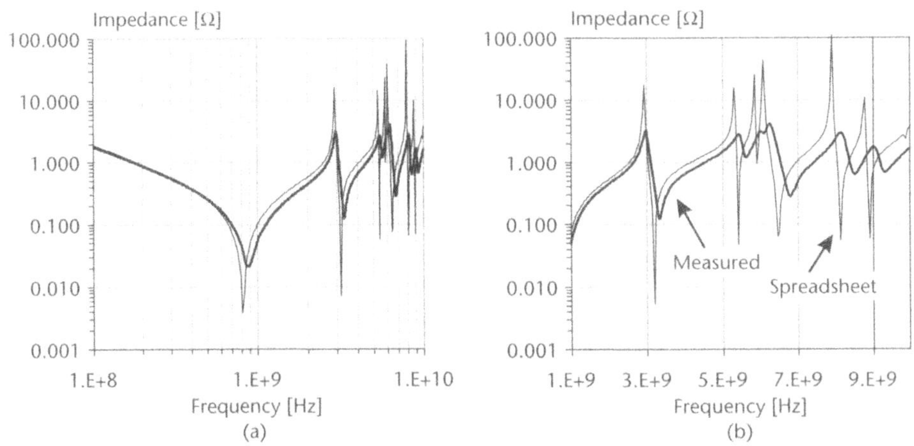

Figure 2.3 (a) Self-impedance of a 5.21× 2.86 cm (2.05 × 1.13 inch) pair of planes, with a 63-μm (2.49-mils) plane separation. The impedance is simulated using a spreadsheet that evaluated an analytical expression for the plane impedance (lossless). The *m* and *n* summation limits in the impedance expression were set to 40. Part (b) is an enlarged view of (a) showing correlation with the higher order modes.

eters, the model topology, and capturing losses can be found in Section 4.3. Perforations, odd-shaped planes, and cutouts (which can significantly alter the modal resonances) can be handled using adaptive grids (see Section 4.3.2). The number of unit cells is constructed such that each transmission line segment in the model is a small fraction of a wavelength of the highest frequency of interest. Thus, for an upper frequency limit of 10 GHz, the wavelength is about 1.52 cm (600 mils). If each transmission line segment is made no more than about 1/20th of the maximum frequency of interest, we need at least 68 × 34 cells to cover the 5.21 × 2.86 cm plane. The location where the current source is injected and the corresponding voltage magnitude measured will affect the impedance profile both in measurement and SPICE simulation. At frequencies well below the series resonance, all points on the plane are at nearly the same voltage, but at frequencies beyond the first resonance dip, the position on the plane becomes very important. Thus, not only is it necessary to have the required grid granularity for transmission line effects, but a sufficiently fine grid is required to ensure that there is a grid point located at the desired position to land the probe (see Section 2.4.3.9). Here the measurement and simulation point was located in the center of the x-y plane. With sufficient node granularity, it is then possible to move the probe point about the grid to observe how the impedance changes across the x-y plane. In this case, since the probe point was in the center of the x-y plane, the first modal resonance was suppressed. Figure 2.4 shows the agreement between the measured and simulated self-impedance profiles for the plane pair structure. Although the conductive and dielectric losses are included, note that the frequency dependence of the dielectric and loss tangent is not captured by the simulation, resulting in a slight frequency offset.

2.3 MATLAB

MATLAB (for Matrix Laboratory) is software for numerical computations and data visualization. As the name suggests, MATLAB is designed especially for matrix

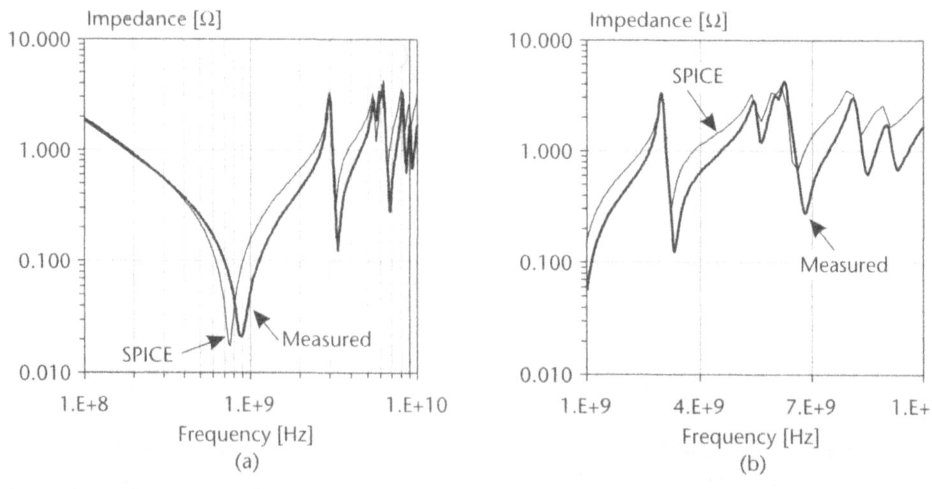

Figure 2.4 SPICE ac simulated self-impedance profile for a plane pair structure using a bedspring model for the planes. Part (b) is an enlarged view of (a) showing correlation with the higher-order modes.

computations (e.g., solving systems of linear equations, computing eigenvalues, and so forth). Since it is a high level language for numerical analysis, numerical code can be developed very compactly. As an example of MATLAB's capabilities as a simulation tool, we solve for the input impedance of a plane pair of the structure shown in Figure 2.1, using the double infinite summation of modal harmonics. This is the same formulation used for the spreadsheet calculation in Section 2.1. The code evaluates the double series summation using two nested for loops with an additional loop used to sweep the frequency. The code is extremely compact and straightforward to implement, and it solves rapidly. Figure 2.5 plots the impedance obtained from MATLAB against measurement results. Virtually identical results are obtained using the spreadsheet method and MATLAB (as expected).

As a second example using MATLAB, we evaluate the plane pair impedance using an expression that includes dielectric losses and conductive loss [1] (see Section 4.2.1.3). The model in [1] is accurate for planes that are much thicker than the skin depth of the conductor. Although the frequency dependence of the dielectric constant and loss tangent are not captured, Figure 2.6 shows that the inclusion of loss significantly improves the correlation.

2.4 Field Solvers

In addition to analytic models and circuit simulators, a field solver is an excellent tool for simulating the behavior of power distribution components and networks. Full-wave field solvers offer the possibility of capturing all electromagnetic effects of a problem from first principles. We can capture all loss mechanisms, distributed effects, and all parasitics. Also, we can observe fields and currents in the structure, permitting a deeper understanding of the underlying physics. However, to take full advantage of the promise of field solvers, we must pay special attention to both the

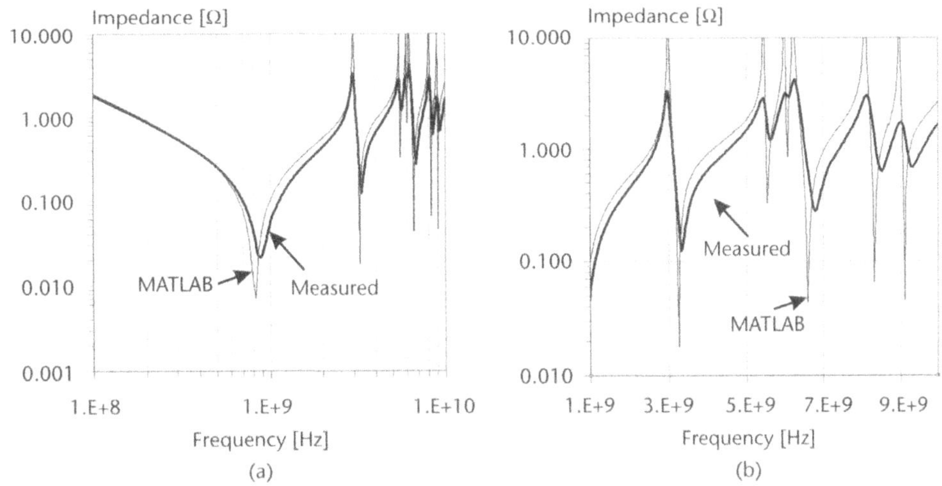

Figure 2.5 Simulated self-impedance profile for a plane pair structure using MATLAB to evaluate a plane-impedance expression that does not include losses. The *m* and *n* summation limits in the impedance expression were set to 40. Part (b) is an enlarged view of (a) showing correlation with the higher order modes.

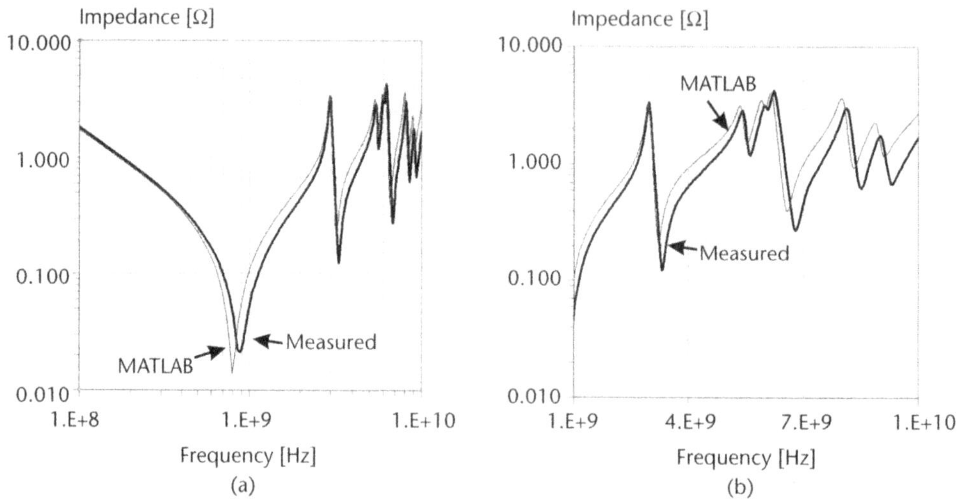

Figure 2.6 Simulated self-impedance profile for a plane pair structure using MATLAB to evaluate an expression for the impedance, which includes dielectric and conductive losses. The *m* and *n* summation limits in the impedance expression were set to 40. Part (b) is an enlarged view of (a) showing correlation with the higher order modes.

problem geometry and materials (i.e., what we are simulating) and the problem setup (i.e., how we are simulating). For example, in order to accurately describe the problem geometry, we must properly account for details such as manufacturing tolerances and cutout faceting. We must also properly account for the conductor and dielectric material properties to ensure accurate results (e.g., electrical conductivity, dielectric constant, or loss tangent). Proper problem setup requires that the user have a solid understanding of the tool and its nuances. Boundary conditions, port definitions, and meshing are examples of elements of the problem setup that need to be carefully considered.

In the next section, we review the different types of field solvers that can be used for simulating elements of the PDN in the frequency domain. Although general tool usage is outside the scope of this book, a list of many potential simulation pitfalls is discussed in the subsequent sections.

2.4.1 Classifications

There are a number of ways to classify field solvers. For the practicing engineer, it may be important to select a field solver based on the type of geometry we are trying to solve. We could also choose a solver based on the numerical method employed and solution domain (i.e., frequency and time domain) in order to understand the advantages and disadvantages depending on the type of problem we are trying to solve.

2.4.1.1 Geometrical Classification

Field solvers can be roughly grouped into three geometrical classes of field-solver code. The first takes the 2D cross-section and solves for the transverse field distributions. These tools are useful for calculating trace impedances and coupling coefficients in microstrip or stripline board routes, for example, or any geometry with a uniform cross-section in the longitudinal direction. They solve rapidly because only the 2D cross section needs to be discretized. Ansoft Maxwell 2D is an example of a commercial 2D field solver.

The second class of code meshes the surface of planar metals. Although this is still a 2D problem, if vias are introduced to establish connection vertically between metal plane layers, then the code is sometimes called 2.5D. The via can then be handled by look-up tables or can be modeled as a lumped element, for example. These tools allow for an arbitrary number of homogenous dielectric layers with patterned planar metal on the conductive layers. Sonnet EM from Sonnet Software and Ansoft SIwave are examples of commercial 2.5D solvers. Figure 2.7 plots the simulated impedance profile of the structure shown in Figure 2.1, using Ansoft SIwave and Ansoft HFSS (a 3D field solver discussed later). In this example, the solve time was reduced by orders of magnitude, using the 2.5D solver without sacrificing accuracy. Note that the structure of Figure 2.1 has a fairly thin dielectric compared to the wavelength at the highest frequency of interest; thicker dielectrics may not be well suited to a 2.5D solver due to field nonuniformity in the Z direction. For most PCB applications, this limitation does not pose a problem.

The final geometrical class meshes a 3D volume. This class of tools is useful for solving for vias transitions, discontinuities in multilayer PCBs, capacitor mounting structures, and so forth. The solution time is much longer compared to 2D and 2.5D solvers, but the 3D solver has the benefit that it can be used to analyze a broad range of problems. Ansoft HFSS and CST Microwave Studio are examples of 3D field solvers.

2.4.1.2 Quasi-Static Versus Full-Wave

As an example to illustrate the difference between quasi-static and full-wave solvers, consider the case of a simple parallel plate capacitor consisting of two circular

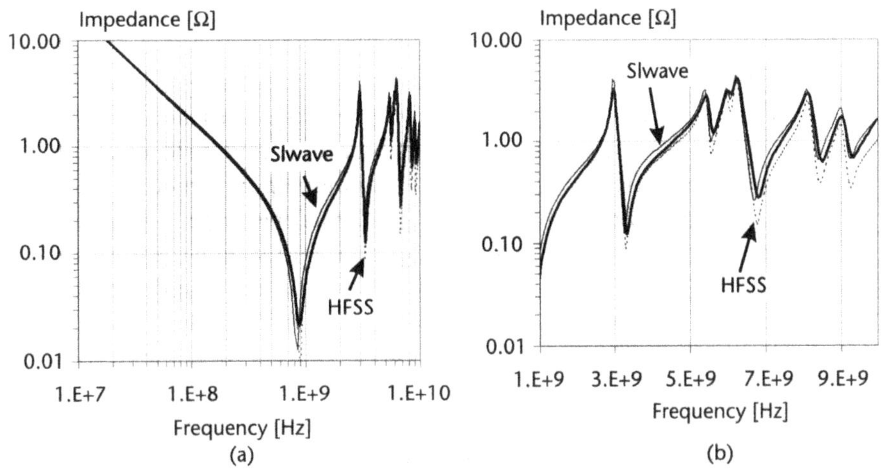

Figure 2.7 Simulated impedance profile of the structure shown in Figure 2.1, using Ansoft SIwave and Ansoft HFSS. The common dark trace on both plots shows measured results. Part (b) is an enlarged view of (a) showing correlation with the higher-order modes.

plates separated by a distance d. Under static conditions, energy will only be stored in the electric field, and no current will be flowing through the device. Using Poynting's theorem for electric energy storage, we can calculate the capacitance as follows:

$$W_{E0} = \frac{1}{2}(qV) = \frac{1}{2}\left(\frac{\varepsilon A}{d}\right)V^2 = \frac{1}{2}(C)V^2 \tag{2.1}$$

where C is found to have the usual expression for the parallel plate capacitance. However, as the frequency increases, the time-varying voltage will give rise to a magnetic field that results in magnetic energy storage, or inductance. We can see this relation from Ampere's law:

$$\nabla \times H = j\omega\varepsilon E \tag{2.2}$$

Similar to (2.1), we can calculate the inductance associated with the first-order magnetic-energy storage, W_{M1}. The frequency at which we can ignore this first-order inductance term can be found by requiring that the energy storage of the zero-order capacitance term be much greater than the energy storage of the first-order inductance, that is,

$$W_{E0} >> W_{M1} \tag{2.3}$$

Inserting the appropriate expressions for the magnetic and electric energy components and simplifying, we obtain:

$$l << \lambda \tag{2.4}$$

This relation requires that the length of the device, l, be much smaller than a wavelength. If this condition is satisfied, then we can assume *static* or at least *quasi-static* conditions will prevail. If this condition is not met, then we are forced to solve Maxwell's equations without the quasi-static assumption. This type of field solver is called a *full-wave* field solver. A full-wave field solver is capable of solving Maxwell equations at any given frequency. It solves for electric and magnetic fields together so that the interaction between electric and magnetic fields is handled properly. Full-wave solvers are important when one must accurately account for displacement currents, electromagnetic radiation, and field coupling. For example, using a quasi-static solver to analyze the plane pair structure shown in Figure 2.1 would only provide the low-frequency capacitive slope. The displacement current through the dielectric would not be captured, nor would the modal resonances, since the standing waves created by the plane boundaries would not be included. Due to the decoupling between electric and magnetic fields under quasi-static conditions, a quasi-static field solver is quicker and can solve much bigger problems than a full-wave solver. There are other advantages to full-wave solvers; for example, full-wave solvers do not require assumptions about the current return path, whereas with quasi-static analysis the return paths are dependent on the placement of current sinks and sources. Ansoft Q3D Extractor is an example of a commercially available quasi-static field solver.

2.4.1.3 Numerical Formulations

Electromagnetic field solvers use various numerical methods to directly solve Maxwell's equations. Most field-solvers work by subdividing the geometry into basic cells or elements where the size of each element is some fraction of a wavelength. Then the task is to find the fields on each cell or the field at the junction of the cells. The final solution requires that the field contribution from each unit cell be summed.

The most popular 3D frequency-domain methods are the finite element method (FEM), finite difference time domain (FDTD), and method of moments (MOM) formulations. For example, Ansoft HFSS uses the FEM formulation and CST Microwave Studio's time-domain solver uses the finite integration technique (FIT). FEM is a general solution that can be applied to a wide range of geometries. The 2.5D field solvers predominantly rely on MOM. Any number of numerical methods can be used for 2D solvers, including MOM and FEM.

Frequency-domain solvers work by discretizing the problem geometry, building a matrix, and inverting the matrix to arrive at a solution. Time-domain solvers work by discretizing the problem geometry and exciting the geometry with an impulse function. Then, with the information about the time response at a particular location, the frequency-domain response can be obtained using a fast Fourier transform (FFT) process, for example. Time-domain solvers can be useful when broadband frequency data is needed. The approach has drawbacks for high-*Q* structures with closely spaced resonances. The high-*Q* condition requires long run times for the energy to decay, and the closely spaced resonances require many time samples for the Fourier transform process to resolve. For a thorough and systematic treatment of the strengths and weaknesses of different field-solver numerical methods, see [2]. Figure 2.8 plots the simulated impedance profile of the structure shown in Figure 2.1, using Ansoft HFSS and CST Microwave Studio's time-domain solver. Although both solvers are 3D, using the FIT reduces the solve time by orders of magnitude-compared to the frequency-domain method.

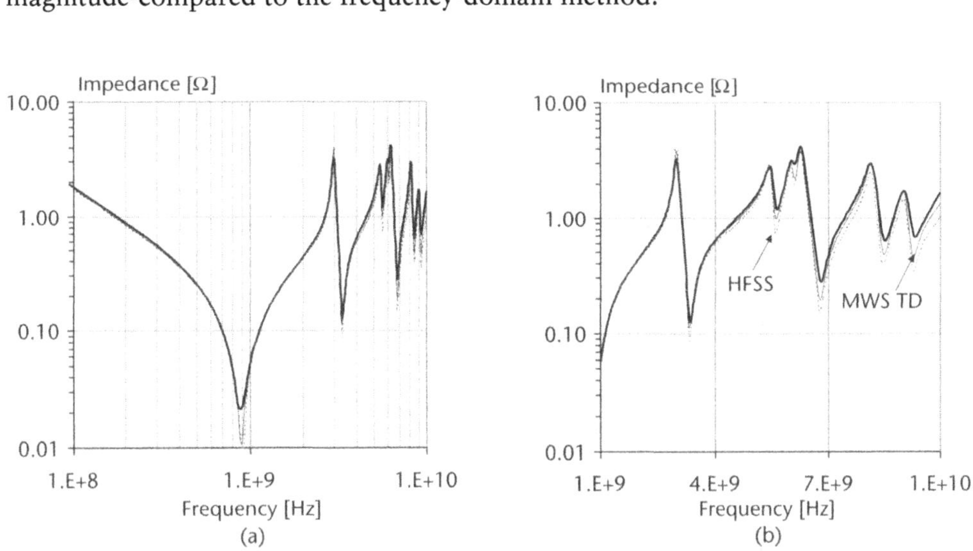

Figure 2.8 Simulated impedance profile of the structure shown in Figure 2.1, using Ansoft HFSS and CST Microwave Studio's time-domain solver (MWS TD). The common dark trace on both plots shows measured results. Part (b) is an enlarged view of (a), showing correlation with the higher-order modes.

In subsequent sections we focus on the FEM formulation using primarily Ansoft HFSS to discuss convergence and explore several sources of simulation inaccuracy. Overall, we have found a 3D full-wave FEM solver to be an excellent tool for simulating many components of the PDN, such as small PCB's, capacitor mounting structures, planar test fixtures, and via structures. Larger structures such as full boards, for example, can be better suited to time-domain formulations that have less computational storage requirements than 3D FEM solvers. 2.5D FEM solvers can also be used to solve large geometries. The choice of Ansoft HFSS is a somewhat arbitrary selection in that there are many capable commercial field-solver packages available.

2.4.2 Convergence

Modern FEM solvers use an iterative process called an *adaptive analysis* in which the mesh is automatically refined in critical regions. The process starts with a solution based on a coarse initial mesh. The mesh is then refined in areas of high error density. This process continues until the mesh has converged at the highest frequency of interest. Since the spatial resolution of the mesh will be determined by the highest frequency of interest, the mesh will also be valid for lower frequencies; however, the converse is false. That is, a low-frequency mesh will not be valid at higher frequencies. Convergence is tracked by changes in the magnitude and phase of the *S*-parameters at the ports. The solution has converged when the Delta-*S* (defined as the change in the *S*-parameters on two successive passes) is lower than a user-selected value. Different stopping criteria may also be specified for specific entries in the *S*-matrix (i.e., magnitude and/or phase of any of the *S*-parameter matrix elements). From there, the optimized mesh is used to solve for the *S*-parameters over the selected frequency range.

2.4.2.1 Solution Time

Solvers must fill and invert a matrix in order to arrive at a solution. (FEM matrices tend to be larger but more sparse, compared to MOM.) The solution time is approximately proportional to N^2, where N is the matrix size. This process must be completed at each discrete frequency point, although other methods can be employed to reduce the total solution time, such as interpolating sweeps (see Section 2.4.3.6). Boundary conditions can also impact the solution time, such as approximating free space using perfectly matched layers (see Section 2.4.3.7). Time-domain solvers solve much more rapidly in general, since there is no matrix to fill and invert. Actually, these field solvers have a solution time that grows more linearly with the problem size. However, solution time can grow considerably when seeking to capture closely spaced or sharp resonances in the frequency domain, requiring more time samples in the time domain.

2.4.2.2 System Limitations on Convergence

Considerations for Delta-*S* selection include the available compute resources and the required solution accuracy. For example, if S_{11} is estimated to be –20 dB at 5 GHz for a particular structure, with Delta-S set to 0.02, this implies a –34 dB-convergence, that is, –20Log (0.02) = –34 dB. The magnitude of a –20 dB S_{11} is 0.1, which means that the solution inaccuracy is less than 20% (i.e., 0.02/0.1). Simulating PDN devices has unique challenges with respect to S-parameter accuracy targets compared to, say, microwave devices. S_{21} values for PDN components are significantly lower compared to microwave devices, because PDN components are characterized at much lower frequencies where dielectric, conductive, and radiation loss are not overwhelming. Thus the default accuracy settings for field solvers may not necessarily be appropriate or sufficiently accurate for PDN components. Also, in general, PDN components are much larger than microwave devices. The low Delta-S requirements coupled with the large structure size can push memory requirements beyond computational resource limits. These limits are exceeded, because as Delta-S is made progressively smaller, a correspondingly finer mesh discretization is required. To maintain this fine mesh over a large volume requires an increasing number of tetrahedrals and unknowns to be determined. Creating a fine uniform mesh can assure the desired accuracy target is satisfied; however, for large structures, this produces very big matrices. By examining the convergence process log, one can determine how much memory and disk space is required in each successive mesh-refinement step. As memory resources are strained, disk-swapping increases, which often increases the solution time dramatically.

2.4.2.3 Verifying Convergence

Convergence can be verified by plotting the maximum magnitude change in the S-parameters as a function of the number of passes (i.e., mesh-refinement steps). Also, the maximum phase change, should also be verified for convergence. By plotting the magnitude and phase change we help avoid false convergence (i.e., by visually identifying local minima). Another approach is to plot the S-parameter values directly as a function of the number of passes; we would expect to see the S-parameters approach a final value with each subsequent pass. Figure 2.9 plots the convergence of the maximum magnitude and phase Delta-S for the plane pair structure

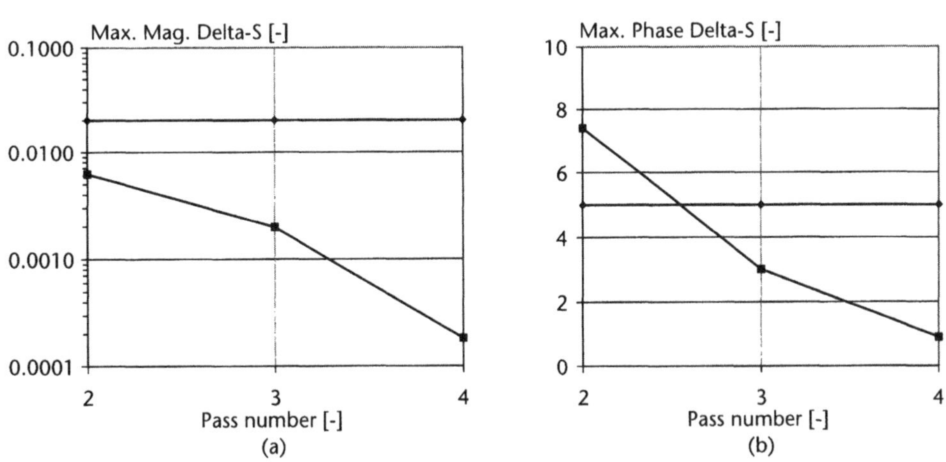

Figure 2.9 (a) Convergence of the maximum magnitude Delta-S. (b) Convergence of the maximum phase Delta-S.

illustrated in Figure 2.1. Maximum magnitude Delta-*S* was set to 0.02 and maximum phase Delta-*S* was set to 5° (shown as a horizontal line on both plots). The convergence criteria are reached only on the fourth pass, because a minimum of two consecutive converged passes was required. The phase converged on the fourth pass, whereas the magnitude converged on the third pass. Figure 2.10 plots the magnitude and phase of the four *S*-parameters as a function of the number of passes. Here, all four entries, for both magnitude and phase in the *S*-matrix, are required to be smaller than the maximum Delta-*S*. We find that the phase of S_{11} drives the adaptive process to require additional passes. The ports are not symmetrical due to the plane pair being located closer to one of the ports.

2.4.2.4 Mesh Seeding to Speed Convergence

Oftentimes the meshing process can be slow using automatic mesh refinement; therefore, it is in the best interest of the user to assist in the process by using mesh seeding. Mesh seeding is a general term to describe manual manipulation of the mesh using different methods. A good application of mesh seeding, which appears in PDN designs, is on microstrip or stripline structures. These structures are used, for example, as dedicated power and ground connections to noise-sensitive circuits, such as PLLs, or as lead-in and lead-out traces on test structures to characterize passive devices, such as decoupling capacitors. The model typically consists of nothing more than a narrow trace over a relatively wide, solid plane. In these structures, the fields will be confined to a narrow region surrounding the trace; whereas a few trace lengths away, there will be very little current flowing on the plane. In these situations, mesh seeding can be used to force a finer mesh on the trace and on the plane directly underneath the trace to speed convergence by helping the mesher locate the area of highest field variation.

There are a number of techniques that can be used to perform seeding. One technique is to simply restrict the maximum length of the cell size on the surface or volume of an object to some small fraction of a wavelength. In the case of a microstrip

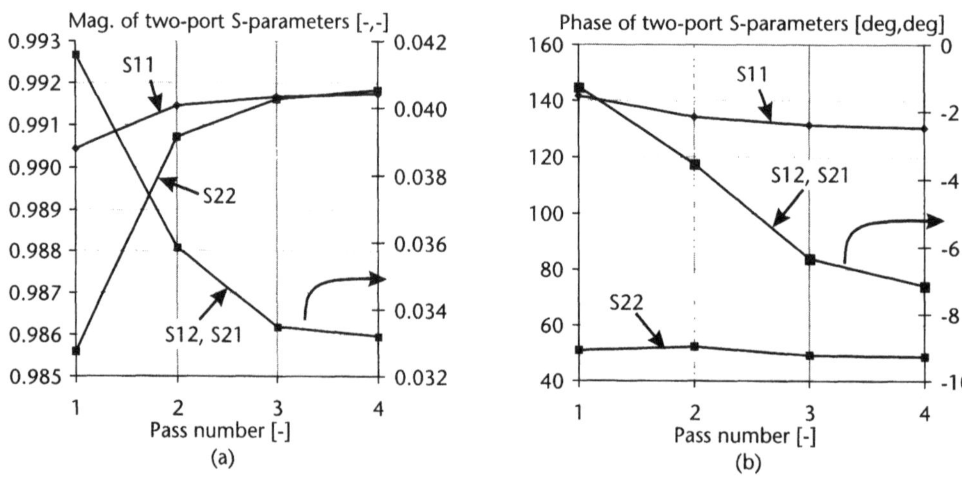

Figure 2.10 (a) Convergence of the magnitude of the four *S*-parameters. (b) Convergence of the phase of the four *S*-parameters.

or stripline, seeding the trace with a maximum cell length may not result in a sufficiently dense mesh on the plane underneath the trace. One way to accomplish this is to create a shadow object on the place that is the same width and length of the trace. This object can then be seeded with a maximum cell length to ensure a dense mesh is created on the plane underneath the trace. For example, Figure 2.11 shows the converged mesh for a 2.54-mm- (100-mils-) long microstrip. The shadow object has created a fine mesh underneath the trace where the field concentration is highest.

Figure 2.12 plots the convergence of the maximum magnitude Delta-*S* for a 2.54-mm- (100-mils-) long microstrip, before and after seeding. Delta-*S* is set to 0.005 in both cases. Without seeding, the convergence criteria are reached only after 10 passes, because a minimum of two consecutive converged passes is required. By seeding the shadow object with a mesh that has a tetrahedral edge length less than 5 mils, the convergence criteria is reached in 6 passes.

2.4.3 Sources of Simulation Inaccuracies

Many impediments can be encountered in the pursuit of obtaining accurate simulation results. The following sections provide a collection of potential sources of inaccuracy. It is certainly not intended to be a complete simulation checklist, but rather a selection of issues that are known to produce inaccurate or erroneous results if not carefully supervised by the user.

2.4.3.1 Geometrical Approximations

Tools approximate curved surfaces or boundaries using arcs with a finite number of straight line segments or facets. A via barrel is an example of a structure that must be approximated with facets. Selecting a large number of facets on a cylindrical via, for example, will increase the size of the mesh and hence the solution time. However, if we limit the number of facets, we will introduce inaccuracy due to this geometrical approximation. Plane cutouts are another example of a smooth surface that needs to be approximated with facets. The number of facets required for accurate modeling depends on the accuracy requirements and also the importance of the

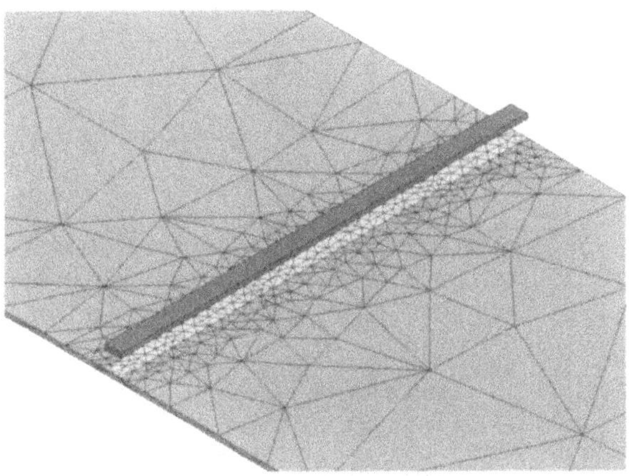

Figure 2.11 Plot of converged mesh for a 2.54-mm- (100-mil-) long microstrip using a shadow object.

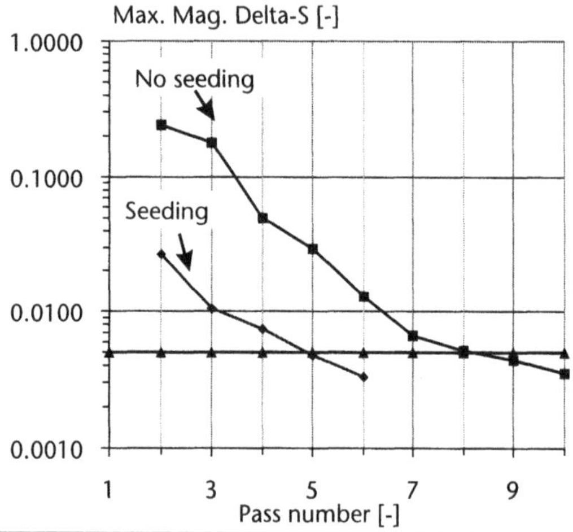

Figure 2.12 Convergence of the maximum magnitude Delta-*S* for a microstrip trace with and without seeding.

object in influencing the overall electrical behavior. For example, although the antipad cutout is only one part of the entire plane pair structure, and the antipad itself can be small in terms of a wavelength, their effect can be very important to accurately capturing the plane impedance [3]. This is particularly true if the antipads of neighboring power-ground vias overlap (see Section 3.4 for more on this topic). The via pair in Figure 2.1 does not have overlapping antipads and therefore we expect their impact on the overall self-impedance profile to be not significant. Figure 2.13 shows the self-impedance profiles using antipads with 4 facets and 30 facets. As expected, the impact of the approximating the antipad is not significant, although there is a slight underestimation of the overall inductance. If the antipads were overlapping, the self-impedance would be more sensitive to the number of antipad facets.

Another accuracy concern with cutouts is that the mesher might not do a good job of meshing around these cutouts. When we control the number of facets on a polygon, we control the resolution of the hole and therefore the resolution of the mesh (to some degree). Figure 2.14 shows the initial mesh generated on the plane

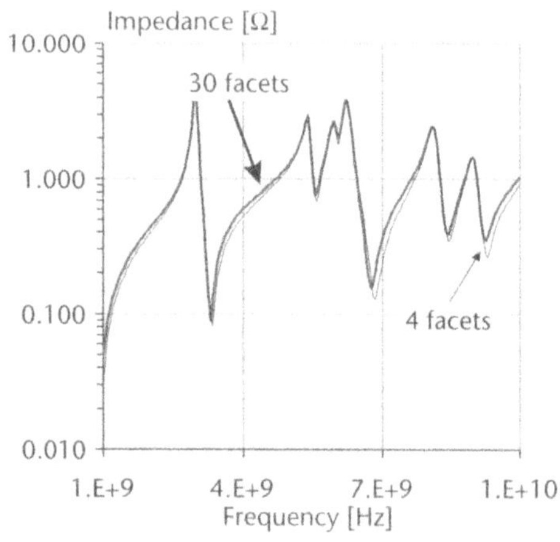

Figure 2.13 Simulation results for a plane pair structure with antipad cutouts approximated with 4 facets and 30 facets.

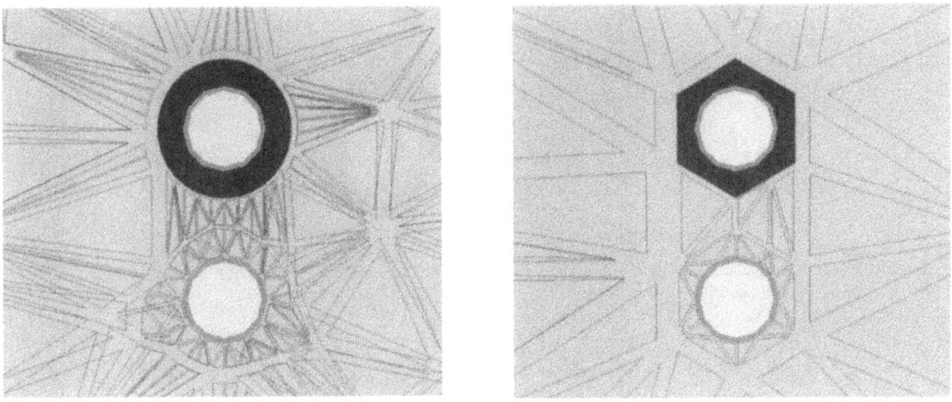

(a) (b)

Figure 2.14 Initial mesh generated on the plane pair structure with the antipads modeled with (a) 30 facets and (b) 6 facets. The region in the center of the plane surrounding the antipad and vias is shown. The tetrahedrals are rendered at 80% of their original size.

pair structure shown in Figure 2.1, with the antipads modeled with 30 sides and 6 sides. The impact of the antipad faceting on the initial mesh is evident.

Another technique for controlling the mesh in critical regions is to use *dummy objects* and/or seeding. Dummy objects are 2D or 3D objects that will influence the mesher but do not modify the electrical properties of the model. The new boundaries that they create will be detected by the meshing algorithm. The dummy object itself can be used to influence the mesher, or the dummy object can be seeded with a mesh that has the required granularity. As an example, the structure shown in Figure 2.14(b) was seeded using a rectangular dummy object. The initial mesh surrounding the cutout region is shown in Figure 2.15. The initial mesh is finer than the mesh dictated by the six sides of the antipad cutout. Other options to ensure a good mesh around cutouts include using an increased mesh frequency and/or imposing a small maximum tetrahedral length on the plane mesh. These last two approaches would fill the box with a very dense mesh, which is not very efficient.

As a final note, even though it is possible to rely completely on automatic mesh refinement, the user should oversee the mesh generation process to ensure a high-quality mesh. Ideally, the mesh should capture the field changes using small, equilateral tetrahedrals. This will produce the most accurate approximation of fields. Ansoft HFSS release 10, for example, allows the user to control the aspect ratio of the tetrahedral. However, it is difficult to satisfy aspect ratio demands if the aspect ratio value is set close to 1, because an arbitrary shape cannot be filled with only equilateral triangles. For smooth shapes, using aspect ratios slightly greater than 1 is a good choice.

2.4.3.2 Solving Inside

At dc, the current is distributed throughout the cross-section of a conductor in such a way as to minimize the resistance. As the frequency increases, the current inside the conductor starts to redistribute to the surface due to skin effects to minimize the

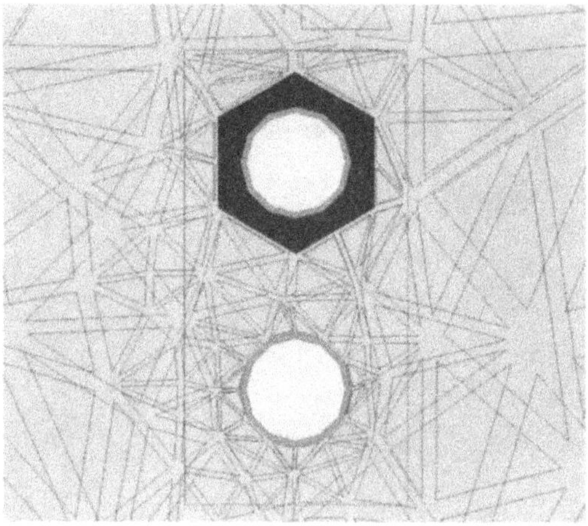

Figure 2.15 Mesh resulting from a rectangular dummy object seeded with a maximum length restriction. The outline of the dummy rectangular region is visible.

inductance. At very high frequencies (where the skin depth is much smaller than the conductor thickness) or with perfect conductors, the fields do not penetrate the conductor but reside on the surface of the conductor. The transition region between the dc and high frequency limit can be important to capture in order to accurately simulate the frequency dependence of resistance and inductance. To fully capture these effects, the interior of the objects needs to be volumetrically meshed. This increases the solve time dramatically, compared to solving for fields on the surface only. Many field solvers use surface impedance concepts to capture current redistribution effects without meshing the interior of an object. On perfect conductors, the electric field is normal to the surface, and the tangential component is zero. Using a finite conductivity boundary condition, the tangential component of the electric field can be calculated, simulating the currents flowing inside the object. As an example, Figure 2.16 plots the self-impedance of the structure shown in Figure 2.1, which is simulated by solving inside and using the surface impedance approximation. Both methods do a good job of capturing the measured impedance profile, primarily

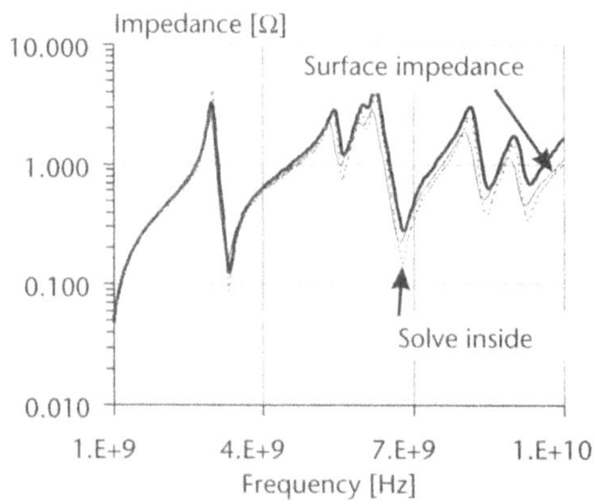

Figure 2.16 Simulated self-impedance of a plane pair with a 63-μm- (2.4-mil-) thick dielectric when solving inside the conductor and not solving inside. The dark trace is measured results.

because the fields are concentrated within the dielectric, between the conductive planes.

However, as the dielectric layer is made progressively thinner, the fields between the planes are forced further into the copper planes. In this situation, solving inside the metal may not do a good job of simulating the field distribution inside the planes. The structure shown in Figure 2.1 had a dielectric thickness of about 63 μm (2.4 mils). Figure 2.17 plots the self-impedance for the same structure but with an 8-μm-thick dielectric. The losses are found to be grossly underestimated when solving inside the metal. The reason for this discrepancy is that in order to capture the field penetration into the planes when solving inside, a very fine vertical mesh is required. By using a sufficiently dense mesh when solving inside, these two approaches will converge. However, this is difficult to achieve in practice due to the large aspect ratio of the plane (i.e., plane width/length compared to the plane thickness). This effect is particularly important to consider when simulating thin-laminate, power/ground plane pairs.

2.4.3.3 Material Properties

Another source of simulation inaccuracy is not including frequency-dependent material properties. For example, for a large number of commonly used PCB and package materials, the relative dielectric constant, ε_r, drops and the dielectric loss tangent, tan_δ, increases with increasing frequency. (Note that frequently ε_r is referred to as Dk and that tan_δ is referred to as Df.) Often, for the sake of simplicity, simulations will assume that the dielectric material has a fixed dielectric constant and loss tangent. For example, Figure 2.18 plots the simulated self-impedance for the plane pair structure illustrated in Figure 2.1, using a fixed dielectric constant and loss tangent. The fixed values were extracted (see Section 4.5.5.2 for more on this procedure) from the measurement data shown in Figure 2.19 at 1 MHz. Using the low-frequency material properties, the parallel resonance, as well as the higher order modes, in the simulated data are shifted to a lower frequency as a result of the

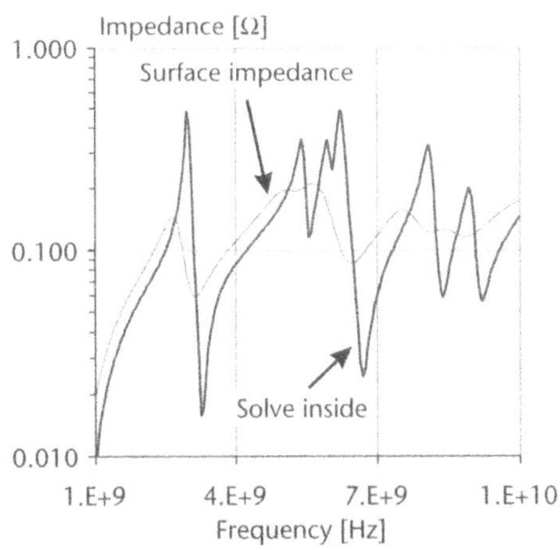

Figure 2.17 Simulated self-impedance of a plane pair with an 8-μm-thick dielectric when solving inside the conductor and not solving inside.

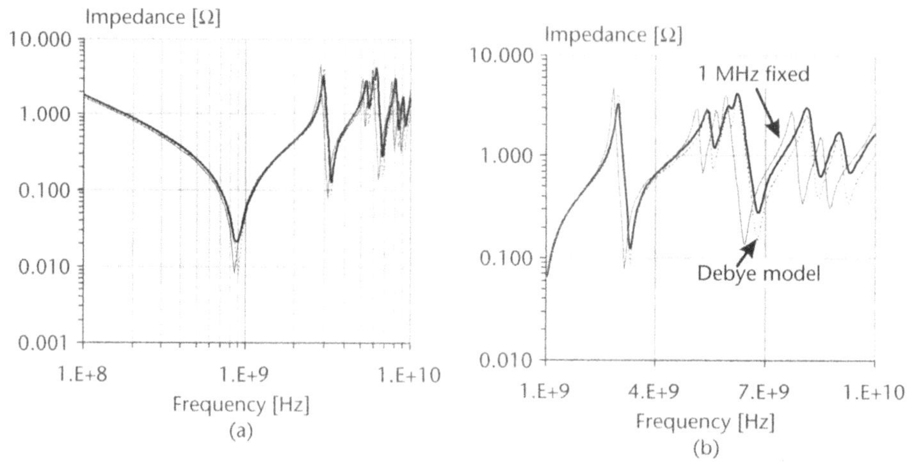

Figure 2.18 Simulated self-impedance of a plane pair using a fixed dielectric constant and loss tangent at 1 MHz. The common dark trace on both plots shows measured results. Part (b) is an enlarged view of (a) showing correlation with the higher-order modes.

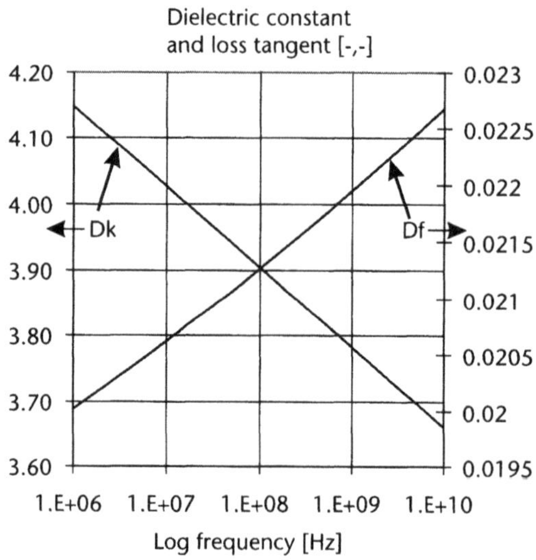

Figure 2.19 Extracted dielectric constant and loss tangent from the measured plane pair test structure.

higher dielectric constant. This is consistent with our expectations, since the second modal resonance and the dielectric constant are inversely proportional; namely

$$f_{res} = \frac{2}{2w_x\sqrt{\varepsilon_0\varepsilon_r\mu_0}} \tag{2.5}$$

where w_x is the long dimension of the plane. The simulated loss is also observed to be slightly underestimated using the low-frequency material properties. Figure 2.20 plots the simulated self-impedance for the plane pair structure shown in Figure 2.1,

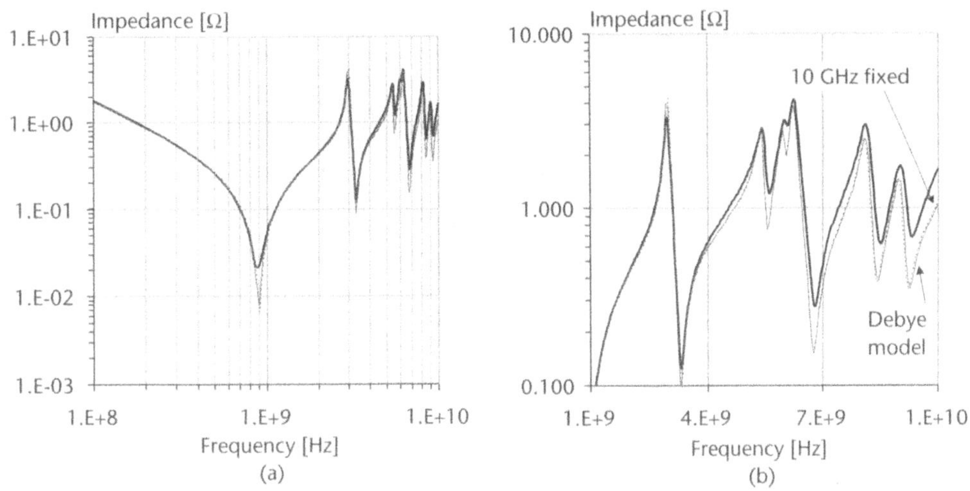

Figure 2.20 Simulated self-impedance of a plane pair, using a fixed dielectric constant and loss tangent at 10 GHz. The common dark trace on both plots shows measured results. Part (b) is an enlarged view of (a) showing correlation with the higher-order modes.

using a fixed dielectric constant and loss tangent at 10 GHz. Aside from a discrepancy near 10 GHz, the Debye model and 10-GHz fixed value are in excellent agreement, and both simulations show good correlation with the measurement data. The modal resonances are better correlated using the 10-GHz fixed value, because the resonances occur at higher frequencies.

2.4.3.4 Geometrical Inaccuracies

Manufacturing tolerances of PCBs and packages can have a significant impact on the accuracy of simulation results. Accurate simulations of vias, antipads, planes, and decoupling capacitors often require further details about their physical construction beyond the details specified or obtained from the vendor. Sample cross-sectioning can go a long way toward reducing this uncertainty. For example, antipad size is one example of a parameter that can have a significant impact on the simulated plane impedance (see Section 3.4), but often it is not tightly controlled. Copper surface roughness is another example of a physical parameter that can have a significant impact on the simulation results and is often unknown. Figure 2.21 plots the self-impedance of the plane pair structure shown in Figure 2.1 with and without surface roughness. The surface roughness obtained from cross-sectioning the sample (see Figure 2.22) was found to be about 1.69 μm and 1.85 μm rms for the upper and lower planes, respectively. The nominal plane thickness was 33.0 μm (1.3 mils). The additional loss mechanism is observed to dampen the modal resonances; this provides a fit to the measurement data that is improved in some areas but worsened in others (e.g., the first inductive upslope after the series resonance has worse agreement when including surface roughness). The roughness is included in the field solver by decreasing the conductivity of the plane layers by a function that depends on the surface roughness and skin depth.

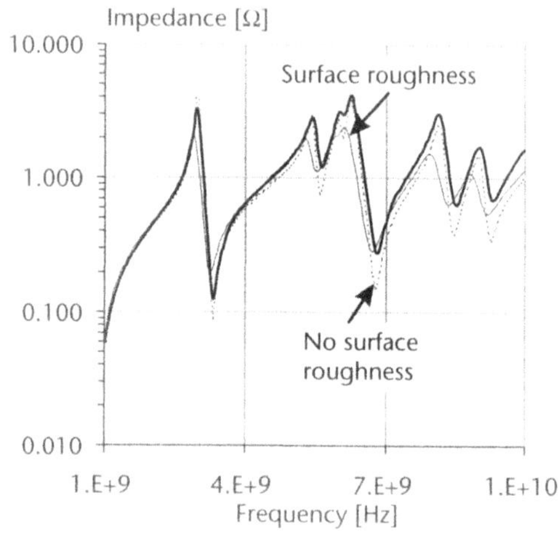

Figure 2.21 Simulated self-impedance of a plane pair showing the impact of including surface roughness on the correlation and loss.

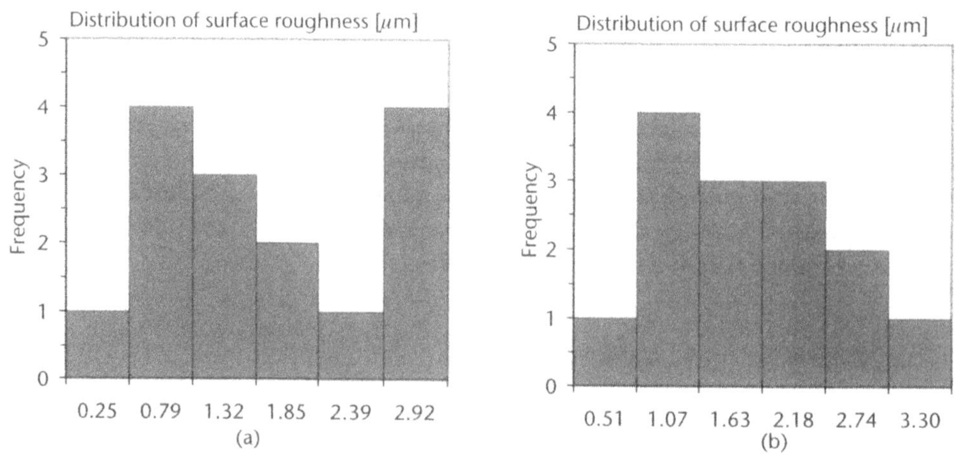

Figure 2.22 Measured surface roughness for the upper plane (a) and lower plane (b). The field solver used the rms value of 1.69 μm and 1.85 μm for the upper and lower planes, respectively.

2.4.3.5 Port Definitions, Impedance Considerations, and Port Sizing

Ports are used to excite the structure and measure the results. For 3D FEM solvers so-called *wave ports* or *lumped ports* are frequently utilized. The decision to use one over the other depends primarily on whether or not the port is located internal or external to the problem. In many PDN applications considered in this book, a lumped port is more appropriate, because the measurement point and is not conveniently located on the face or boundary of the structure. For example, it is easier to use a lumped port for the plane pair structure in Figure 2.1, because the measurement is located at the vias. However, lumped ports provide no information about the impedance or phase velocity of the structure to which the port is connected. Therefore, it is not possible to perform de-embedding using lumped ports. De-embedding can be a useful technique when we need to relocate the reference point closer to the device under test (DUT). For example, measurement test structures for passive devices, such as decoupling capacitors, can use lead-in and lead-out traces to connect to the DUT. De-embedding can then be used to move the reference point to the DUT, removing the influence of the lead-in and lead-out traces. Wave ports provide the flexibility of de-embedding. Wave ports rely on a separate 2D solver on each port to determine what the field excitation should look like and the impedance and phase velocity of the port. Once this is computed, a line of uniform length is extended sufficiently from the port to launch the desired mode. The information about the impedance and phase velocity are used to move the launch back to the wave port face and to move it beyond, into the structure, to provide de-embedding. Alternatively, wave ports can be used to extend the length of a uniform transmission line. For example, instead of solving for the *S*-parameters for a 25.4-cm (10-inch) microstrip trace, one can solve for a 2.54-mm (100-mil) microstrip trace and extend the length in either direction as a postprocessing step to yield a 25.4-cm (10-inch) transmission line. Figure 2.23 shows the S_{12} plot of a 2.54-mm (100-mil) microstrip that was extended to 1.27 cm (500 mils) using two approaches: (1) by de-embedding within the field solver using wave ports and (2) by concatenating five, 2.54-mm (100-mil) sections of transmission line using MATLAB to multiply the ABCD matrices of each section. Fairly good agreement is obtained up to 10 GHz using these equivalent approaches. The discrepancy may have to do with numerical noise or error associated with the calculation of the complex propagation constant at the port. Using either of these techniques can greatly decrease the solve time.

Wave ports calculate the impedance of the port using a few different techniques. For example, using Z_{vi}, we find the impedance by calculating the voltage and current at the wave ports. The voltage can be calculated using the following expression of Faraday's Law:

$$\nabla \times E = \int E \cdot d\ell = \int_1^2 E \cdot d\ell + \int_2^1 E \cdot d\ell = V_{12} + V_{21} = -j\omega\mu H = 0 \qquad (2.6)$$

Then the current can be calculated as follows:

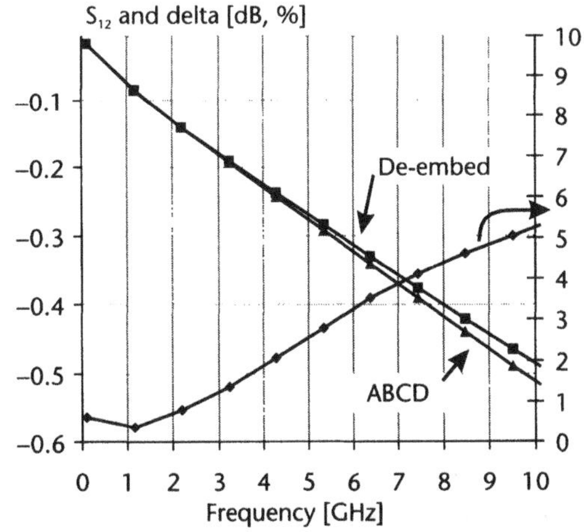

Figure 2.23 Plot of a simulated 2.54-mm (100-mil) microstrip that was extended to 1.27 cm (500 mil) using two approaches: de-embedding and by multiplying the ABCD parameters.

$$\nabla \times H = J + j\omega\varepsilon E = J \tag{2.7}$$

In case of a microstrip, the line integral of the electric field is usually defined from the ground plane to the signal trace and back. The implicit assumption in (2.6) is that the structure is TEM such that the flux passing through the surface enclosed by the line integral is zero. This is not the case with a microstrip. The same argument applies to the calculation of the current for a microstrip; that is, the electric field through the line contour is not zero. For a pure TEM line, such as a stripline, we can calculate the impedance directly from the product of V and I using (2.6) and (2.7). For a microstrip or non-TEM structure, using definitions of the impedance that incorporate the average power (e.g., Z_{pv} and Z_{pi}), we can obtain a more reliable definition of the impedance.

The outer edge of a wave port is defined to have a perfect E boundary. Therefore, the port is defined within a waveguide. This permits the calculation of the natural field patterns or modes. If the structure itself is enclosed within a metal boundary (e.g., coaxial cables or waveguides), then this automatic assignment is not a problem, as the fields will be contained within those boundaries. However, for open structures, such as microstrips, the surrounding fields can be disturbed by improper sizing of the port. Figure 2.24 shows the E-field distribution for a 100-μm-wide (4-mil-wide) microstrip over a 100-μm-thick (4-mil-thick) dielectric, using two different sizes of wave ports. Improper sizing of the wave port creates artificial coupling to the side walls, which will decrease the port impedance.

Figure 2.25 plots the simulated port impedance for the microstrip shown in Figure 2.24 and a stripline using the three different definitions for port impedance. The microstrip plot shows that for scale factors (defined as the width/height of the port divided by the trace width) greater than six, the delta in the impedance is less than 5%. (Delta is defined as the percentage change in the impedance with each progressive step increase in the scale factor.) The stripline plot converges more rapidly as a function of port size: for ports with a scale factor greater than four, the delta in the impedance is less than 2%. Both plots show that the port must be adequately sized in order to obtain the asymptotic impedance. Note also the vertical scale difference between Figure 2.25(a) and Figure 2.25(b); microstrip impedance is much more sensitive to improper port sizing.

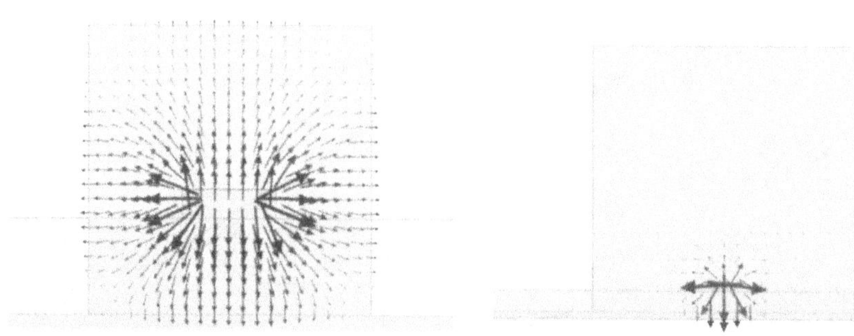

(a) (b)

Figure 2.24 Electric-field distribution for a 100-μm-wide (4-mil-thick) microstrip over a 100-μm-thick (4-mil-thick) dielectric. (a) The port width and height are 3 times the trace width. (b) The port width and height are 10 times the trace width.

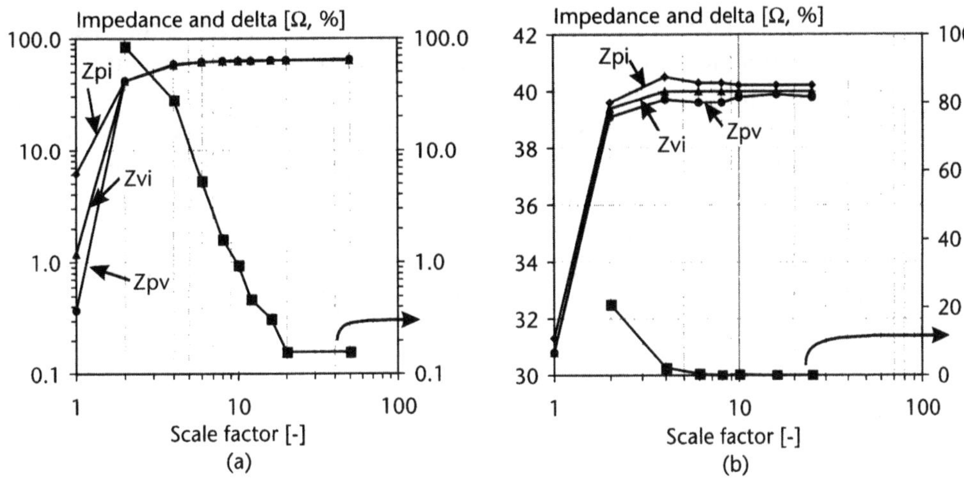

Figure 2.25 Simulated port impedance for a microstrip (a) and a stripline (b) as a function of the scale factor. The stripline has the same dimensions as the microstrip; the top plane is symmetrically placed 100 μm (4 mils) from the trace.

Lumped ports are generally easier to size. The default boundary is perfect H on all edges that do not come in contact with the metal or with another boundary condition. Edges in contact with a conductor are perfect E. The lumped port impedance is normalized to a constant user defined impedance. As discussed previously, wave ports compute the field distribution at the port plane, then it numerically adds some line length to the port and applies the computed excitation field. By the time the excitation arrives at the original port plane, the fields are in the best possible configuration. However, in a lumped port the fields are applied brute force at the port plane. Consequently, it will take time and distance for the fields to form the correct mode. Thus, there can be different field components in the vicinity of the lumped port that are not part of the desired mode. Figure 2.26 shows the vectors of the surface current magnitude of a microstrip simulated using wave ports and lumped

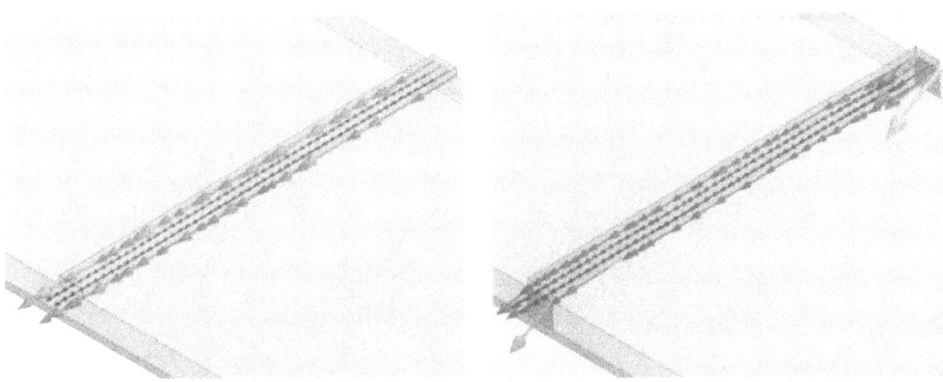

(a) (b)

Figure 2.26 Vector plot of the surface current magnitude for a microstrip using (a) wave ports and (b) lumped ports. The tails of the vectors are extended to better illustrate the trajectory. The size and shading of the arrow is proportional to the vector magnitude.

ports. The vectors are plotted along the microstrip trace length. The formation of *evanescent modes*, or modes that are formed around a discontinuity (i.e., the ports) but do not propagate down the transmission line, are observed at the lumped ports but not on the wave ports. Any current flowing perpendicular to the direction of propagation are not part of the normal quasi-TEM mode. Evanescent modes can introduce additional coupling in the simulated structure. For example, the evanescent modes associated with the lumped ports located at either end of the test vias, used to simulate the structure in Figure 2.1, can couple to the nearby power and ground planes. This additional coupling is nonphysical, since a measurement probe launch would likely not couple to the planes.

2.4.3.6 Interpolation Sweeps

Discrete simulations perform a full solution at every frequency using the optimized mesh. The solve time is the single frequency solve time times the number of frequency points. However, if the problem is a broadband problem and the solution parameters vary rapidly over the frequency range, then many frequency points must be calculated. One option for dealing with this is to use interpolating sweeps, which aim to calculate a solution at a minimum number of discrete points. Then a polynomial expression is used to fit the solution points with a maximum specified error tolerance. The authors have encountered scenarios, discussed below, where the broadband interpolating sweep can introduce inaccuracy by the presence of interpolating spikes or by omitting resonances altogether. Figure 2.27 plots the interpolated self-impedance of the plane pair, shown in Figure 2.1, against measurement. The interpolating sweep, performed in HFSS release 10, is found to miss the resonances entirely above the series resonance.

At particular frequency points a separate discrete sweep can be performed to check the accuracy of the interpolating sweep. If the broadband nature of the sweep is causing difficulties for the interpolation algorithm, the single interpolating sweep can be subdivided into smaller sweeps. Furthermore, by partitioning the interpolating sweep in this way, the user may verify that the endpoints of each sweep meet and that the overall trend of the combined sweep appears reasonable. In Figure 2.28, the wideband interpolating sweep in Figure 2.27 was subdivided into three narrower

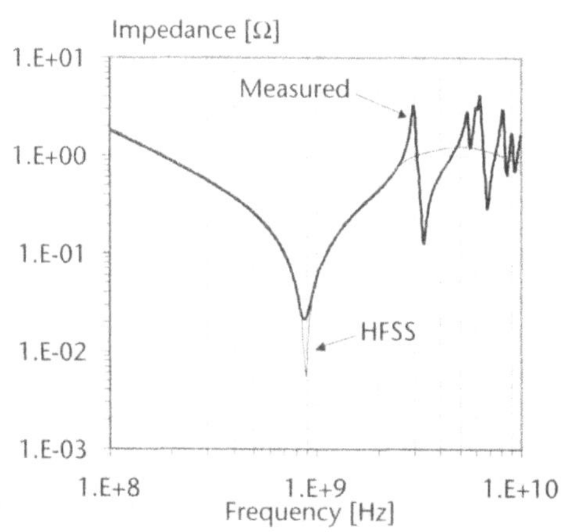

Figure 2.27 Plane pair simulated and measured using a single interpolating sweep.

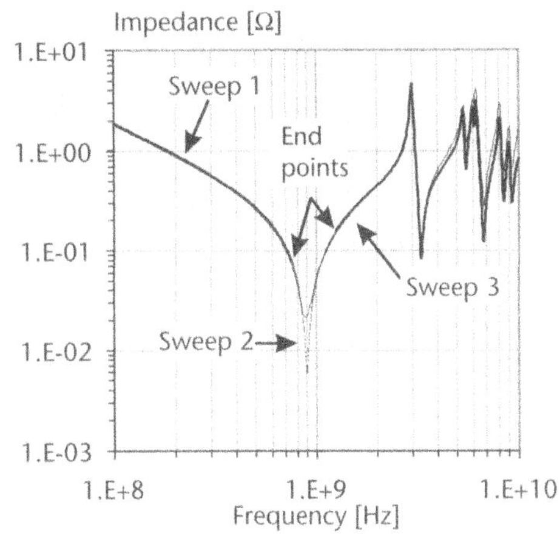

Figure 2.28 Plane pair simulated with subdivided interpolating sweeps. The three interpolating sweeps are overlaid on the measurement data.

sweeps, namely, 50–750 MHz, 750–1,250 MHz, and 1,250 MHz–10 GHz. In this example, by partitioning the problem in three partitions, not only is the problem resolved with the interpolating sweep algorithm, but we get confirmation that the results are reasonable, since the endpoints of the individual sweeps align.

2.4.3.7 Simulating Open Boundaries

When the FEM formulation is used to analyze a problem that radiates energy, fields that are unbounded must be truncated to limit the computational domain. Boundary conditions can be used to simulate an environment where the fields continue to propagate to infinity with no reflections. The absorbing boundary condition (ABC) or the perfect matched layer (PML) method can be used to simulate an open boundary. However, the ABC cannot yield zero reflection for all angles of incidence; therefore it cannot be placed too close to the object. In order to reduce the reflection caused by the ABC, the box must be offset from the structure, increasing the solve time. PML can absorb all the outgoing fields without any reflection, so PML can be placed very close to the structure. However, PML relies on lossy medium layers to absorb the outgoing fields that must be discretized, increasing the solve time. ABCs should be located at least a quarter wavelength of the highest frequency of the solution from the radiating source. There are some cases when little radiated energy is expected, and the boundary may be located closer than one-quarter wavelength. For example, consider the plane pair structure shown in Figure 2.1. The plane pair separation is quite small compared to the problem size, so we are able to locate the ABC fairly close to the plane edge while still accurately capturing the plane self-impedance. Figure 2.29 plots the self-impedance of the plane pair for different radiation boundary offsets. The maximum frequency of interest here is 10 GHz, which translates to a free space quarter wavelength of 7.6 mm (300 mil). Figure 2.29 shows that only when the ABC is placed against the plane edge do we see a significant change in the impedance profile. The plane resonances are generated by standing wave reflections due to plane-edge boundaries. By placing the radiation boundary against the plane edge, we absorb these standing wave reflections and

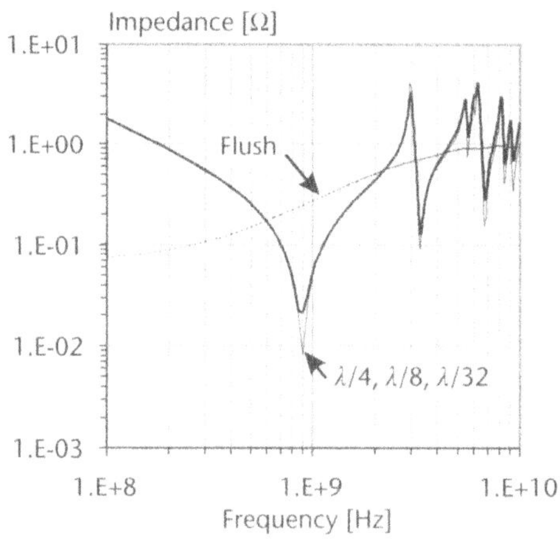

Figure 2.29 Self-impedance of a power-ground plane pair. Open-plane edge boundaries are simulated using an ABC offset by different distances.

suppress the modal resonances. Since these boundary conditions produce near-zero reflection, they can also be used to simulate the presence of perfect plane-edge termination [4]. As a footnote, virtually identical results were obtained using the PML in this case, as long as the boundary was not flush with the plane edge.

2.4.3.8 Basis Functions

Common to all numerical methods in field solvers is the use of *basis functions*. The unknown solution (fields) is expanded in terms of known expansion functions with unknown coefficients. Then the coefficients are determined such that the sum of the basis functions approximates the unknown solution. A perfect representation of a function would require a complete set of basis functions. For most problems we must truncate the number of basis functions, because the number of unknowns can quickly overwhelm the computer memory resources. For example, in HFSS, the solver interpolates field quantities from both nodal values at vertices and on edges using 20 unknowns per tetrahedron. A low-order mode is available that instead interpolates the field quantities using the nodal values at vertices only, reducing the number of unknowns to six per tetrahedron. Using only six unknowns per tetrahedron solves much more rapidly, but can also introduce inaccuracy.

As an example, Figure 2.30 shows a test structure for measuring the impedance of a 0306 mounted decoupling capacitor along with the equivalent 3D model. The structure consists of two pairs of through vias connected to plane layers 2 and 3. A pair of test points (TP39P and TP39N) connects to the same plane layers. A shorting bar is in place of the actual decoupling cap to measure the loop inductance. S_{21} of the structure is measured using picoprobes (see Figure 2.31) connected to an Agilent 4396A VNA with a frequency range spanning from 100 kHz to 1.8 GHz. The structure is simulated using different order basis functions to examine the impact on correlation. Figure 2.32 plots the measured and simulated loop inductance and impedance. The simulated loop inductance using a zero order basis function has a 22% discrepancy with measurement at 100 MHz, whereas the first order basis function has only a 7% discrepancy at the same frequency. The magnitude plot, shown in

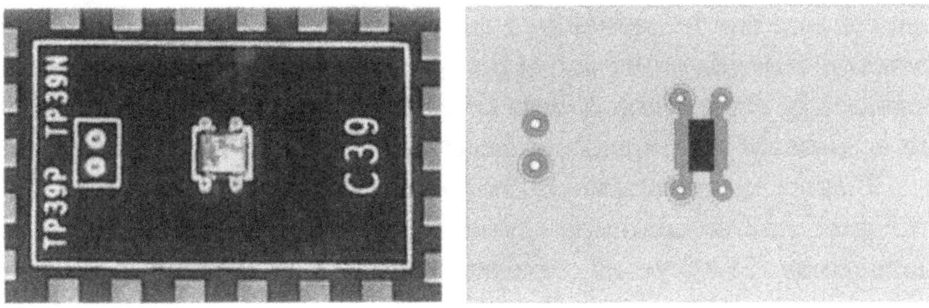

(a) (b)

Figure 2.30 (a) Decoupling capacitor test structure and (b) its equivalent 3D model. The size of the structure is roughly 9 mm × 15 mm.

Figure 2.31 1,000-μm picoprobes connected to the test structure from opposite sides.

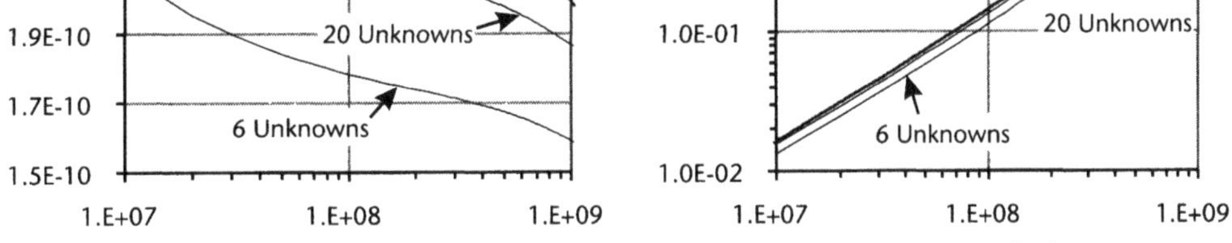

(a) (b)

Figure 2.32 Measured and simulated (a) loop inductance and (b) impedance magnitude using different basis functions.

Figure 2.32(b), does not reveal as great a difference, because the data is riding on a linear (i.e., ω) slope. The ω dependency is removed for the inductance plot, so the difference is larger (visually). The computation penalty for the enhanced accuracy is a solve time that increased by a factor of 10. The impact of using a zero order basis function depends on the size of the structure and the number of tetrahedrons. With a sufficiently dense mesh, a zero order basis function has very little impact on accuracy, since the additional tetrahedrons are used to describe the field variation.

Further improvements in the correlation between measurement and simulation in Figure 2.32 would require cross-sectioning the test structure to obtain the actual dimensions. This level of correlation was not undertaken in this particular study.

2.4.3.9 Probe and Port Location

The exact x-y coordinates of the measurement probe position and simulated port position can be significant in obtaining good correlation between measurement and simulation. For example, PCBs can and do exhibit position-dependent impedance profiles. Figure 2.33 shows the eigenmode solution for the structure shown in Figure 2.1. Whereas the field solvers discussed thus far rely on an external source of energy to excite the problem, eigenmode solvers rely on the stationary field configuration to determine the resonant frequencies of the geometry. A plane pair can be considered a waveguide-type structure with eigenmodes. Figure 2.33(a) plots the eigenmodes alongside the measurement data; the measurement probe location is in the center of the plane. The correlation between the measurement data and eigenmodes is poor because this particular measurement location suppresses the first modal resonance. Figure 2.33(b) plots the eigenmodes alongside the results from Ansoft HFSS, with no conductive or dielectric losses. The absences of losses in the simulation make the resonances sharper and easier to correlate with the eigenmode results. The port is located in the corner of the plane where there is little mode cancellation. Excellent correlation is observed between the HFSS and the eigenmode solver.

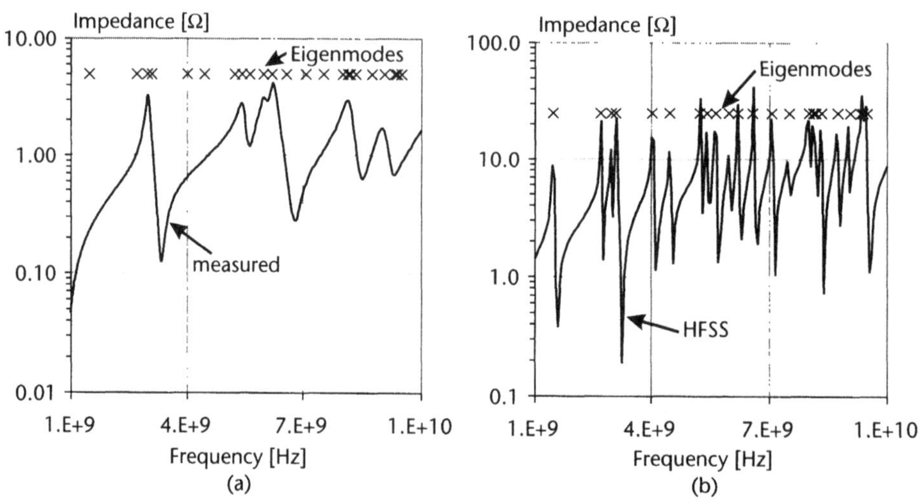

Figure 2.33 (a) Plot of the eigenmodes against measurement data for the structure shown in Figure 2.1. The measurement probes are centered in the plane. (b) Plot of the eigenmodes against Ansoft HFSS. The ports are located at the corner of the plane.

References

[1] Xu, M., H. Wang, and T. Hubing, "Application of the Cavity Model to Lossy Power-Return Plane Structures in Printed Circuit Boards," *IEEE Trans. on Advanced Packaging*, Vol. 23, No. 1, February 2003.

[2] Swanson, D. G., and W. J. R. Hoefer, *Microwave Circuit Modeling Using Electromagnetic Field Simulation*, Norwood, MA: Artech House, 2003.

[3] Miller, J. R., and I. Novak, "Modeling the Impact of Power/Ground Via Arrays on Power Deliver," *Proceedings of the 14th Topical Meeting on Electrical Performance of Electronic Packaging*, October 2004.

[4] Novak, I., "Reducing Simultaneous Switching Noise and EMI on Ground/Power Planes by Dissipative Edge Termination," *IEEE Trans. on CPMT, Components, Packaging, and Manufacturing Technology,* Vol. 22, No. 3, August 1999, pp. 274–283.

CHAPTER 3

Characterization and Modeling of Vias

3.1 Introduction

Vias form part of the path in PDN carrying the current from source to load. They stitch common planes together vertically and can also provide a current return path for signals. A via consists of a barrel, pad, and antipad. The barrel is a conductive cylinder that allows electrical connection between PCB or package layers. The pad is used to connect the via barrel to a signal or supply trace. Nonfunctional pads can also exist on layers where no trace connection is made. Antipads are clearance holes between the pad and the surrounding plane. The most common type of via, called a through via, extends from the top of the stackup to the bottom layer. For PCBs, through vias are formed by drilling a hole through the PCB and plating it with copper. This plating forms the wall of the via barrel, which are typically hollow. A via constructed in this manner is illustrated in Figure 3.1 and called a plated-through-hole (PTH) via.

3.2 Via Partial Inductance

The inductance, not the capacitance, of vias is usually the primary concern in PDN applications because it increases the impedance between the source and load. Consequently, it is very useful to have expressions to calculate the inductance of a via. In this section, a complete derivation of via inductance from energy relations is presented, outlining all the key steps and assumptions [1]. Based on this derivation, the concept of partial inductance is reviewed and several useful expressions for partial inductance are presented. The accuracy of these expressions is then evaluated by comparing these formulas to the field solution.

From Poynting's theorem the energy stored in a magnetic field can be written as

$$W_M = \frac{1}{2}\iiint \vec{B}\cdot\vec{H}dV \tag{3.1}$$

From circuit theory the energy stored in an inductor is given by

$$W_M = \frac{1}{2}Li^2 \tag{3.2}$$

Equating these two definitions and solving for the inductance gives

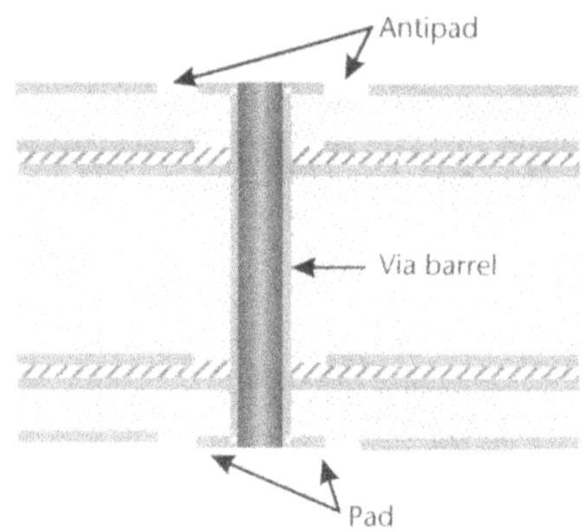

Figure 3.1 Illustration of a PTH via cross-section connecting to two plane layers.

$$L = \frac{1}{i^2} \iiint \vec{B} \cdot \vec{H} dV \tag{3.3}$$

The vector magnetic potential A can be defined such that

$$\vec{B} = \nabla \times \vec{A} \tag{3.4}$$

Substituting (3.4) into (3.3) gives

$$L = \frac{1}{i^2} \iiint \left(\nabla \times \vec{A}\right) \cdot \vec{H} dV \tag{3.5}$$

A standard vector identity is

$$\left(\nabla \times \vec{A}\right) \cdot \vec{H} = \nabla \cdot \left(\vec{A} \times \vec{H}\right) + \vec{A} \cdot \left(\nabla \times \vec{H}\right) \tag{3.6}$$

Inserting the above expression into (3.5) gives

$$L = \frac{1}{i^2} \iiint \nabla \cdot \left(\vec{A} \times \vec{H}\right) dV + \frac{1}{i^2} \iiint \vec{A} \cdot \left(\nabla \times \vec{H}\right) dV \tag{3.7}$$

Applying the divergence theorem and substituting Ampere's law in differential form (under quasi-static conditions) gives

$$L = \frac{1}{i^2} \oint_S \left(\vec{A} \times \vec{H}\right) dS + \frac{1}{i^2} \iiint \vec{A} \cdot \vec{J} dV \tag{3.8}$$

Since the volume integration is over all space, the surface enclosing this volume is at infinity. Applying the boundary condition that the magnetic field intensity **H** must be zero at infinity, the left hand term vanishes. Note that if the current itself were assumed to return at infinity, the magnetic field could not be zero at infinity. The volume current density **J** is zero outside of the conductor, therefore, the bounds of integration for the right hand integral are over the volume bounded by the conductor. The vector magnetic potential **A** for a volume current is defined as

$$\vec{A} = \frac{\mu}{4\pi} \iiint \frac{\vec{J}}{\vec{R}_{ij}} dV \tag{3.9}$$

where $\mathbf{R}_{ij}$ is the distance vector from the line element $d\mathbf{l}$ to the field point. Substituting (3.9) into (3.8) yields

$$L_{ij} = \frac{\mu}{4\pi i^2} \int_{V_i} \int_{V_j} \left(\frac{\vec{J}_i}{\vec{R}_{ij}} \cdot \vec{J}_j \right) dV_i dV_j \tag{3.10}$$

This is the most general formula for calculating the self and mutual inductance for a given problem definition. Using (3.10), the inductance can be calculated for two loops, i and j (see Figure 3.2). If the loops are assumed to have a uniform current density $\mathbf{J}_i$ and $\mathbf{J}_j$, and constant cross-sectional area, a_i and a_j, the following substitutions can be made: $dV = dl_i dS_i$ and $J_i = i_i/a_i$, and (3.10) can then be rewritten as

$$L_{ij} = \frac{1}{a_i a_j} \frac{\mu}{4\pi} \oint_i \int_{a_i} \oint_j \int_{a_j} \frac{\vec{dl}_i \cdot \vec{dl}_j}{\vec{R}_{ij}} da_i da_j \tag{3.11}$$

For loops that are thin filaments, the current density disappears for places off the line contour and Neumann's formula is obtained

$$L_{ij} = \frac{\mu}{4\pi} \oint_i \oint_j \frac{\vec{dl}_i \cdot \vec{dl}_j}{\vec{R}_{ij}} \tag{3.12}$$

By rewriting the integrations over the lengths as summations over segments of the loop the *partial inductance* concept is introduced [2]

$$L_{ij} = \sum_{k=1}^{K} \sum_{m=1}^{M} \frac{\mu}{4\pi} \int_{b_k}^{c_k} \int_{b_m}^{c_m} \frac{\vec{dl}_k \cdot \vec{dl}_m}{\vec{R}_{km}} \tag{3.13}$$

where loop i is divided into K segments and loop j is divided into M segments. The line integral is then calculated for each segment over the geometry of each segment

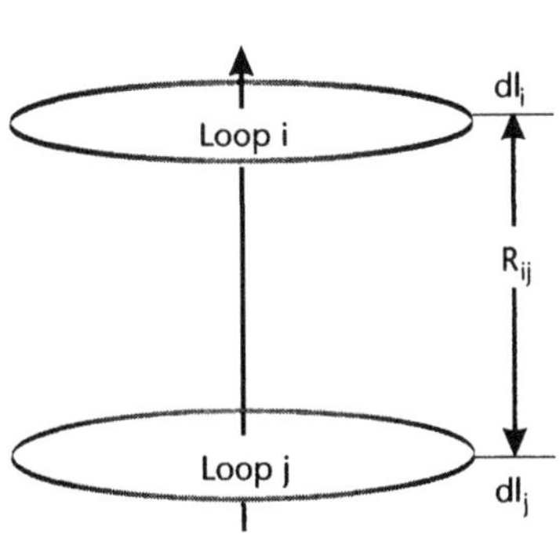

Figure 3.2 Two coupled loops, *i* and *j*.

in free space. This integration will be well defined for a particular problem definition and produces a *unique* partial inductance value for a given segment geometry.

Closed-form expressions for the partial inductance can be obtained from Neumann's formula. Rosa derived an expression for the partial mutual inductance between two thin filaments of length h and separation d from Neumann's formula [3]. The geometrical relation of the filaments is shown in Figure 3.3.

Inserting the problem definition into (3.13), the integral to be solved is

$$L_{ij} = \frac{\mu}{4\pi}\int_0^h\int_0^h \frac{\vec{dy}\cdot\vec{dy}'}{\sqrt{(y-b)^2+d^2}} \tag{3.14}$$

Solving the double integration in (3.14) produces

$$L_{ij} = 5.08h\left[\ln\left(\left(\frac{h}{d}\right)+\sqrt{1+\left(\frac{h}{d}\right)^2}\right)-\sqrt{\left(1+\left(\frac{d}{h}\right)^2\right)}+\left(\frac{d}{h}\right)\right] \tag{3.15}$$

where the units of (3.15) are in pH and the dimensions are in mils (1 mil is equal to 0.0254 mm or 25.4μ). The partial *self*-inductance can be calculated using (3.15) with the separation of the filaments equal to the conductor radius. If the filament height is much greater than the filament separation, (3.15) can be shown to reduce to the following expression for partial inductance

$$L_{ij} = 5.08h\left[\ln\left(\left(\frac{2h}{d}\right)-1\right)\right] \tag{3.16}$$

Note that both (3.15) and (3.16) provide the external inductance of the via. At high frequencies this is a good approximation due to skin effects. However, for solid vias at low frequencies the internal inductance needs to be added to the external inductance.

Figure 3.4 plots the partial-self inductance calculated using (3.15) and (3.16) for a via that is 0.254 mm (10 mil) in diameter as a function of via height. These results are compared to 3D quasi-static field solution using Ansoft Q3D Extractor 7. Figure 3.4(b) shows that (3.16) produces nonphysical results (i.e., negative inductance val-

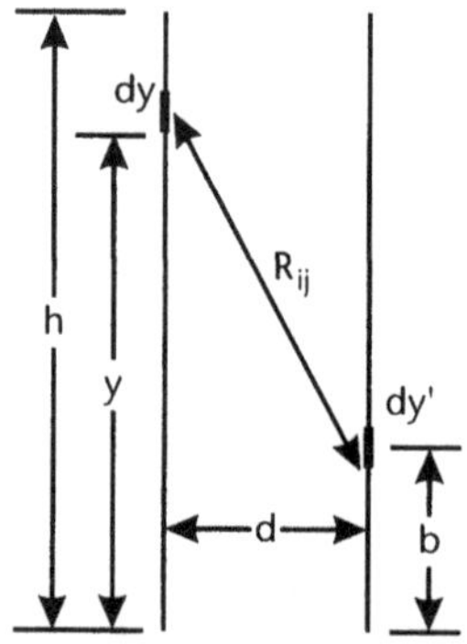

Figure 3.3 Two filamentary conductors.

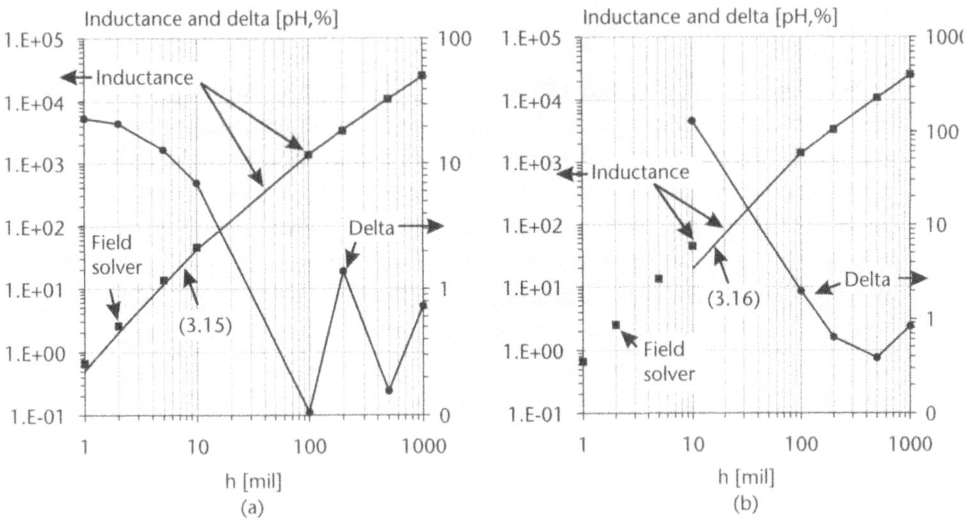

Figure 3.4 Via partial self-inductance as a function of via height for a via diameter of 0.254 mm (10 mils): (a) compares field solution to (3.15) and (b) compares field solution to (3.16).

ues) when the aspect ratio of the conductor approaches unity. Figure 3.4(a) shows excellent agreement between (3.15) and the field solution over a wide range of conductor aspect ratios. The largest discrepancy is observed for short conductors when the port definitions and current redistribution effects in the field solver become increasingly important in order to obtain accurate partial self-inductance values.

In this section, equations for calculating the partial inductance of a stand-alone via have been developed. In real-world PCB designs there are often large numbers of vias to account for; the challenge of modeling a complex via network is discussed in the following section.

3.3 Via Loop Inductance

Vias do not exist in isolation. There are planes to which they connect and neighboring vias to which they are coupled. Current flowing through a via will often find its current return on a neighboring via. This current path defines the loop inductance of the via network. Figure 3.5 illustrates a simple power-ground via arrangement. If the separation, s, between the vias is small compared to the via length, then the loop inductance, L_{loop}, is

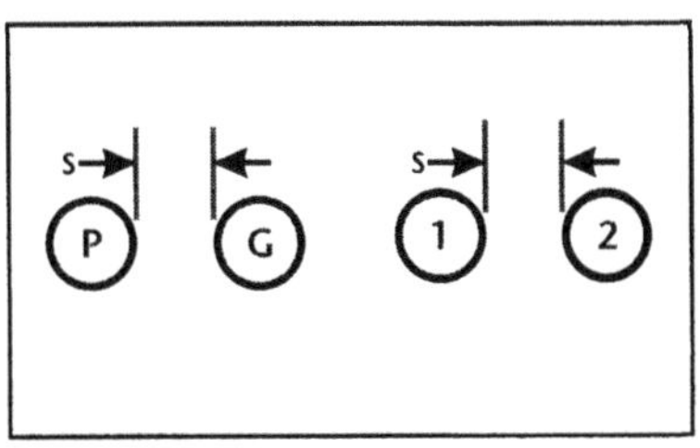

Figure 3.5 Power-ground via arrangement.

$$L_{loop} = L_{11} + L_{22} - M_{12} - M_{21} \tag{3.17}$$

where the L_{nn} is the partial self-inductance for each via and M_{nm} is the partial mutual inductance between the vias. In large, complex boards with thousands of vias, capturing the partial self- and partial mutual inductance terms for all vias is a massive computational undertaking. For example, if a via network has n vias (with equal numbers of power and ground vias), the loop inductance equation will have n^2 terms. However, the number of elements can be reduced to n terms by ignoring the mutual terms and only capturing the self-inductance terms. This is a reasonable approximation of the loop inductance, if the mutual-inductance terms are small compared to the self-inductance terms. Applying this simpliciation to the simple two-via model and assuming that the two vias are identical in construction, (3.17) simplifies to

$$L_{loop} = 2L_{11} = 2L_{22} \tag{3.18}$$

The inaccuracy introduced by this simplification can be estimated by rewriting (3.17). The error in the loop inductance introduced by ignoring the mutual inductance is described by the last term in the following expression:

$$L_{loop} = 2L_{11}(1 - M_{12}/L_{11}) \tag{3.19}$$

Thus for this simple two-via network, as long as the mutual inductance is small compared to the self-inductance, (3.19) can be approximated by (3.18).

Equation (3.15) can be used to calculate both the partial self- and partial mutual inductance terms between vias. As mentioned previously, the partial self-inductance can be calculated with the separation of the filaments, d, equal to the conductor radius. The same expression can be used to calculate partial mutual inductance by instead setting d equal to the center-to-center distance between the vias. Figure 3.6

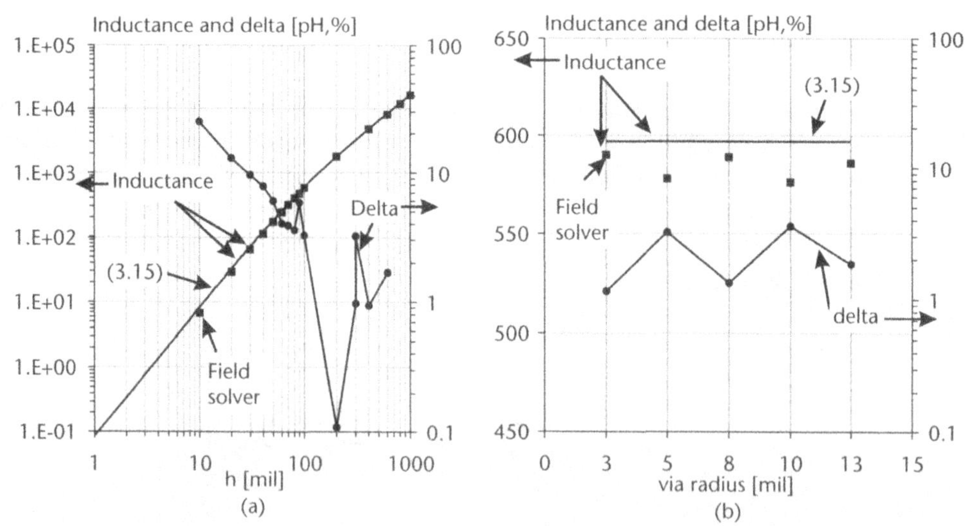

Figure 3.6 (a) Via partial mutual inductance as a function of via length, h. The vias are 0.254 mm (10 mils) in diameter. (b) Via partial-mutual inductance as a function of via radius. Here the via length is fixed at 2.54 mm (100 mils).

compares the partial mutual inductance obtained from (3.15) and using Q3D Extractor 7 for the via pair shown in Figure 3.5 with a center-to-center distance of 0.762 mm (30 mils). Good agreement is obtained using (3.15) over a wide range of conductor aspect ratios. In spite of (3.15) not being an exact expression for the mutual inductance of two parallel cylindrical vias, the expression is found to be quite accurate for vias having an appreciable cross section compared to the center-to-center distance. This remains true as long as the via length is great compared to the center-to-center distance.

With good agreement obtained between field solution and (3.15), for both partial self- and partial mutual inductance calculations, the error in the loop inductance as a result of ignoring the partial-mutual inductance can be found by calculating the percentage difference between (3.18) and (3.19), that is,

$$Error = \frac{1}{\left(\dfrac{L}{M} - 1\right)} \tag{3.20}$$

Figure 3.7 plots (3.20) using self- and mutal-inductance values obtained from a field solver and (3.15) over a range of via aspect ratios. This plot permits an assessment of the importance of mutual inductance for via loop inductance calculations. For example, from Figure 3.7, for a pair of vias 2.54 mm (100 mil) long and spaced 1.27 mm (50 mils) apart, the loop inductance would be overestimated by about 45%. Figure 3.7(b) plots the approximate and full equations for loop inductance alongside the percentage error. Equation (3.15) was used to obtain the self- and mutual-inductance values.

The analysis methodology for a simple two-via network can be generalized to analyze the more complex via networks found on typical PCB designs. To do this, we calculate the loop-inductance equation for a single current-source via surrounded by multiple current-return vias. Figure 3.8 shows the equivalent partial

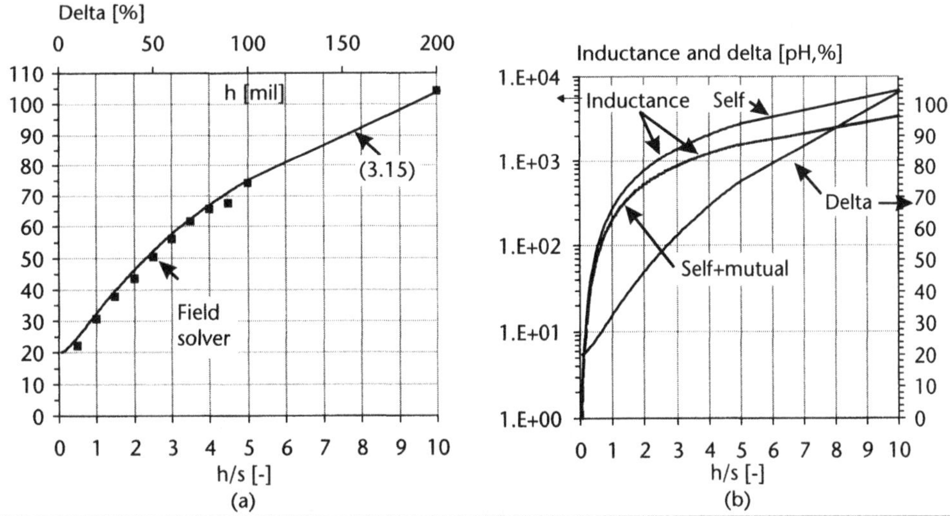

Figure 3.7 (a) Error introduced by ignoring the partial-mutual inductance over a range of via aspect ratios. (b) Approximate and full equations for loop inductance alongside the percentage error. The vias are identical with a via diameter of 0.254 mm (10 mils).

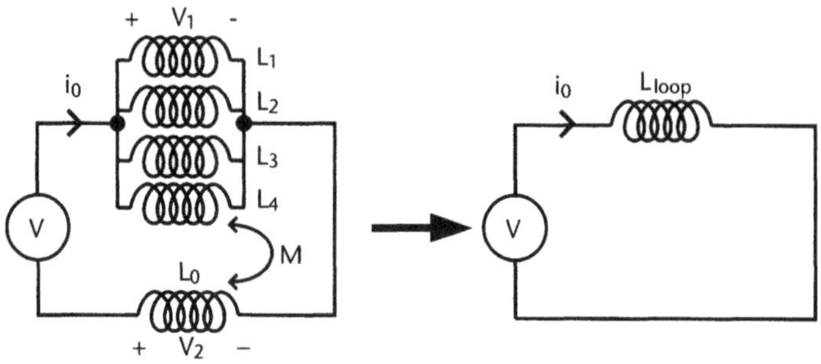

Figure 3.8 Equivalent partial inductance and inductance circuits for the arrangement shown in Figure 3.9.

inductance and inductance circuits for the via arrangement illustrated in Figure 3.9. The repeating pattern is simplified to the basic unit cell, which consists of a single power via surrounded by four current-return vias. This via arrangement is typical of what is found to connect power and ground to a packaged device, such as an ASIC or CPU. This arrangement is used because it minimizes the loop inductance between power and ground vias.

For the mesh with partial inductances, Kirchhoff's current law requires that the total current through the return vias equals the current through the source via

$$i_0 = i_1 + i_2 + i_3 + i_4 \tag{3.21}$$

Likewise, for the mesh with partial inductances, the voltage across the mesh, V_1, can be written as follows

$$\begin{aligned}
V_1 &= L_1 \frac{di_1}{dt} + M_{12}\frac{di_2}{dt} + M_{13}\frac{di_3}{dt} + M_{14}\frac{di_4}{dt} - M_{10}\frac{di_0}{dt} \\
V_1 &= M_{21} \frac{di_1}{dt} + L_2\frac{di_2}{dt} + M_{23}\frac{di_3}{dt} + M_{24}\frac{di_4}{dt} - M_{20}\frac{di_0}{dt} \\
V_1 &= M_{31} \frac{di_1}{dt} + M_{32}\frac{di_2}{dt} + L_3\frac{di_3}{dt} + M_{34}\frac{di_4}{dt} - M_{30}\frac{di_0}{dt} \\
V_1 &= M_{41} \frac{di_1}{dt} + M_{42}\frac{di_2}{dt} + M_{43}\frac{di_3}{dt} + L_4\frac{di_4}{dt} - M_{40}\frac{di_0}{dt}
\end{aligned} \tag{3.22}$$

Designating the mutual inductance between the neighboring return current vias as M_n, the mutual inductance between opposite current return vias as M_o, the mutual

Figure 3.9 A current source via surrounded by four current-return vias.

inductance between return current vias and source via as M_s, and the self-inductance of all of the vias as L then (3.22) simplifies to

$$\begin{aligned} V_1 &= L\frac{di_1}{dt} + M_n\frac{di_2}{dt} + M_o\frac{di_3}{dt} + M_n\frac{di_4}{dt} - M_s\frac{di_0}{dt} \\ V_1 &= M_n\frac{di_1}{dt} + L\frac{di_2}{dt} + M_n\frac{di_3}{dt} + M_o\frac{di_4}{dt} - M_s\frac{di_0}{dt} \\ V_1 &= M_o\frac{di_1}{dt} + M_n\frac{di_2}{dt} + L\frac{di_3}{dt} + M_n\frac{di_4}{dt} - M_s\frac{di_0}{dt} \\ V_1 &= M_n\frac{di_1}{dt} + M_o\frac{di_2}{dt} + M_n\frac{di_3}{dt} + L\frac{di_4}{dt} - M_s\frac{di_0}{dt} \end{aligned} \tag{3.23}$$

The voltage across the source current via is

$$V_2 = -M_s\frac{di_1}{dt} - M_s\frac{di_2}{dt} - M_s\frac{di_3}{dt} - M_s\frac{di_4}{dt} + L\frac{di_0}{dt} \tag{3.24}$$

For the mesh with the equivalent loop inductor

$$V = V_1 + V_2 = L_{loop}\frac{di_0}{dt} \tag{3.25}$$

Collecting terms and solving simulataneous voltage equations, the resultant formula is

$$L_{loop} = \frac{1}{4}(5L + 2M_n + M_o - 8M_s) \tag{3.26}$$

Figure 3.10 plots the four partial inductance terms in (3.26) for the via arrangement illustrated in Figure 3.9 across a range of via pitches using both Q3D Extractor 7 and (3.15). The error between the field solver and (3.15) is less than about 10% across the range of via pitches.

Going a step further and assuming that all of the mutual inductances are the same, (3.26) is reduced to

$$L_{loop} = \frac{5}{4}L\left(1 - \frac{M}{L}\right) \tag{3.27}$$

and if the partial mutual inductance is small compared to the self-inductance, then the loop inductance can be approximated as

$$L_{loop} = \frac{5}{4}L \tag{3.28}$$

Figure 3.11 plots (3.26), the loop inductance formula with no simplification about the coupling, against (3.27), the loop inductance formula assuming equal coupling between all of the vias. Specifically, all of the partial mutual term in (3.27) was assumed to be equal to M_s, the mutual inductance between return-current vias

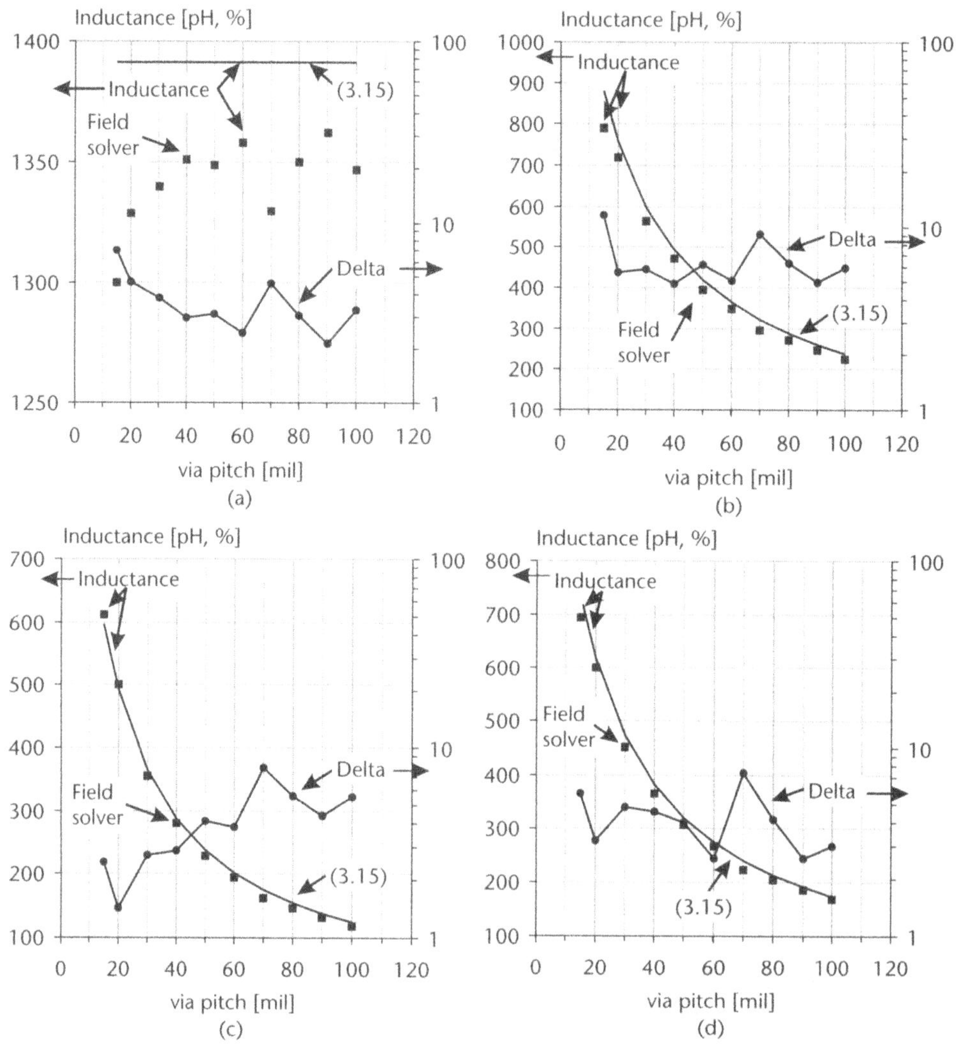

Figure 3.10 Calculated and simulated partial-self- and mutual-inductance terms for a current-source via surrounded by four current-return vias. (a) Plots the self-inductance of all of the vias, L. (b) Plots the mutual inductance between return current vias and source via, M_s. (c) Plots the mutual inductance between opposite current return vias, M_o. (d) Plots the the mutual inductance between the neighboring return current vias, M_n. The vias were 2.54 mm (100 mils) long with a diameter of 0.254 mm (10 mils).

and source via. The plot shows less than 15% error is incurred in the loop inductance by calculating the loop inductance using (3.27). The greatest error is found when the vias are very closely spaced and the partial-mutual inductance is a significant fraction of the partial-self inductance.

Figure 3.12 plots (3.26) against (3.28): the full and approximate equations for loop inductance, respectively, alongside the percentage error. This is the same type of plot as that shown in Figure 3.7 for a via pair. Notice that even though there are four return-current vias instead of one, the loop inductance is overestimated by

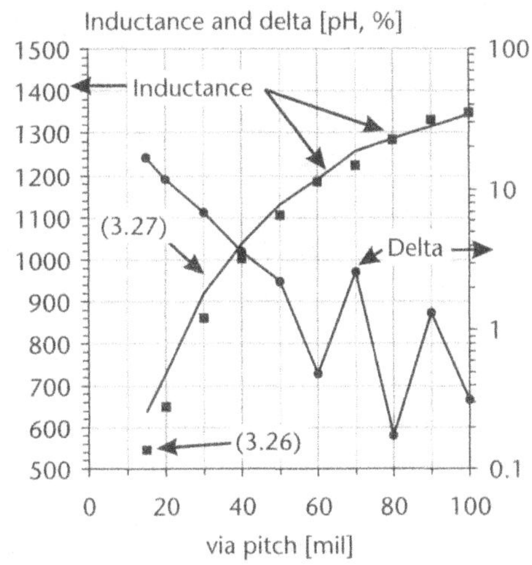

Figure 3.11 Loop inductance calculated using (3.26) and (3.27). The partial inductance terms were obtained from field solution. The vias are 2.54 mm long (100 mils) with a diameter of 0.254 mm (10 mils).

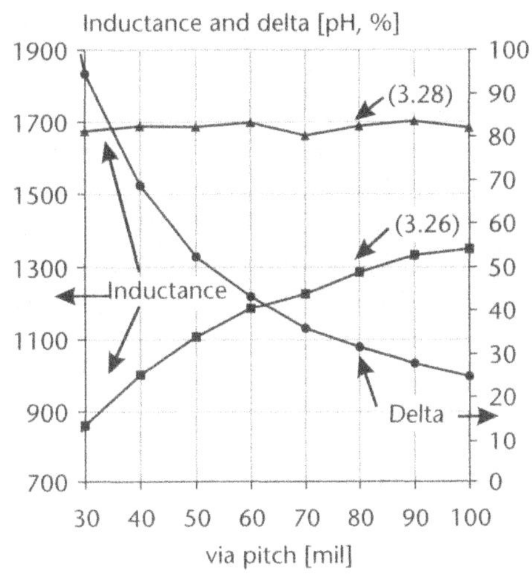

Figure 3.12 Approximate and full equations for loop inductance alongside the percentage error. The partial-inductance terms were obtained from field solution.

approximately the same percentage. This overestimate is due to the fact that the loop inductance in both cases can be approximated by the following expression:

$$L_{loop} = \frac{(n+1)}{n} L\left(1 - \frac{M}{L}\right) \tag{3.29}$$

where n is the number of return current vias. For example, for $n = 1$ we obtain (3.19) and for $n = 4$ we obtain (3.27). Consequently, the error in ignoring the partial mutual inductance for n number of vias is given in (3.20).

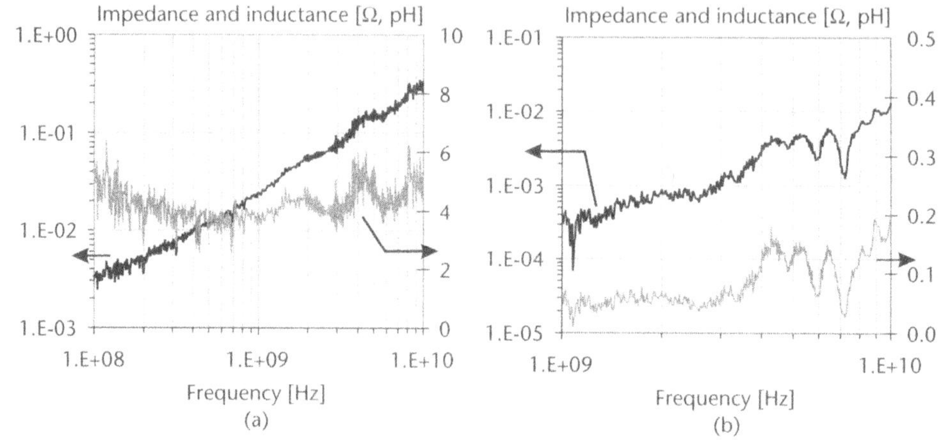

Figure 3.13 Impedance and inductance measurement of shorts, separated by (a) 20 mils and (b) 150 mils.

In summary, to be completely accurate, all of the partial self- and partial mutual terms between all of the vias must be captured. Accuracy can be traded for reduced modeling complexity by eliminating the coupling between the vias and just capturing the partial-self inductance terms in the current loop. This section examined the error introduced by this simplification. For via pairs and via checkerboard patterns, it was shown that the loop inductance equation can be approximated by (3.29) and the error in ignoring the partial mutual inductance in the circuit model can be determined using (3.20). The partial self- and partial mutual inductance terms can then be calculated using (3.15). An excellent correlation was shown between the field solutions and formulas for vias, both in terms of partial inductance and loop inductance values. These formulas are correlated to the measured loop inductance of via test structures in the next section.

3.3.1 Measurement Correlation

When attempts are made to measure the inductance of a single, stand-alone via, the measuring instruments/probes create a closed loop around the via. To reduce the contribution of this uncertainty, a pair of vias can be connected in a loop by a plane. Through calibration, the portion of loop inductance created by the finite probe pitch is removed. As long as the coupling between the probes and plane sides is negligible, the measured loop inductance will be a good approximation of the loop inductance of the via pair alone (provided that the horizontal plane inductance is negligible compared to the via inductance).

Figure 3.13 shows the uncertainty associated with the removal of the probe-tip inductance. First the calibration was done on the wafer substrates. During the short calibration, the probes were placed on adjacent strips on the substrate, 20 mils apart. After calibration, the probes were put back onto the same shorting features (20 mils apart). With everything ideal, you would expect a zero impedance reading. (Remember, these strips are isolated, so the two probes are shorted but not connected.) The residual calibration error produces an impedance reading corresponding to a 4-pH equivalent inductance. When we place the two probes on shorting strips 150 mils apart, we get the impedance plot of Figure 3.13(b), which reduces the extracted inductance to a fraction of a picoHenry.

The measurement structure, illustrated in Figure 3.14, consisted of two via pairs connecting to a common plane. As shown in Figure 3.15, picoprobes were landed on the same via pair loop, using the two-port shunt through measurement technique described in Chapter 5. Measuring the different combinations allows for four unique measurements, namely $L_{loop1\text{-}2}$, $L_{loop3\text{-}4}$, $L_{loop2\text{-}4}$, and $L_{loop2\text{-}3}$. Figure 3.16 shows the measured impedance and extracted loop inductance for the four combinations. The loop inductance was estimated by finding the best fit to the impedance profile using a fixed inductance value. Table 3.1 lists the best fit loop inductance from measurement alongside the partial inductance obtained from (3.15). Equation (3.15) was used to determine both the partial self- and partial mutual inductance values, and then the loop inductance was calculated using (3.17). The agreement across the four measurements is better than 10%. Improved correlation may be possible by including the partial inductance of the plane and the parasitics of the probe. In addition, any coupling between the plane and the probe could be included.

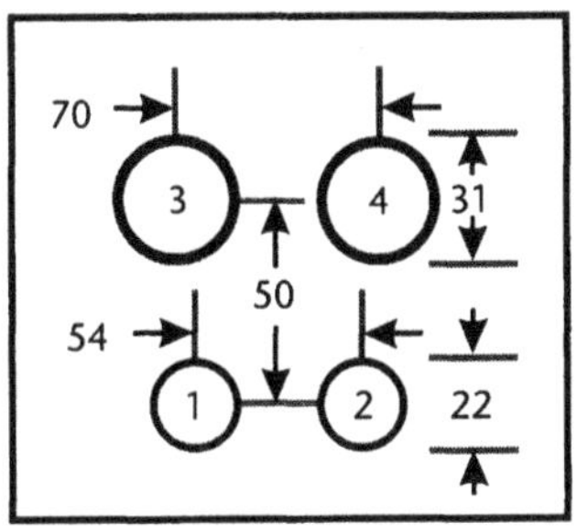

Figure 3.14 Top view of measurement structure. The dimensions are in mils. The via height is 2.29 mm (90 mils).

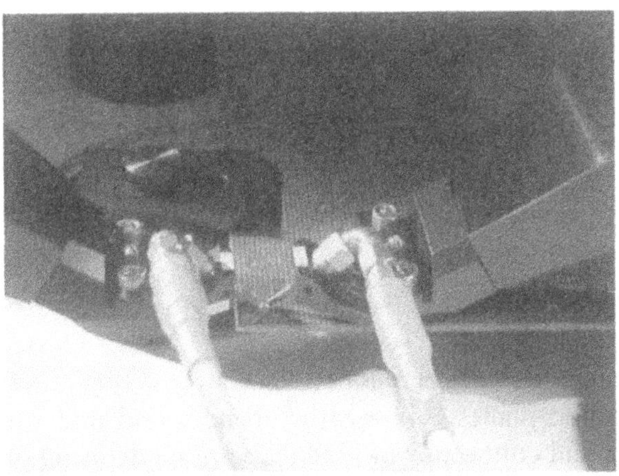

Figure 3.15 Measurement setup showing the picoprobes, micromanipulators, cables, and test board used for measuring the loop inductance.

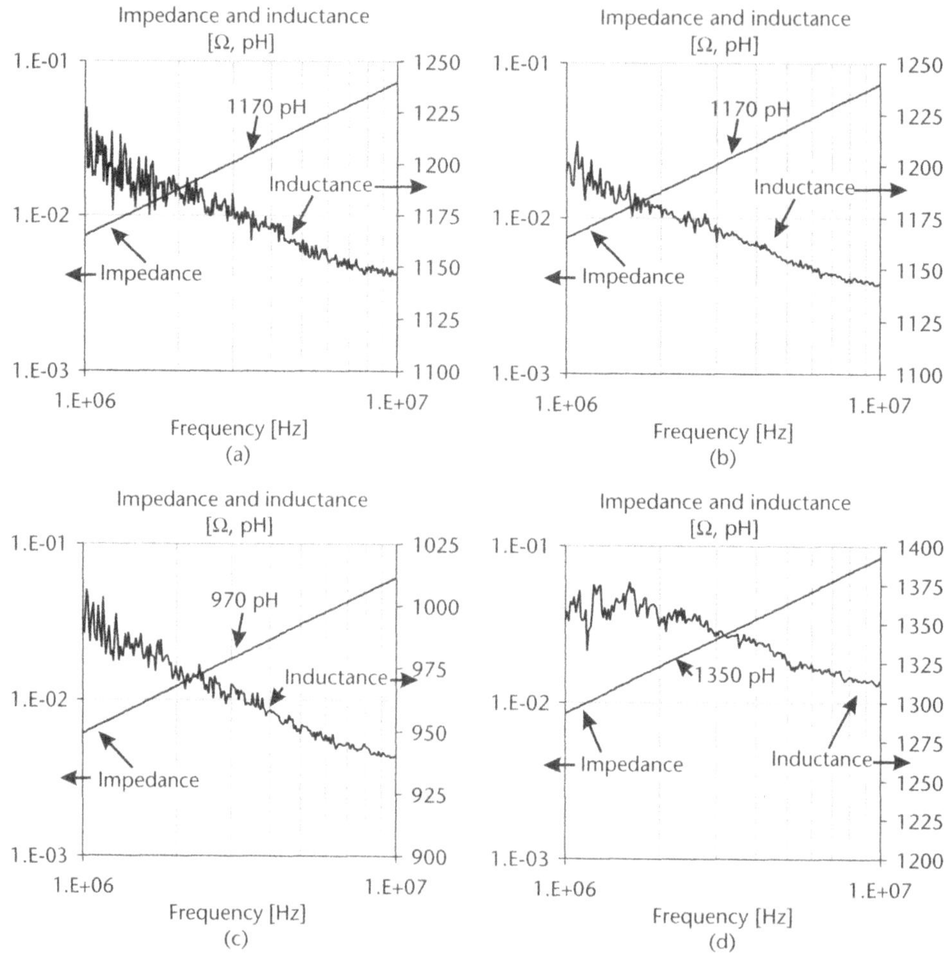

Figure 3.16 (a–d) Measured impedance and extracted loop inductance for the four via-pair combinations.

Table 3.1 Tabulated Data from Measurement and (3.15) for the Four Via Pairs

Via Pair	*Measurement [pH]*	*(3.15) [pH]*	*Delta [%]*
12	1,170	1,201	2.7
34	1,170	1,090	6.8
24	970	1,048	8.0
32	1,350	1,356	0.4

3.4 Via Arrays

This section examines the impact of via arrays on the electrical characteristics of power/ground planes. Via arrays can be found underneath a package where the power/ground connections to the PCB are made using multiple power and ground vias. Board designers often incorporate multiple parallel vias to minimize the inductance incurred in connecting packages to the board's power planes. With increasingly tight ball pitches, this results in a very dense pin or ball grid array field. The antipads associated with these via connections will perforate the plane region underneath the package, often to such an extent that only a thin web of metal exists underneath the package. Smaller antipad diameters can help to mitigate this effect, but multilayer boards require a certain minimum antipad diameter to be manufacturable. As a consequence of these perforations, the power plane pair impedance can be significantly altered by the via array [4]. It has also been shown that the effective inductance of a plane pair increases with the density of the via array [5, 6].

The following three sections examine the impact of these arrays on impedance. In the first section, the impedance of a test board consisting of an 8 × 8 array of vias is measured. These results are then compared to full-wave electromagnetic field solution using Ansoft HFSS v9.2. In this second section we examine the simulation environment and the validity of various simplifying assumptions (e.g., losses). In the third section, the model is parameterized, and by sweeping some of the physical parameters of the array, such as antipad diameter and dielectric thickness, we can examine the impact of the array's physical dimensions on the power plane impedance.

3.4.1 Measuring Via Arrays

A number of test boards were designed and fabricated that all had in common an 8 × 8 via array of checkboard power and ground vias centered on a 6.35-cm (2.5-inch) square plane pair. Figure 3.17 shows a top view of the array illustrating the staggered arrangement of the power/ground vias. The openings in the power-ground plane pair are created by overlapping antipads. Here, the antipad diameter is 1.473 mm (58 mils) and the via center-to-center distance is 1.27 mm (50 mils), respectively. Each via pair can be uniquely identified using the row and column identifiers shown on the left and top edge of the array in Figure 3.17. For example, (A1,B1) corresponds to the via pair located in the bottom left hand corner. The matrix is

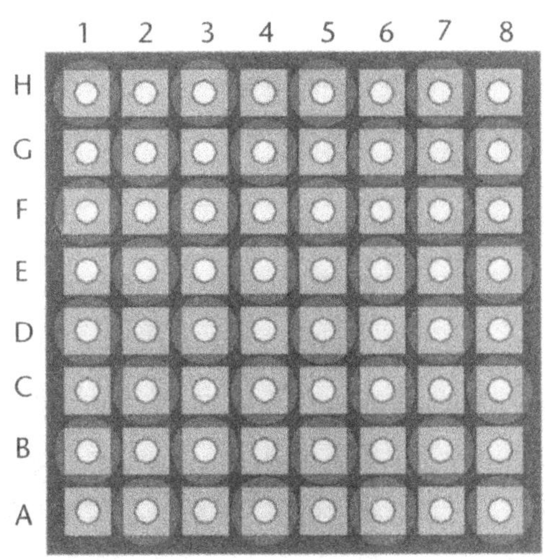

Figure 3.17 Via matrix arrangement. The array is centered on a 6.35-cm^2 (2.5-inch^2) plane pair that is not shown.

fully symmetric; thus, (A1,B1) has the same geometry and environment as (G8,H8). Figure 3.18 shows a cross section of the structure. The via diameter is 0.559 mm (22 mils) and the PTH have a wall thickness of about 33μ (1.3 mils). The dielectric thickness between the plane pair is 8μ.

An HP 4396B vector network analyzer (VNA) was used to measure the self-impedance magnitude and phase from 100 kHz to 1.8 GHz. With the board secured on its side, semirigid coaxial probes were used to contact the via pairs from opposite sides of the board (see Figure 3.19).

The power/ground via pairs were measured as a function of pair location. Real and imaginary S_{21} measurement data were processed and the corrected self-impedance magnitude and phase was obtained. In this arrangement, the parasitics associated with the via are in series with the source and load impedances. If the via impedance is small compared to the 50Ω environment, its contribution can safely be ignored. Further details about this measurement technique can be found in Section 5.4.1.

Figure 3.20 plots the impedance measured on the diagonal of the matrix from the corner (A1,B1) to the center (D4,E4). Also shown for reference is the measured

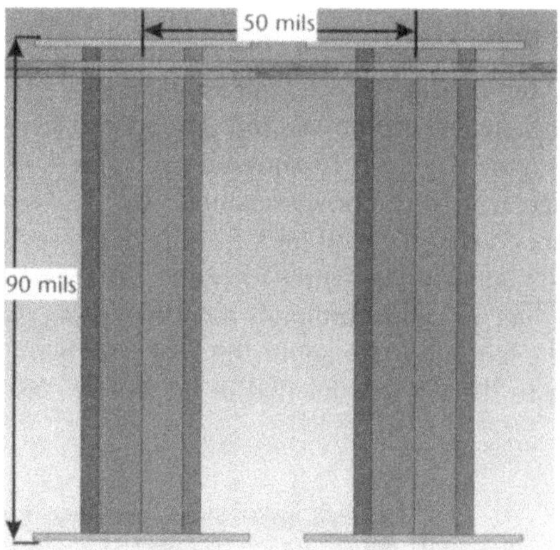

Figure 3.18 Side view of the test structure showing a pair of vias, the plane pair, and via pads. The copper weight is 1 oz.

Figure 3.19 Measurement configuration showing the semirigid coaxial probes.

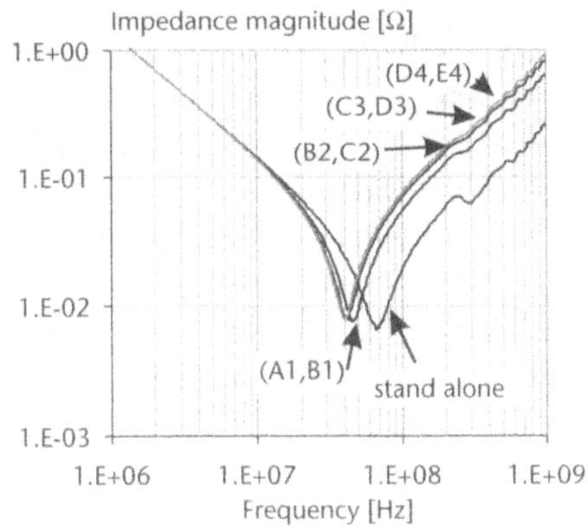

Figure 3.20 Measured impedance profiles at via pairs on the diagonal, starting with (A1,B1) progressively moving towards the center, (D4,E4).

impedance at a stand-alone via pair that is outside of the array surrounded by a solid plane pair. Up to about 10 MHz, the locations show similar self-impedance profiles, as the impedance is dominated by the static capacitance of the plane pair. Above 10 MHz, the impedance profiles diverge signifcantly; the plane perforation pushes the series resonance lower and increases the plane's equivalent inductance. The largest impedance change within the array is observed one step in from the corner location. Thereafter, the impedance changes are smaller. In fact, the final two measurement pairs near to the center are overlapping on this scale.

Figure 3.21(a) plots the measured equivalent inductance ($L_{eq} = \text{Im}(Z)/\omega$), at the via pairs, moving diagonally from the corner via pair to the center of the array. The measured equivalent inductance at the stand-alone via pair is shown as a reference. The inductance is found to be a strong function of location within the array. Figure 3.21(b) plots the ratio of the equivalent inductances at the via pairs, measured along the diagonal, to the inductance at the stand-alone via pair. The equivalent inductance within the array is found to vary by as much as a factor of 4.

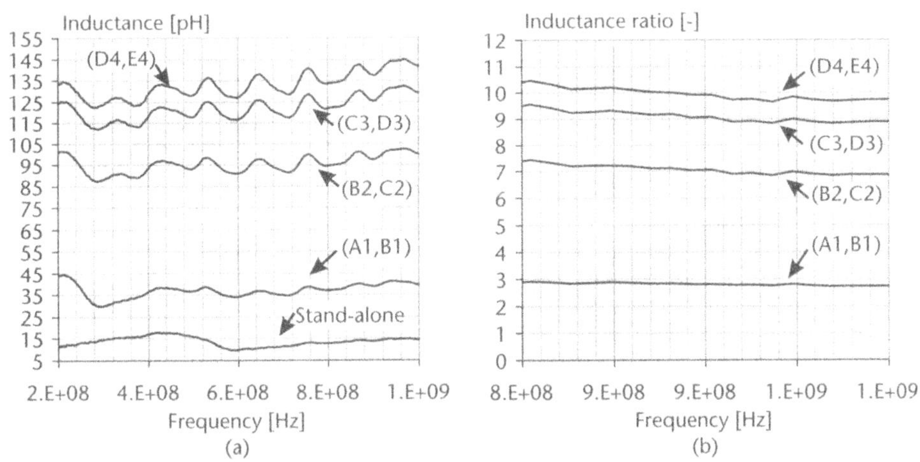

Figure 3.21 (a) Measured equivalent inductance at via pairs on the diagonal starting with (A1,B1) and progressively moving toward the center (D4, E4). (b) Ratio of the equivalent inductance at via pairs along the diagonal of the array relative to the stand-alone via pair.

By measuring rows AB, BC, CD, and DE across each column (1–8) and taking advantage of the symmetry of the structure, it is possible to obtain a plot of the impedance and inductance of the entire array. Figures 3.22(a) and 3.22(b) plot the impedance and equivalent inductance, respectively, of the array at 500 MHz. Moving in from the array perimeter, the impedance magnitude and equivalent inductance are found to increase sharply toward the array center. The highest impedance, in the center of the array, is found where the plane pair is heavily perforated on all sides by the antipad cutouts. The lowest impedance is found at the corners of the array where the plane is solid on two sides.

3.4.2 Modeling Via Arrays

A 3D full-wave electromagnetic field simulator can be used to extract the S-parameters of the entire via array. The 50Ω lumped gap ports were defined at each of the via pad pairs, on the top and bottom of the test board, and the two-port S-parameters were calculated for each pair. An absorbing boundary was used to simulate the open nature of the problem and to ensure that negligible energy is reflected back toward the structure. In order to properly capture the antipad overlap, the antipad was modeled with 30 facets, very closely approximating the true area overlap. Real and imaginary S_{21} solver data was then processed, and the corrected impedance magnitude and phase information was obtained. Figures 3.23(a) and 3.23(b) compare the impedance measured and simulated at the corner and center of the array, respectively. The dielectric materials in these simulations were assigned a loss tangent of 0.02. Overall, excellent agreement is obtained between the measurement and simulation. The resonance near 170 MHz, more pronounced in the corner of the array, was not captured by the simulation and remains to be understood.

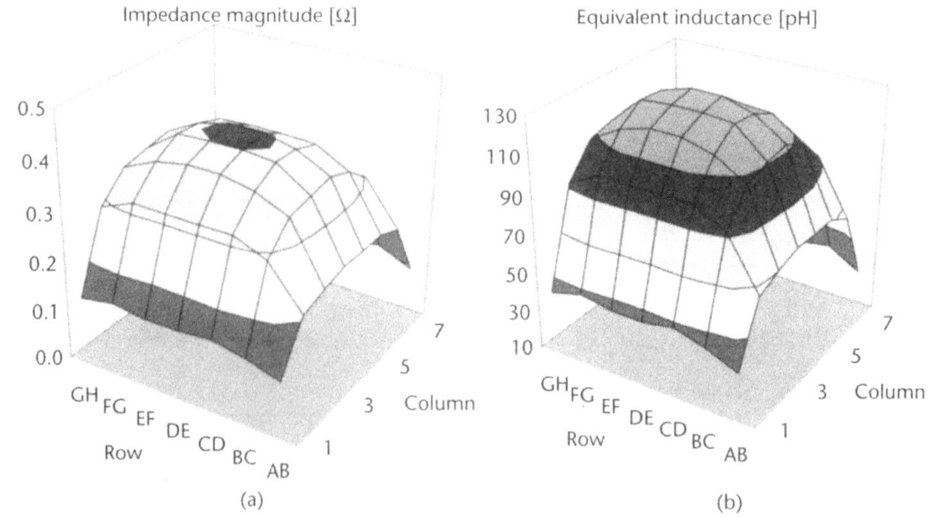

Figure 3.22 (a) Plot of measured impedance magnitude of the entire array at 500 MHz. (b) Plot of the equivalent inductance of the entire array at 500 MHz.

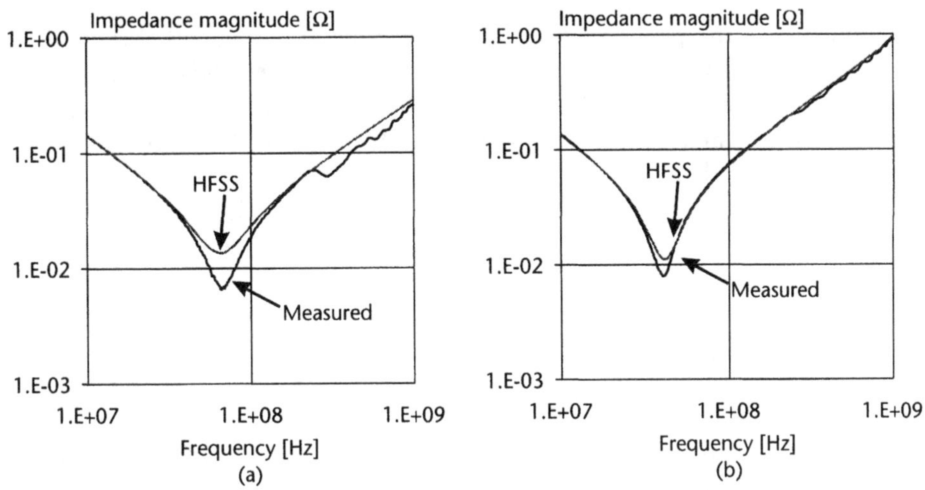

Figure 3.23 (a) Measured and simulated impedance data at a via pair located at the corner of the array (A1,B1). (b) Measured and simulated impedance data at a via pair located at the center of the array (D4,E4).

3.4.3 Parameterizing Via Arrays

In this section, two physical parameters of the array, antipad diameter and dielectric thickness, are parameterized, so that the impact on the power plane impedance can be examined.

3.4.3.1 Antipad Diameter

By parameterizing the antipad diameter in the field solver, it is possible to examine the impact of antipad overlap on the impedance and equivalent inductance as a function of location within the array. Figure 3.24 plots the equivalent inductance of the locations (A1,B1) and (D4,E4) as a function of antipad overlap. In this geometry, an antipad overlap of zero means that the antipad diameter is 1.27 mm (50

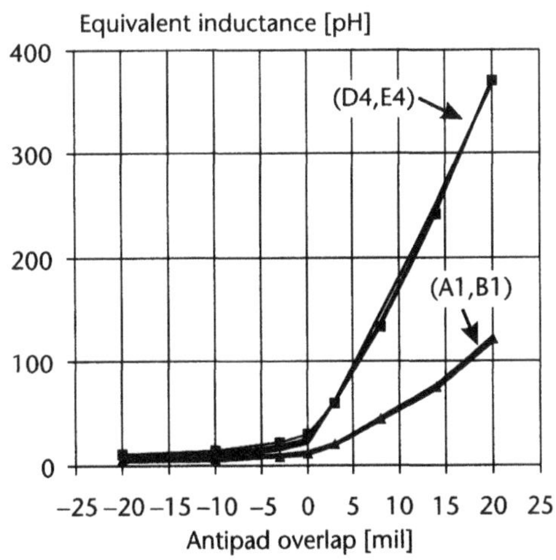

Figure 3.24 Simulated equivalent inductance at the corner via pair (A1,B1) and (D4,E4) as a function of the antipad overlap.

mils). For the two pair locations, the inductances, at 250 MHz, 500 MHz, 750 MHz, and 1 GHz, were compared. At all frequencies, the plots show that the inductance at the two locations increases sharply when the antipad overlap is positive. A positive value for the antipad overlap implies that the via center-to-center spacing is such that there is an overlapping opening in the power and ground planes. On a linear-linear scale, the equivalent inductance at all frequencies points overlap. Figure 3.25 shows the same graph as Figure 3.24, except on a log-linear scale. This graph more clearly shows that in the case of negative overlap, the inductance does increase as the antipad diameter increases, although not as sharply as when there is positive antipad overlap.

Figure 3.26 plots the ratio of the inductance at (D4,E4) over the inductance at (A1,B1), at 250 MHz, 500 MHz, 750 MHz, and 1 GHz. The plot shows that with positive antipad overlap the center via pair locations will have much higher inductance than the perimeter vias. With negative antipad overlap, the ratio of the inductances of the center vias to the perimeters vias asymptotically approaches one. At lower frequencies (e.g., 250 MHz), the ratio approaches one more slowly, probably due to current spreading.

3.4.3.2 Dielectric Thickness

In the previous section, it was found that across a wide range of frequencies, the equivalent inductance is a strong function of antipad diameter, particularly in the case of positive antipad overlap. In addition, it was found that the antipad overlap helps to determine how the impedance varies across the array. These simulations were performed with a fixed dielectric thickness of 8μ. In this section we examine the impact of dielectric thickness on these conclusions.

Figure 3.27 plots the simulated equivalent inductance at the corner via pair (A1,B1) and (D4,E4) as a function of the antipad overlap for the 8μ-, 24μ-, and 100μ-thick dielectrics. The plot is generated for 500 MHz because it exhibited the

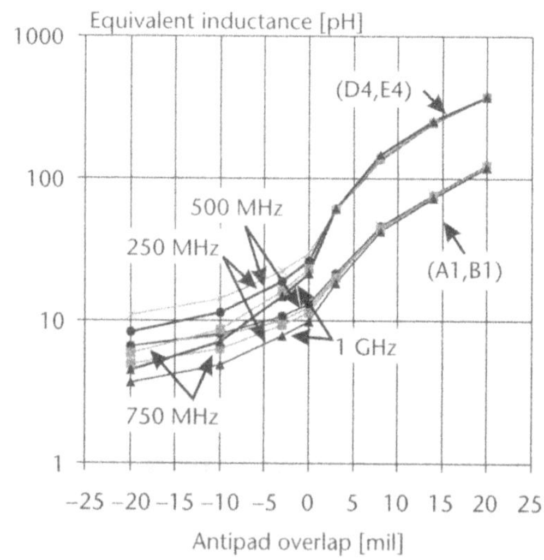

Figure 3.25 Simulated equivalent inductance at the corner via pair (A1,B1) and (D4,E4) as a function of the antipad overlap on a log-linear scale.

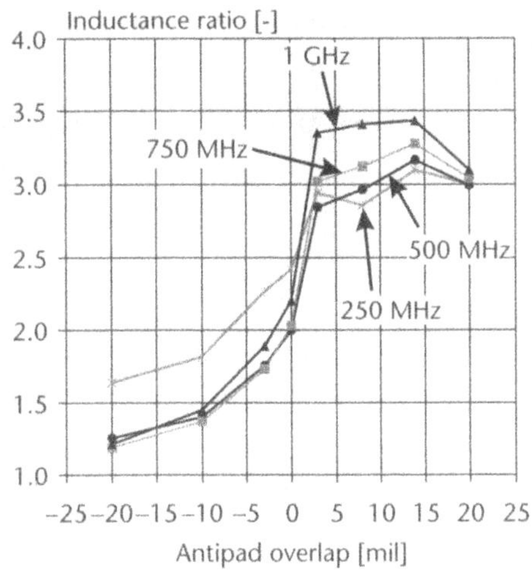

Figure 3.26 Ratio of the inductance at (D4,E4) to (A1,B1) at 250 MHz, 500 MHz, 750 MHz, and 1 GHz as a function of antipad overlap.

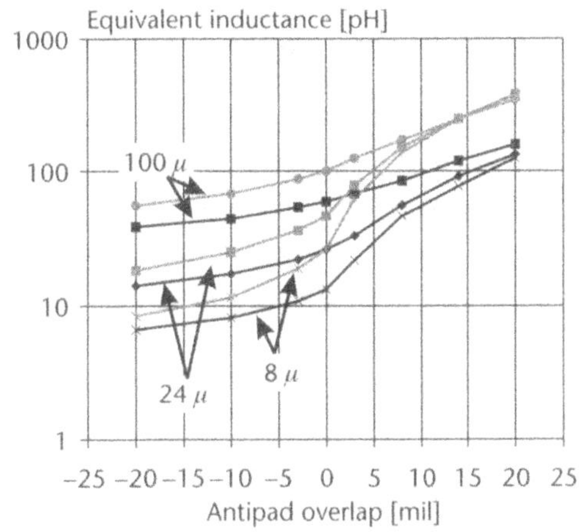

Figure 3.27 Simulated equivalent inductance at the corner via pair (A1,B1) and (D4,E4) as a function of the antipad overlap for 8μ-, 24μ-, and 100μ-thick dielectrics. Dark traces were simulated at (D4,E4), and light traces were simulated at (A1,B1).

smoothest impedance profile. Higher amplitude resonances are observed at other frequency points for the 24μ- and 100μ-thick dielectrics, due to the lower loss of the power plane pair and reduced damping. Looking at Figure 3.27, we see that for a given negative antipad overlap the dielectric thickness will increase the inductance at both locations in the array. This indicates that for the range of dielectric thicknesses studied, the plane inductance dominates for small antipads. As the antipad overlap increases, the inductance profiles converge for the center and perimeter locations; this conversion indicates that the antipad inductance dominates the plane inductance.

Figure 3.28 plots the ratio of the equivalent inductances of the 24μ- and 100μ-thick dielectrics to the 8-μm-thick dielectric at (A1,B1) and (D4,E4). Again, these results are plotted at 500 MHz only. This plot shows graphically what was

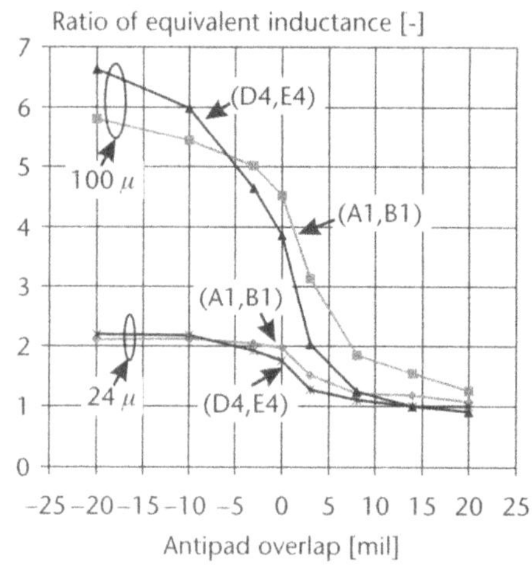

Figure 3.28 Ratio of equivalent inductance of the 24μ- and 100μ-thick dielectric to the 8-μm-thick dielectric at (A1,B1) and (D4,E4).

stated above. Namely, for positive antipad overlap the dielectrics yield approximately the same equivalent inductance. One conclusion that can be drawn from this is that increasing the antipad size reduces the benefits of using thinner dielectrics to reduce the power plane impedance. Notice that the ratio is slightly larger than one for the corner location with the 100μ dielectric. This indicates that, with a 100μ-thick dielectric, the plane inductance is becoming a fraction of the antipad inductance. With even thicker dielectrics the profile in Figure 3.28 would become flatter as plane inductance dominates across a wider range of antipad diameters.

Since the plane inductance increasingly dominates as the antipad size shrinks, we expect an inductance ratio that is proportional to the difference in the loop area introduced by the larger plane separation. For example, moving from a 24μ-thick dielectric to a 100μ-thick dielectric should increase the inductance by about a factor of 4. Figure 3.28 shows an increase of about a factor of 3. This discrepancy is a consequence of the inductance profiles as a function of frequency (for a given antipad size) being nonmonotonic at certain frequencies due to resonances, making it difficult to make a comparison. The comparison point of 500 MHz was chosen because the inductance profile was well behaved across the range of dielectrics and antipad sizes.

In Figure 3.27, it was shown that thinner dielectrics exhibit more impedance variation across the array as the antipad overlap increases. Figure 3.29 examines the sensitivity of the impedance variation to the dielectric thickness. The graph plots the ratio of the inductance at (D4,E4) to (A1,B1) at 500 MHz as a function of antipad overlap and dielectric thickness. The plot shows that thinner dielectric materials will demonstrate more impedance variation as the antipad size is increased, while using a thick dielectric will minimize this variation. (Here *thin* and *thick* dielectrics are relative to the antipad radius.) One way of interpreting this trend is that the electromagnetic fields in the dielectric material will not be perturbed by the antipad cutout, providing its radius is small relative to the thickness of the dielectric.

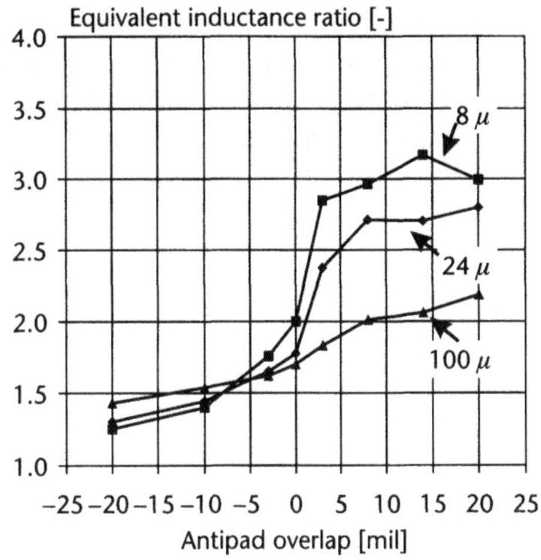

Figure 3.29 Ratio of the inductance at (D4,E4) to (A1,B1) at 500 MHz as a function of antipad overlap.

3.4.4 Via Array Summary Points

In summary, the impedance across the via array varies with location; vias on the outside of the array have the lowest impedance, and the impedance increases sharply towards the array center. This variation can be reduced by using thicker dielectrics and decreasing the antipad size.

Simulation results showed that the impedance is a strong function of antipad size. Positive antipad overlap increases the impedance dramatically for both the center and perimeter vias. The impedance is reduced and a more uniform impedance is maintained across the array when antipads in power/ground via arrays do not overlap.

Negative antipad overlap causes the plane inductance to dominate the antipad inductance over a large range of dielectric thicknesses. However, antipad diameter helps shape the impedance profile even with negative antipad overlap. The antipad size matters less and less as the thicker dielectrics are used. With positive antipad overlap, the differences between the dielectric thicknesses are less important as the antipad inductance dominates. Thus, the benefits of using thinner dielectrics to reduce the power plane impedance will be tempered by the relative proportion of antipad inductance.

References

[1] Miller, J. R., I. Novak, and T. Chou, "Calculating Partial Inductance of Vias for Printed Circuit Board Modeling," *Proceedings of the Electrical Performance of Electronic Packaging Conference*, October 21–23, 2002, pp. 123–126.

[2] Ruehli, A., "Inductance Calculations in a Complex Integrated Circuit Environment," *IBM J. Res. Dev.*, Vol. 16, No. 5, 1972, pp. 470–481.

[3] Rosa, E., "The Self and Mutual Inductance of Linear Conductors," *Bulletin of the National Bureau of Standards*, Vol. 4, 1908, pp. 301–344.

[4] Hyungsoo, K., et al., "Analysis of Via Distributions Effect on Multi-Layered Power/Ground Transfer Impedance of High Performance Packages," *Proceedings of the 11th Topical Meeting on Electrical Performance of Electronic Packaging*, October 2002, pp. 171–178.

[5] Yang, Z., et al., "Impact and Modeling of Anti-Pads on Power Delivery System," *Proceedings of the 12th Topical Meeting on Electrical Performance of Electronic Packaging*, October 2003, pp. 117–120.

[6] Miller, J. R., I. Novak, and J. Delap, "Characterizing and Modeling the Impact of Power/Ground Via Arrays on Power Plane Impedance," *Proceedings of DesignCon 2005*, Santa Clara, CA, January 31–February 3, 2005.

CHAPTER 4

Characterization and Modeling of Planes and Laminates

4.1 Introduction

Planes with dielectric layer separation, also called the power-bus, serve several purposes in PDNs: they carry dc current from source to load, they connect bypass capacitors horizontally to active devices, and many times they also provide a return path for signals. The nature and depth of modeling and characterization depends on our interests and (ultimate) goals. If we are worried about the dc voltage drop on a high-current but otherwise well-filtered supply rail, our main focus may be limited to the equivalent dc resistance and voltage-drop profile. For lower current multi-GB IO rails the dc voltage drop may matter less, but we may need to characterize and model the high-frequency return-path function of the plane over a wide frequency band. If our task is to characterize the material properties of the conductive planes and dielectric laminates, we may not care much for the practical limitations of connecting geometry in the real usage, but we might want to model and capture the pure material properties as accurately as possible. Finally, if our focus is electromagnetic compatibility, we probably need to characterize primarily the high-frequency resonance peaks.

There are several commercial tools available to simulate PDN planes (see Chapter 2). Tools which use the finite difference time domain (FDTD) method (see, e.g., [1]) solve for the structure's response in the time domain, and obtain the frequency domain response by translating the result with fast Fourier transform. FDTD solutions are known to have time-efficient execution for large problems, but the translation to the frequency domain is limited by the equidistant time-sample requirement of the transformations. Equidistant time samples result in a linear scale in the frequency domain. PDNs may require characterization in the frequency domain over several decades of frequency. The linear scale either ignores low-frequency details or requires an unrealistically large number of points, many of which at the high end of the frequency range are not needed. This is because PDN components necessarily come in a limited range of quality factor (Q), which is more readily suited for logarithmic description.

While FDTD does not lend itself to simple implementation by the end user, analytical plane models based on the cavity resonances (see, e.g., [2]) are easy to program in spreadsheet programs or MATLAB for rectangular shapes; solutions based on the *Partial Element Equivalent Circuit* (PEEC) (see, e.g., [3]) are easy to program in SPICE.

4.2 Analytical Plane Models

For parallel plane pairs separated by a uniform dielectric material, analytical impedance expressions are available describing the self- and transfer impedances between rectangular ports. Analytical expressions are available for simple plane shapes, such as rectangular, triangular, or circular. Of these options, the rectangular shape is the most widely used, both directly and as a building block, to construct irregular plane shapes.

4.2.1 Analytical Models for Rectangular Plane Shapes

The parameters are defined in Figure 4.1. Figure 4.1(a) shows a top view of the plane structure, defining the size of the rectangular plane shape as w_x by w_y. There are two ports on the plane pair. Port i is at coordinates (x_i, y_i), and port j is at coordinates (x_j, y_j). Figure 4.1(b) defines the vertical geometry and the material properties. We assume a uniform plane spacing of h and a uniform relative dielectric constant of ε_r. The upper and lower planes have thicknesses of t_u and t_l, and conductivities of σ_u and σ_l, respectively.

In PCBs with regular lamination processes, as well as in organic packages, the two planes separated by a dielectric layer may represent either a core, or a prepreg with the planes belonging to cores above and below. If located on the same core, the nominal plane thicknesses and conductivities are probably, but not necessarily, the same. If located on different cores, or if a build-up process is used to create the PCB or package, the two plane layers may be nominally different. One of the planes may be of copper with regular thickness, such as 18 μm (0.5 oz) or 35 μm (1 oz), while the other plane could be a 70-μm (2 oz) or 140-μm (4 oz) copper plane or a similarly thick aluminum layer for heat-spreader purposes and to help to distribute large currents.

4.2.1.1 Lossless Cavity Model

If we assume that the plane separation, h, is small compared to any wavelength of interest, the field can be considered to be constant along the z-axis; this results in a

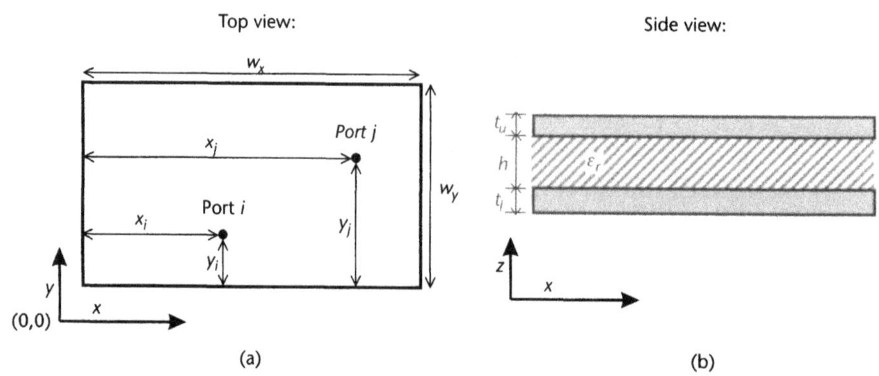

Figure 4.1 Rectangular parallel plane pair, separated by a uniform layer of dielectric material. Part (a) shows the top view and (b) defines the vertical geometry. Not to scale.

2D waveguide cavity with open boundaries. Waves traveling horizontally between the planes experience full reflection at the open boundaries; this reflection gives rise to a 2D series of modal resonances at frequencies where the w_x or w_y dimension equals integer multiples of the half wavelength. This structure has been analyzed for planar array antennas [4], for planar microwave circuits [5], and more recently for PDN applications.

In Figure 4.1, the generic transfer impedance between ports *i* and *j* for a lossless structure is given by:

$$Z_{ij}(\omega) = j\omega\mu h \sum_{n=0}^{\infty}\sum_{m=0}^{\infty} \frac{\chi_{mn}^2}{w_x w_y \left(k_{xm}^2 + k_{yn}^2 - k^2\right)} f(x_i, y_i, x_j, y_j) \tag{4.1}$$

where

$$f(x_i, y_i, x_j, y_j) = \cos\left(\frac{m\pi x_i}{w_x}\right)\operatorname{sinc}\left(\frac{m\pi t_{xi}}{2w_x}\right)\cos\left(\frac{n\pi y_i}{w_y}\right)\operatorname{sinc}\left(\frac{n\pi t_{yi}}{2w_y}\right)\cdot$$
$$\cos\left(\frac{m\pi x_j}{w_x}\right)\operatorname{sinc}\left(\frac{m\pi t_{xj}}{w_x}\right)\cos\left(\frac{n\pi y_j}{w_y}\right)\operatorname{sinc}\left(\frac{n\pi t_{yj}}{2w_y}\right) \tag{4.2}$$

$$k = \omega\sqrt{\varepsilon\mu} = \omega\sqrt{\varepsilon_r \varepsilon_0 \mu_0} = \sqrt{\varepsilon_r}\,\frac{\omega}{c} \tag{4.3}$$

$\omega = 2\pi f$ is the angular frequency;
μ is the permeability of dielectric ($\mu = \mu_0 = 4\pi 10^{-7}$);
c is the speed of light.

$$k_{xm}^2 = \left(\frac{m\pi}{w_x}\right)^2,\; k_{yn}^2 = \left(\frac{n\pi}{w_y}\right)^2$$

$$\chi_{mn} = 1 \text{ for } m = 0 \text{ and } n = 0;\; \sqrt{2} \text{ for } m = 0 \text{ or } n = 0;\; 2 \text{ for } m \neq 0, n \neq 0$$

(x_i, y_i), (x_j, y_j) are the coordinates of the ports;
(t_{xi}, t_{yi}), (t_{xj}, t_{yj}) are the port dimensions.

The analytical expression of (4.1) can be solved for any pair of arbitrarily located points on the planes, either self-impedance or transfer impedance, and is well suited for programming in spreadsheets with macro capabilities to evaluate the summations.

If the port dimensions are much smaller than the smallest wavelength of interest, the *sinc* functions disappear in (4.2) leaving the four *cosine* terms. Furthermore, if the two ports are at the same location such that the self-impedance of the plane is measured, then (4.2) simplifies to:

$$f(x_{ij}, y_{ij}) = \cos\left(\frac{m\pi x_{ij}}{w_x}\right)^2 \cos\left(\frac{n\pi y_{ij}}{w_y}\right)^2 \tag{4.4}$$

Though not limited by finite spatial granularity, as it is the case with SPICE-grid plane models, the analytical expression has a double infinite series, which for practical calculations must be truncated. This means that instead of being infinite, we have to use finite $n = N$ and $m = M$ limits. Figure 4.2 illustrates the effect of truncation with the simulated impedance of a plane pair of 25.4 × 25.4 cm (10 × 10 inch) size, 50-μm (2-mil) plane separation and dielectric constant of 4.0. The self-impedance was computed at one of the corners.

There are two major trends to observe as we change the summation limits in (4.1). As it was detailed for instance in [6, 7], at low frequencies, the impedance minima converge very slowly; at high frequencies, after the last impedance peak in the summation, the impedance drops monotonically, as opposed to a rising function that we would expect from the inductive behavior. Since the summation is based on poles, the impedance maxima included in the summation of a low-loss structure are captured properly, regardless of the summation limits. To ensure sufficient frequency coverage for the peaks, the N and M summation limits have to be chosen such that the highest included modal peak is safely above the highest frequency of interest. With $n = N$ and $m = M$ summation limits along the n and m variables, the frequencies of the last captured modal resonances are:

$$f_N = \frac{N}{2a}\frac{c}{\sqrt{\varepsilon_r}}, \quad f_M = \frac{M}{2b}\frac{c}{\sqrt{\varepsilon_r}} \tag{4.5}$$

While peaks automatically provide good convergence, the convergence around the minima is very slow. This is the most critical at the first impedance minimum of the self-impedance profile, because it usually reaches milliohm values in a low-loss PCB or package structure. Figure 4.3 shows the convergence with two plots using the same plane pair that was used for Figure 4.2. Figure 4.3(a) plots the simulated impedance magnitude, enlarged around the first minimum. Figure 4.3(b) plots the extracted frequencies of the impedance minimum from each trace. The summation limit was varied from $N = M = 1$ to $N = M = 20$ in increments of one.

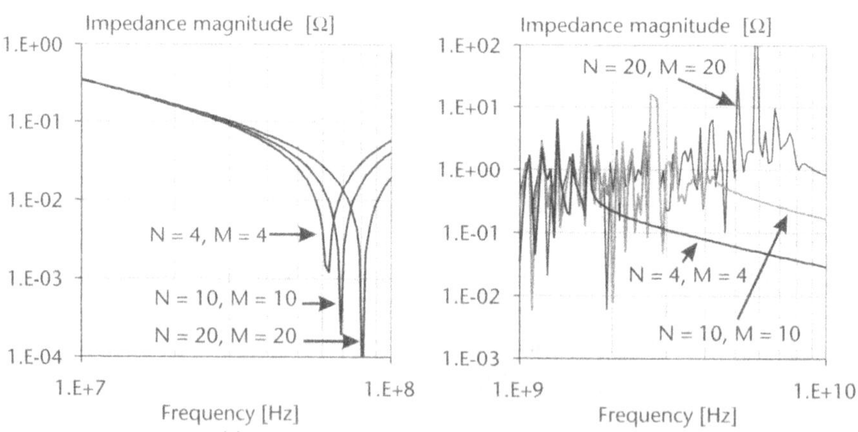

Figure 4.2 Self-impedance magnitude at the corner of a lossless square plane pair: (a) low-frequency response and (b) high-frequency response.

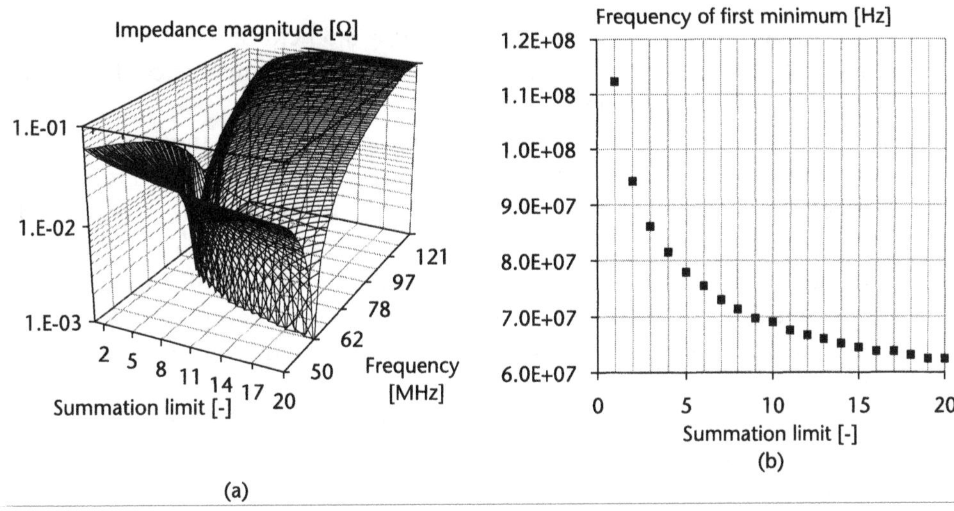

Figure 4.3 Convergence of the first impedance minimum as a function of summation limit: (a) plots the simulated impedance magnitude and (b) plots the minimum-impedance frequency points.

For laminates with a plane separations much larger than the thickness of either of the conductive layers, t_u or t_l, the lossless assumption yields fairly good agreement with measured data. Figure 4.4 compares the measured self-impedance magnitudes at a corner of a square parallel plane pair with simulated data using (4.1). The plane pair was square, with $w_x = w_y = 25.4$ cm (10 inches) and $h = 0.79$ mm (31 mils). The bare two-sided FR4 laminate had 35-μm (1 oz) copper on either side, and it was measured in several sizes: while the measurement test points were attached to one of its corners, the laminate sheet was repeatedly cut into four equal-size sections. First, the laminate was measured as is with its full size. Second, the sheet was cut into four squares of $w_x = w_y = 12.7$ cm (5 inches). Third, one of the smaller squares was fur-

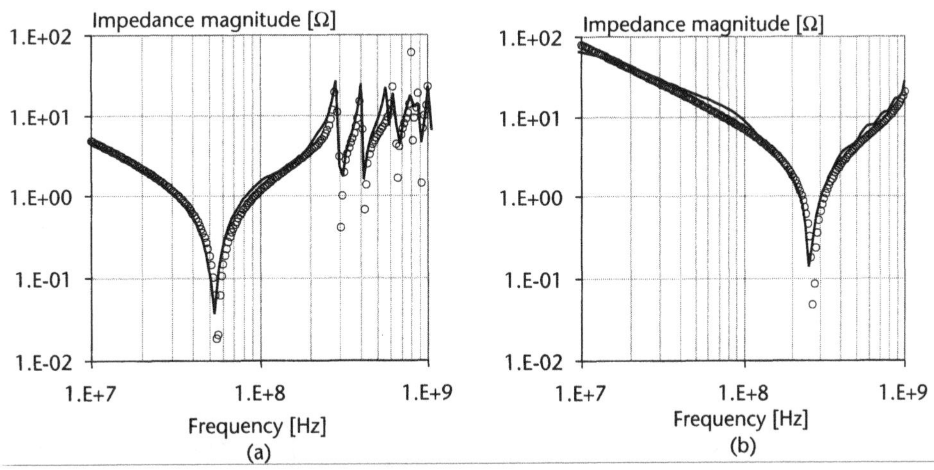

Figure 4.4 Comparison of measured self-impedance at a corner of a square pair of planes to simulated self-impedance using (4.1). Continuous lines: measured data. Small circles: simulated data. Plane separation is $h = 0.79$ mm (31 mils). (a) Plane size is $w_x = w_y = 25.4$ cm (10 inches). (b) Plane size is 6.35 cm (2.5 inches).

ther cut into four equal sections of $w_x = w_y = 6.35$ cm (2.5 inches). Figure 4.4 shows the correlation between measured and simulated self impedances for the (a) full-size and (b) smallest-size squares.

Besides the slow convergence around the minima, another drawback of the analytical solution based on (4.1) is the dual summation. One option to reduce the complexity of the calculations is to make use of the natural symmetry of the rectangular plane shape. By appropriately partitioning the planes, the responses for even and odd modes can be calculated, leading to an overall reduction in computation time [8]. A second possibility to reduce the computational complexity is to model the plane pair as a section of a rectangular waveguide so that one direction is automatically accounted for instead of using a summation. By applying magnetic walls along the opposite sides of length w_y, and applying open termination at the other two sides, the boundary conditions can be captured by a Green's function [9].

Finally, another major limitation of (4.1) is that the analytical expression assumes the structure is lossless. Plane pairs in real PCB exhibit many different types of losses, of which conductive losses and dielectric losses are generally the most prominent loss mechanisms. Several investigators have modified (4.1) to include conductive and dielectric losses. In the following sections, several different lossy formulations are discussed.

4.2.1.2 Light-Losses Cavity Model

One approach to capture losses is to modify the real wave number, k, with a complex wave number given by [4]:

$$k = \beta - j\alpha = k' - jk'' \tag{4.6}$$

where $k' = k$ from (4.3), and

$$k'' = \omega\sqrt{\varepsilon_r \varepsilon_0 \mu_0}\left(\frac{\tan\_\delta}{2} + \frac{\delta}{2h}\right) \tag{4.7}$$

where tan_δ is the loss tangent of the dielectric material, and δ is the skin depth at the frequency of interest, given by:

$$\delta = \sqrt{\frac{1}{\pi f \sigma \mu}} \tag{4.8}$$

The model assumes that there is small dissipation (i.e., $k' >> k''$).

This model accounts for the attenuation but neglects any changes to the phase constant β caused by the nonideal conductor and dielectric substrate. For this reason, strictly speaking, the solution is not causal [10]. The overall attenuation constant, α, is calculated by summing the dielectric loss and conductive loss.

4.2.1.3 Modified Cavity Model Using the Complex Propagation Constant

A second approach to capture losses is to substitute the complex propagation constant for the real wave number, k, in (4.3) [11]. The per-unit-length transmission line parameters are obtained using a lumped-element model of a radial transmission line. The effect of the imperfect dielectric substrate is represented by the shunt conductance, G_d, and the conductive losses are represented by the series impedance, Z_{cu}. The modified cavity model expression and accompanying formula for the propagation constant are

$$Z_{ij}(\omega) = j\omega\mu h \sum_{n=0}^{\infty}\sum_{m=0}^{\infty} \frac{\chi_{mn}^2}{w_x w_y \left(k_{xm}^2 + k_{yn}^2 + \gamma^2\right)} f\left(x_i, y_i, x_j, y_j\right) \tag{4.9}$$

where

$$\gamma = \sqrt{(Z_{cu} + j\omega L)(G_d + j\omega C)} = j\omega\sqrt{\varepsilon\mu}\sqrt{\left(1 - \frac{j(1+j)\delta}{h}\right)(1 - j\tan\_\delta)} \tag{4.10}$$

4.2.1.4 Equivalent Circuit-Based Cavity Model

A third approach to include losses takes the dual infinite series of (4.1) and transforms it into an equivalent circuit [12]. By using parallel resonant circuits and ideal transformers, the transfer impedance between port i and port j is:

$$Z_{ij}(\omega) = \sum_{n=1}^{\infty}\sum_{m=1}^{\infty} \frac{\chi_{mn}^2}{\frac{1}{j\omega L_{mn}} + j\omega C_{mn} + G_{mn}} f\left(x_i, y_i x_j, y_j\right) + \frac{1}{j\omega C_{00} + G_{00}} \tag{4.11}$$

where

$$C_{00} = \frac{\varepsilon_0 \varepsilon_r w_x w_y}{h}, \quad G_{00} = C_{00}\omega\left(\tan\_\delta + \frac{\sqrt{\frac{2}{\omega\mu\sigma}}}{h}\right) \tag{4.12}$$

$$f_{mn} = \frac{\sqrt{\left(\frac{m}{w_x}\right)^2 + \left(\frac{n}{w_y}\right)^2}}{2\sqrt{\varepsilon_0 \varepsilon_r \mu}} \tag{4.13}$$

$$L_{mn} = \frac{h}{\omega_{mn}^2 w_x w_y \varepsilon_0 \varepsilon_r}, \quad C_{mn} = \varepsilon_0 \varepsilon_r \frac{ab}{s}, \quad G_{mn} = C_{mn}\omega_{mn}\left(\tan\_\delta + \frac{\sqrt{\frac{2}{\omega_{mn}\mu\sigma}}}{h}\right) \tag{4.14}$$

The static capacitance of the plane pair is separated (i.e., $m = 0$, $n = 0$) from the higher-order modes in (4.11). Each mode of the structure can be represented as a resonant circuit with the resonant frequency equal to (4.13). The equivalent circuit of (4.11) through (4.14) is shown in Figure 4.5 for two ports.

4.2.1.5 Transmission Plane Model

Finally, a fourth approach for including losses in the cavity model was described in [10]. In this approach, partial differential equations are derived in the frequency domain for the parallel plane structure. This yields expressions for the distributed admittance, $Y(\omega)$, and impedance, $Z(\omega)$, of the plane pair which are then substituted into (4.1). The modified cavity model expression and accompanying formula are

$$Z_{ij}(\omega) = Z(\omega)\sum_{n=0}^{\infty}\sum_{m=0}^{\infty}\frac{\chi_{mn}^{2}}{w_x w_y\left(k_{xm}^2 + k_{yn}^2 + Z(\omega)Y(\omega)\right)} f\left(x_i, y_i, x_j, y_j\right) \tag{4.15}$$

where

$$Z(\omega) = j\omega L + Z_s(w), \quad Y(w) = \frac{1}{h}\left(j\omega\varepsilon_0\varepsilon_r(\omega) + \sigma\right) \tag{4.16}$$

$$L = \mu h, \quad Z_s = \frac{2k_m}{\sigma}\cot(k_m t), \quad k_m = (1-j)\frac{1}{\delta} \tag{4.17}$$

t is the thickness of the top and bottom metal planes.

In (4.16), σ is a bulk conductivity of dielectric and is zero for FR4-type dielectrics. It was included in the derivation because some PDN dielectric materials actually may have nonzero conductivity to suppress noise.

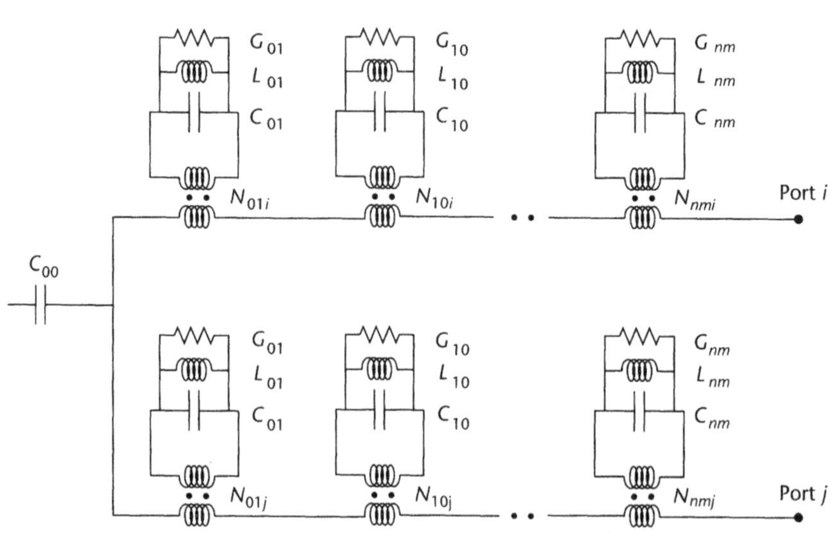

Figure 4.5 Equivalent circuit of the plane transfer impedance based on a cavity resonance model for two ports and $n \times m$ modes.

In this formulation, the complex, relative dielectric constant, $\varepsilon_r(\omega)$, included in (4.16), is defined over a broadband frequency range to ensure that (4.15) is causal. The Debye model is used to capture the frequency dependence of the complex dielectric constant and is included here for reference

$$\varepsilon_r(f) = \varepsilon_r(\infty) + \varepsilon_{rd} F_d(f) \tag{4.18}$$

where

$$F_d(f) = \frac{1}{(m_2 - m_1)\ln(10)} \ln\left[\frac{10^{m2} + jf}{10^{m1} + jf}\right] \tag{4.19}$$

$$\varepsilon_{rd} = -\frac{\tan\delta(f_0)\varepsilon_r(f_0)(m_2 - m_1)\ln(10)}{\text{Im}(\theta)} \tag{4.20}$$

$$\varepsilon_r(\infty) = \varepsilon_r(f_0)\left(1 + \tan\_\delta(f_0)\frac{\text{Re}(\theta)}{\text{Im}(\theta)}\right), \quad \theta = \ln\left[\frac{10^{m2} + jf_0}{10^{m1} + jf_0}\right] \tag{4.21}$$

To employ this model the dielectric constant and loss tangent only need be defined at one frequency point, f_0; $m1$ and $m2$ are the parameters that define the linear portion of real part of the dielectric constant. The imaginary part of the dielectric constant (and thus loss tangent, too) will be linear in a narrower range, inside the $m1$ and $m2$ limits.

4.2.1.6 Cavity Model Simulations

To examine the accuracy of the four different lossy plane impedance formulations introduced above, the plane pair structure shown in Figure 2.1 was simulated. The m and n summation limits were set to 80 for all the simulations. The formulas were coded into MATLAB and the results postprocessed and plotted in Excel. Figure 4.6 plots the four lossy impedance expressions alongside measurement data of the test structure.

The different formulations produce similar results which are fairly well correlated to measurement although with some magnitude offset and frequency offset at higher frequencies. The transmission plane model, Figure 4.6(d), shows very little phase offset due to the inclusion of the frequency dependent dielectric constant in the model.

Plotting the low-frequency behavior, however, reveals several major differences among the formulations. Figure 4.7(a) plots the low frequency behavior of the *lossless* plane expression along side measurement data. Below the series resonances frequency, both impedance plots show a capacitive slope dictated by the static capacitance of the plane pair.

Figure 4.7(b) plots the four lossy plane expressions alongside the lossless plane expression. All but the transmission plane model show differing degrees of low-frequency roll off compared to the lossless case. The low-frequency roll off is due to the onset of skin effects (i.e., the skin depth approaches the thickness of the metal planes). Mathematically, the low-frequency response of these expressions can be

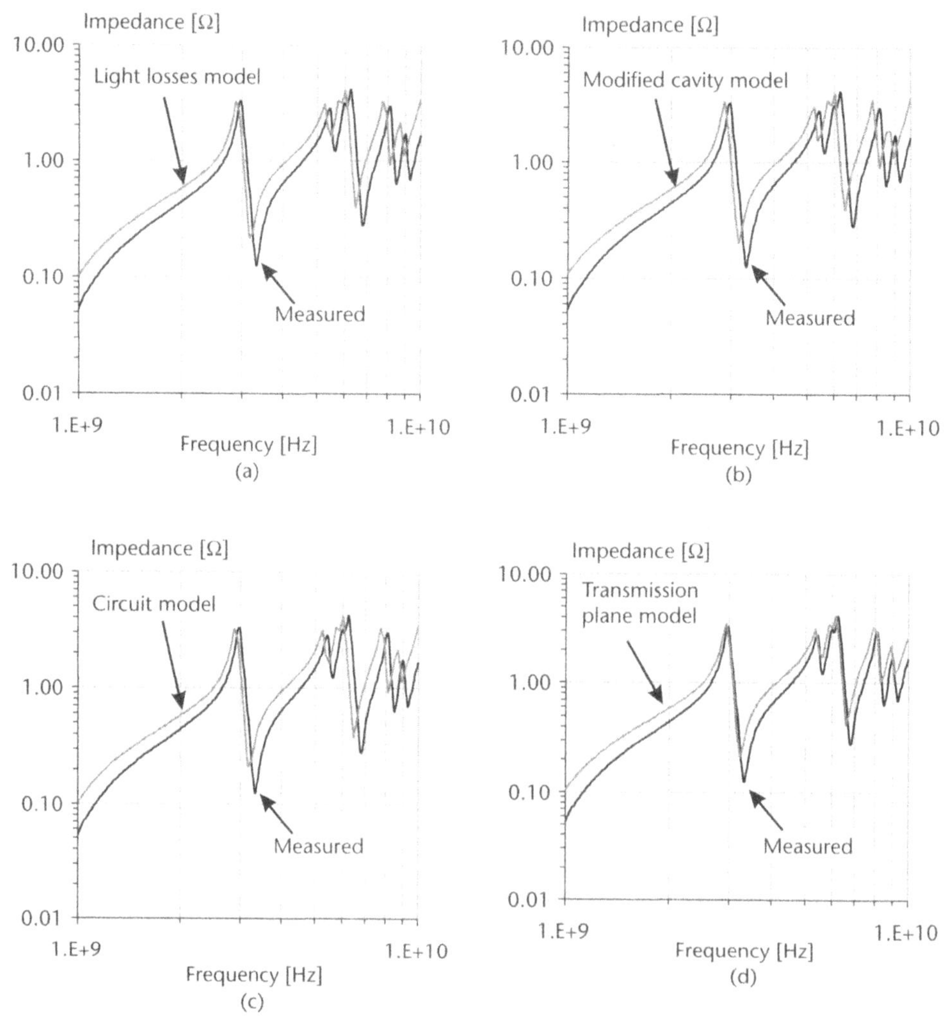

Figure 4.6 Measurement data for the test structure shown in Figure 2.1, plotted alongside the following four lossy impedance expressions: (a) light losses model, (b) modified cavity model, (c) circuit model, and (d) transmission plane model.

understood by examining the $m = 0$, $n = 0$ mode, which should capture the static capacitance of the plane pair. Substituting $m = 0$, $n = 0$ in the lossless case (4.1), the low-frequency impedance of the plane pair is

$$Z_{ij}(\omega) = -j\frac{h}{\omega w_x w_y \varepsilon_0 \varepsilon_r} = \frac{1}{j\omega C_{00}} \tag{4.22}$$

This is the impedance of a lumped capacitor, C_{00}, representing the static capacitance of the plane pair. On the other hand, substituting $m = 0$, $n = 0$ into the modified plane impedance expression which assumes light losses (4.6) through (4.7) yields:

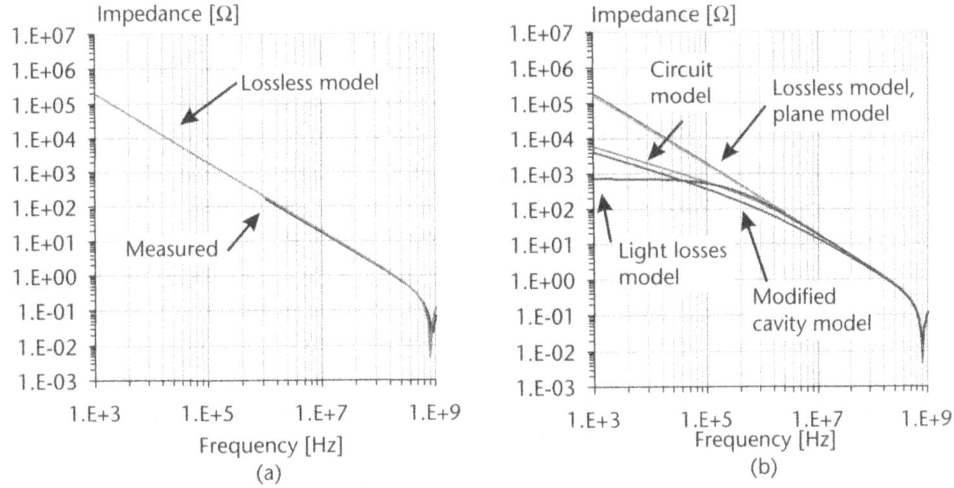

Figure 4.7 Measurement data for the strucutre shown in Figure 2.1, plotted alongside the (a) lossless model and (b) the four lossy models, together with the lossless model. The low-frequency roll off observed in (b) occurs at different frequencies for the different models.

$$Z_{ij}(\omega) = -j\frac{h}{w_x w_y h\varepsilon_0\varepsilon_r\left(1-j\left(\frac{\tan\_\delta}{2}+\frac{\delta}{2h}\right)\right)^2} = \frac{1}{j\omega C_{00}\left(1-j\left(\frac{\tan\_\delta}{2}+\frac{\delta}{2h}\right)\right)^2} \tag{4.23}$$

Equation (4.23) shows that the impedance can only be approximated by (4.22) if the dielectric thickness is much thicker than the skin depth and assuming light dielectric losses. The same requirement holds true for the modified plane impedance expression which uses the complex propagation constant, (4.9):

$$\begin{aligned} Z_{if}(\omega) &= -j\frac{h}{w_x w_y h\varepsilon_0\varepsilon_r\left(1-\frac{j(1+j)\delta}{h}\right)(1-j\tan\_\delta)} \\ &= \frac{1}{j\omega C_{00}\left(1-\frac{j(1+j)\delta}{h}\right)(1-j\tan\_\delta)} \end{aligned} \tag{4.24}$$

The low-frequency impedance of the equivalent circuit based model, which can be found by substituting (4.12) into (4.11) and solving for the $m = 0$, $n = 0$ mode, places a similar requirement on the plane thickness and dielectric loss:

$$Z_{ij}(\omega) = \frac{1}{j\omega C_{00} + G_{00}} = \frac{1}{\omega C_{00}\left(j+\left(\tan\_\delta+\frac{\delta}{h}\right)\right)} \tag{4.25}$$

Finally, the $m = 0$, $n = 0$ mode of the transmission plane model can be evaluated by substituting (4.16) into (4.15):

$$Z_{ij}(\omega)\frac{1}{w_x w_y Y(\omega)} = \frac{1}{j\omega C_{00}} \tag{4.26}$$

This is the only lossy, plane-impedance expression of the four discussed above which yields the same result obtained from the lossless plane impedance expression, (4.22); so it places no apparent limitation at low frequencies on the dielectric thickness or loss tangent. From (4.23) through (4.25), the low-frequency accuracy of the other lossy plane impedance expressions above are a function of the skin depth and dielectric loss. In Figure 4.7(b) the impedance frequency deviates from the linear capacitive slope when the skin depth approaches the thickness of the dielectric.

Simulations were performed to examine the accuracy limitations of the four lossy impedance expressions. In the first simulation, the dielectric loss tangent, tan_δ, was swept from 0.001 to 0.2, while the conductivity was set to a very large value (10^{25}), approximating a perfect conductor. By using a large conductivity, the skin depth will be smaller than the metal thickness at low frequencies so the impact of loss tangent on the impedance expressions can be examined. The dielectric constant, ε_r, was set to an arbitrary value of 4.1 for these simulations. Note that for the transmission plane model, the loss tangent and dielectric constant were set to these values at 1 MHz only; at other frequency points, the parameters vary according to the Debye model to capture the frequency dependence of the loss tangent and dielectric constant. For each value of tan_δ, all four expressions were evaluated and the loss tangent value was extracted as a function of frequency from the impedance. Specifically, the loss tangent was obtained from the phase of the impedance as follows

$$\tan\_\delta = \tan(\theta(\omega)) \tag{4.27}$$

where $\theta(\omega)$ is in radians. The loss tangent was extracted as a function of frequency. The upper frequency limit was chosen such that it was much less than the series resonance frequency, where the loss tangent can be reliably extracted from the impedance. *If* the four expressions placed no limitations on the value of the loss tangent and the expressions were causal, the extracted loss tangent values would match the simulated values across the range of extraction frequencies.

Figure 4.8 plots the percentage difference in the extracted loss tangent value over a range of extraction frequencies and loss tangent values for all four expressions. The light losses formula, in Figure 4.8(a), shows the highest overall difference in the extracted loss tangent value due to the explicit assumption about losses in this formulation (i.e., $k' >> k''$). For loss tangent values less than or equal to 20%, the difference is better than 1%. Figure 4.8(b) and Figure 4.8(c) show that the difference in the extracted loss tangent is negligible over a wide range of loss tangent values. Notice that, for these two plots, the error is increasing as the loss tangent decreases; although the conductivity is high, the δ / h term in the denominator starts to influence the extraction of loss tangent as tan_δ is made progressively smaller. With higher loss tangent values, at higher extraction frequencies, Figure 4.8(b) and Figure 4.8(c) show less difference in the extracted loss tangent value, as the influence of the δ/h term becomes negligible. Finally, Figure 4.8(d) plots the difference in the extracted loss tangent using the transmission plane model; this model shows the

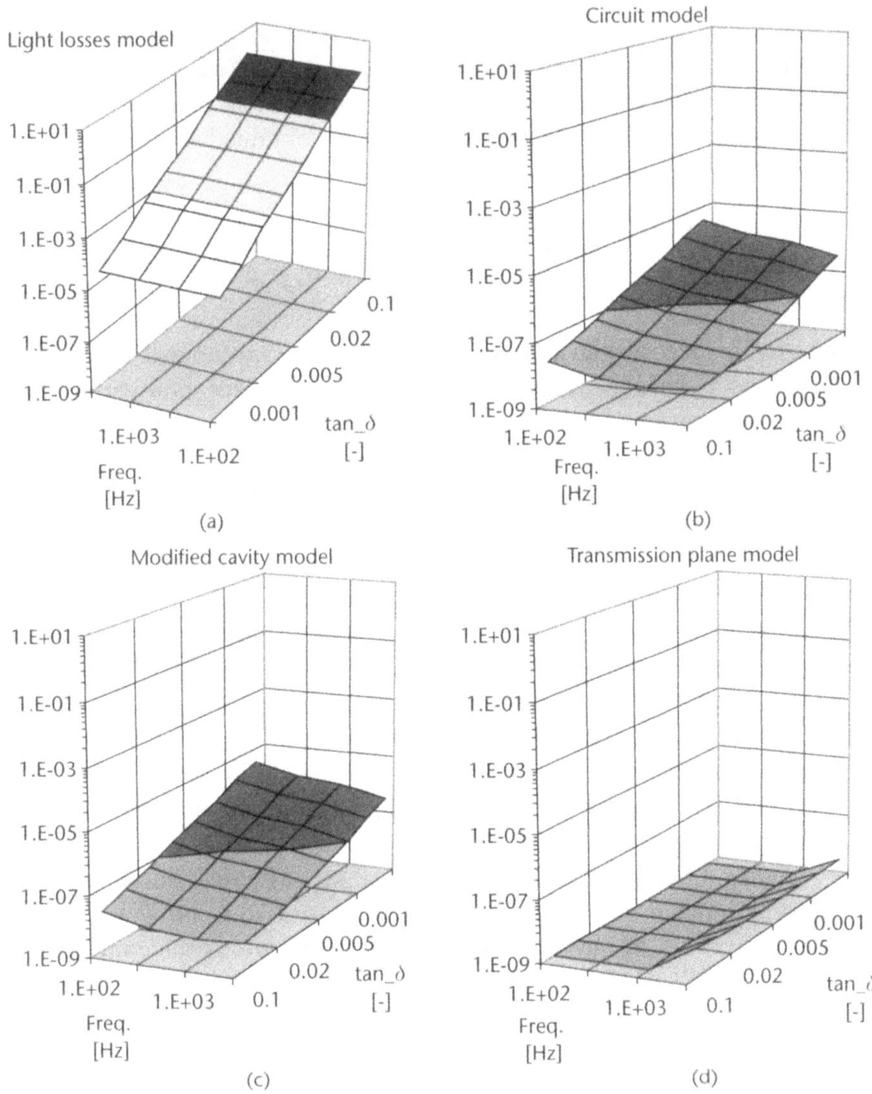

Figure 4.8 Surface plots of the percentage difference in the extracted loss tangent across a range of extraction frequencies and loss tangent values using the (a) light losses model, (b) circuit model, (c) modified cavity model, and (d) the transmission plane model.

lowest overall difference in the extracted loss tangent of the four lossy plane expressions.

In the second batch of simulations, the dielectric loss tangent was fixed to a low value (0.0001) to approximate a lossless dielectric, while the conductivity was swept over a large range. In particular, the base copper conductivity (5.8×10^7) was multiplied by 1, 10^3, 10^6, 10^{12}, and 10^{18}. The large range of conductivity values was chosen to evaluate the limits of the expressions not (necessarily) to represent practical values. By using a low dielectric loss, the impact of the metal conductivity on the impedance expressions can be examined. The dielectric constant, ε_r, was set to an arbitrary value of 4.1 for these simulations. As before, the dielectric constant and loss tangent were only fixed to these values at 1 MHz for the transmission plane model. At each conductivity multiplier, all four expressions were evaluated and the loss tangent and dielectric constant was extracted as a function of frequency from the impedance. The dielectric constant was obtained directly from the imaginary portion of the impedance as follows

$$\varepsilon_r(\omega) = -\frac{h}{\omega\left(\mathrm{Im}(Z(\omega)) - \omega L\right)\varepsilon_0 w_x w_y} \tag{4.28}$$

where L is the inductance of the plane at the series resonance. If the four expressions placed no limitations on the conductor losses and the expressions were causal, we expect the extracted loss tangent and dielectric constant values to match the simulated values across the range of extraction frequencies.

Figure 4.9 plots the difference in the extracted dielectric loss using the four different impedance formulations. Figures 4.9(a) through 4.9(c) show a sharply increasing difference in the extracted dielectric loss as the conductivity is reduced. Also, there is a slight increase in the difference with decreasing frequency. Both of these trends are due to skin effects; only at very high conductivities and/or at high frequencies do we find skin effects be minimized. For the base copper conductivity (i.e., multiplier of 1), the error in the extracted loss tangent is very significant due to skin effects. Figure 4.9(d) plots the difference in the extracted loss tangent using the transmission plane model. This model shows the lowest overall difference in the extracted loss tangent of the expressions. The increase at the lowest conductivity values is quite moderate (0.01%).

Figure 4.10 plots the difference in the extracted dielectric constant using the four different impedance formulations. Figures 4.10(a) through 4.10(c) show very significant differences in the extracted dielectric constant as the conductivity is reduced due to skin effects. Figure 4.10(d) plots the difference in the extracted dielectric constant using the transmission plane model. This model shows no dependence of the dielectric constant on the copper conductivity.

Finally, we present a simple work-around for avoiding the roll-off at low-frequencies due to skin effects observed in most of the lossy impedance expressions. For example, (4.23) shows that the low-frequency roll off becomes significant when the skin depth is approximately equal to or greater than the dielectric thickness, h. If we substitute the following modified expression for the skin depth, δ_{mod}, then we can "clip" the skin depth at low frequencies, thereby avoiding the roll-off while still maintaining the proper skin effect behavior at high frequencies:

$$\delta_{\text{mod}} = \frac{1}{\frac{1}{\delta} + \frac{1}{t}} \tag{4.29}$$

where t is the thickness of the upper and lower planes. As an example, δ_{mod} was substituted into the modified plane impedance expression, which assumes light losses (4.6) through (4.7). Figure 4.11 plots the plane impedance for the structure shown in Figure 2.1 using the light-losses cavity model and using the modified light losses cavity model, which uses (4.29). The inclusion of (4.29) is observed to remove the

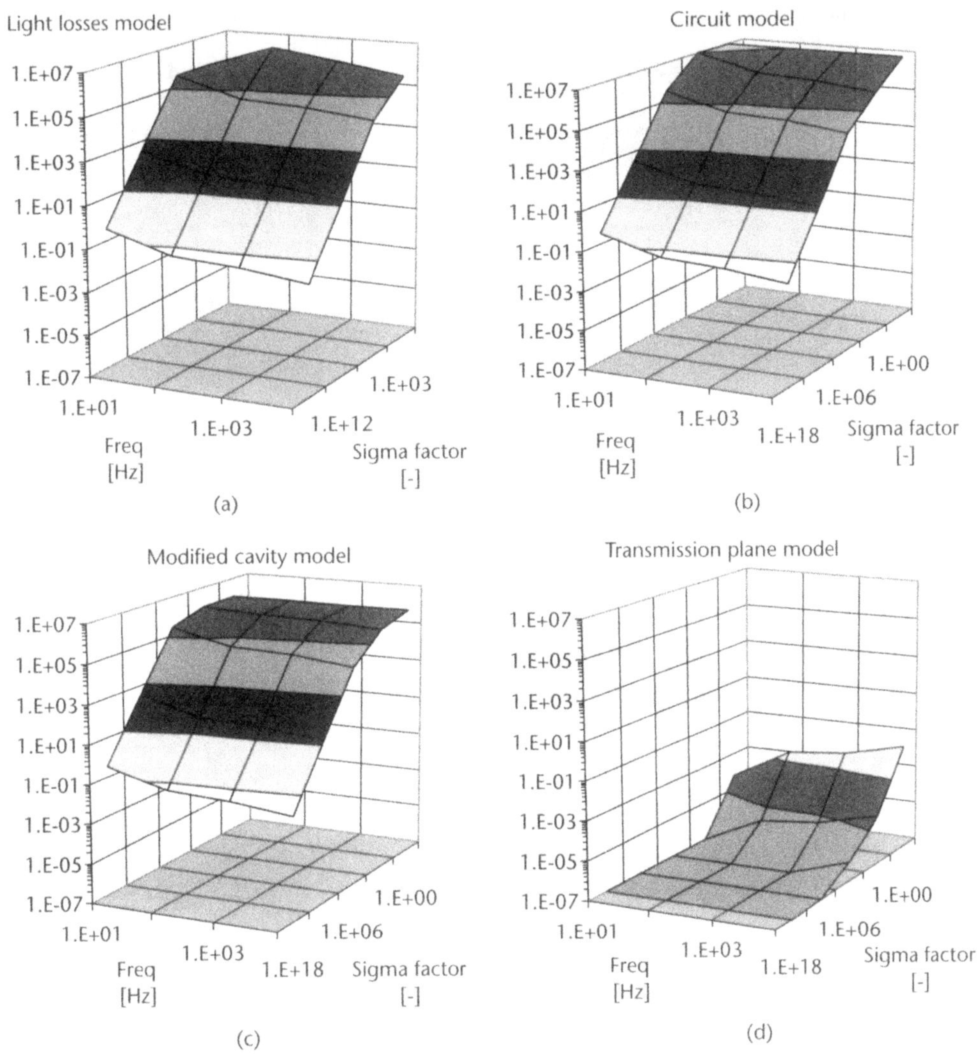

Figure 4.9 Surface plots of the percentage difference in the extracted dielectric loss across a range of extraction frequencies and copper conductivity values using (a) the light losses model, (b) circuit model, (c) modified cavity model, and (d) the transmission plane model. The *sigma factor* is a multiplicative constant applied to base copper conductivity.

low-frequency roll-off due to skin effects. Of course, the limitation on the light losses are not removed by this technique. However, as we observed from the error plots, inaccuracy due to dielectric losses only start being significant at very high tan_δ values.

4.2.2 Analytical Plane Models for Arbitrary Plane Shapes

The analytical plane models are based on the cavity modal resonances; in their original forms as shown above, they are all limited to rectangular plane shapes. The analytical models can still be applied to irregular-shape power-ground plane pairs by using the segmentation method [13, 14]. First we approximate the irregular plane

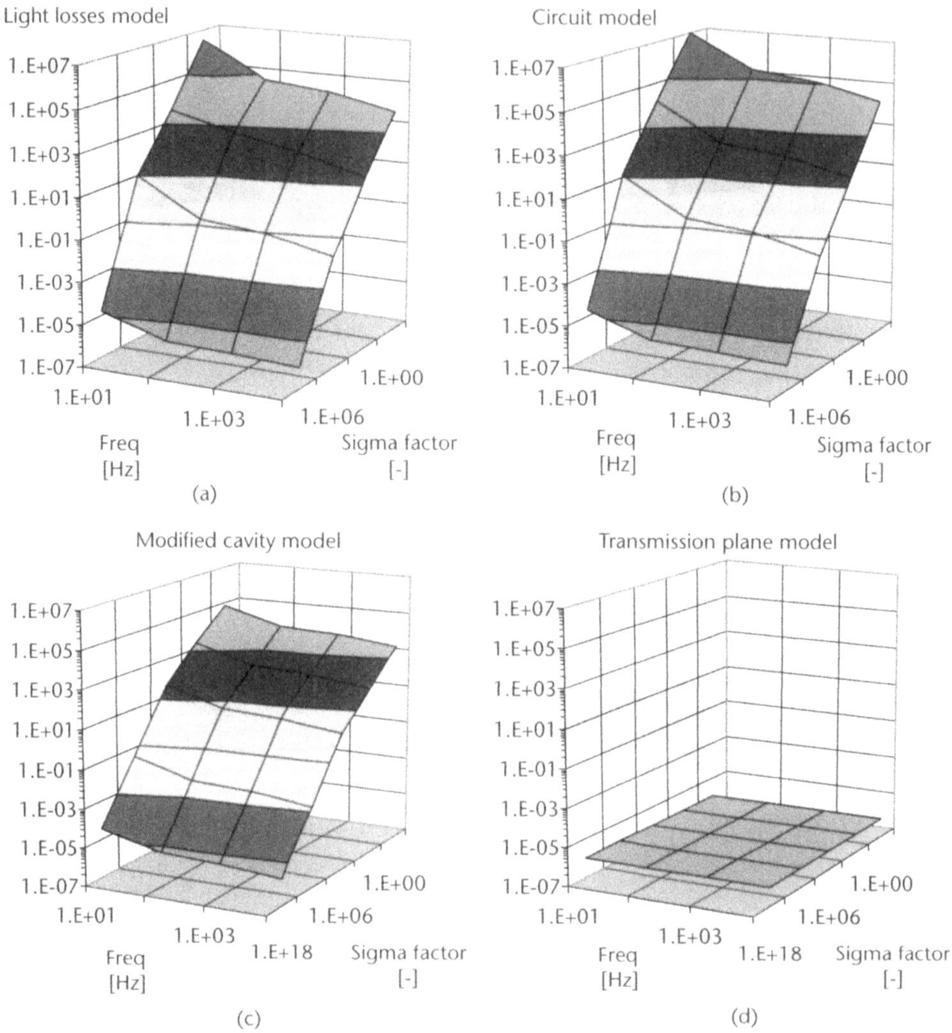

Figure 4.10 Surface plots of the percentage difference in the extracted dielectric constant across a range of extraction frequencies and copper conductivity values using the (a) light losses model, (b) circuit model, (c) modified cavity model, and (d) the transmission plane model. The *sigma factor* is a multiplicative constant applied to base copper conductivity.

shape with the sum of a series of rectangular shapes; these shapes approximate the irregular plane shape with sufficient accuracy. Second, we assign temporary ports along the sides of neighboring rectangular shapes. By enforcing the continuity of voltages and currents at the temporary ports, the impedance matrix of the combined rectangles can be obtained. The process is illustrated with a simple L shape, which is decomposed into two rectangles, as shown in Figure 4.12.

The illustration is based on the fact that along the marked line we can decompose the L shape into two rectangles, marked as α and β on the right. The sum of the two shapes make up the original shape, marked γ. Furthermore, we assume that before the cut, there are p and q ports in segments α and β, respectively. After the cut, two matching sets of temporary ports are added along the cut line to the two segments. To get sufficient accuracy from the segmentation method, the temporary

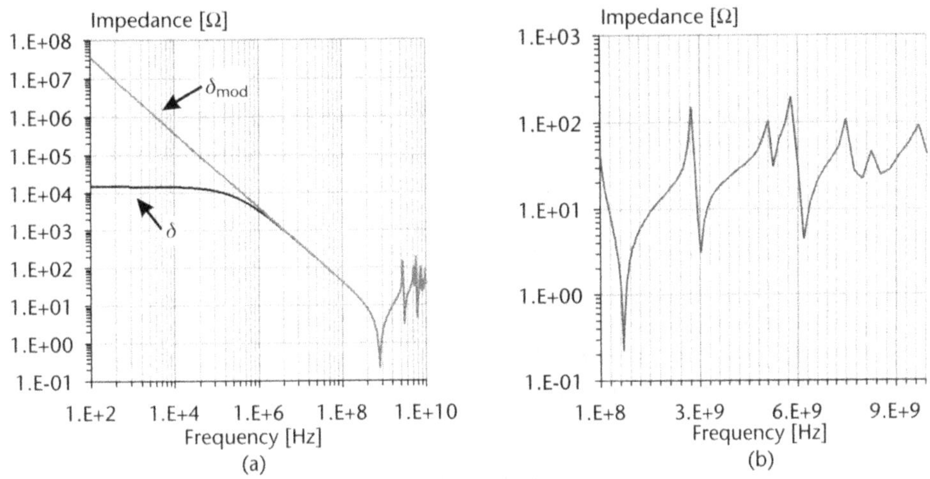

Figure 4.11 Impedance profile obtained using the light losses cavity model with and without (4.29). Part (b) is an enlarged view of (a) showing that the two approaches yield identical high frequency results.

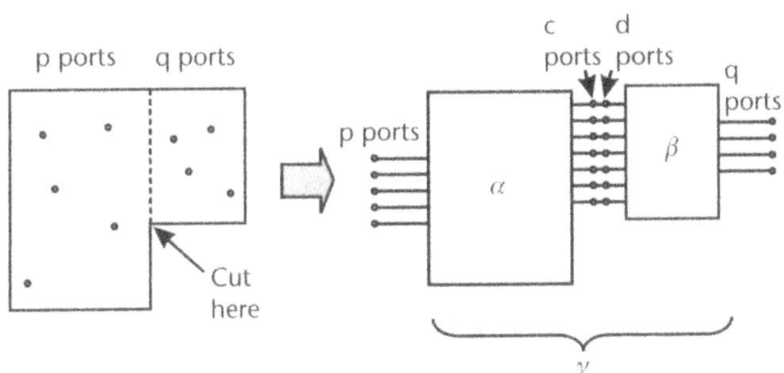

Figure 4.12 Illustration of the segmentation method to calculate the impedance matrix of irregular plane shapes.

ports have to be assigned with a spacing much less than the shortest wavelength of interest; these temporary ports are marked as *c ports* and *d ports* on the figure. The impedance matrices for α, β and γ can be partitioned into sub matrices corresponding to the c, d, p, and q sets of ports:

$$Z_\alpha = \begin{bmatrix} Z_{pp\alpha} & Z_{pc} \\ Z_{cp} & Z_{cc} \end{bmatrix}, \quad Z_\beta = \begin{bmatrix} Z_{dd} & Z_{dq} \\ Z_{qd} & Z_{qq\beta} \end{bmatrix}, \quad Z_\gamma = \begin{bmatrix} Z_{pp\gamma} & Z_{pq} \\ Z_{qp} & Z_{pp\gamma} \end{bmatrix} \tag{4.30}$$

At the temporary ports, voltages and currents must equal on the two sides. From this condition we get:

$$Z_\gamma = \begin{bmatrix} Z_{pp\alpha} - Z_{pc} Z'_{dp} & Z_{pc} Z'_{dq} \\ Z_{qd} Z'_{dp} & Z_{qq\beta} - Z_{qd} Z'_{dq} \end{bmatrix} \tag{4.31}$$

where

$$Z'_{dp} = \left[Z_{cc} + Z_{dd}\right]^{-1} Z_{cp}, \quad Z'_{dq} = \left[Z_{cc} + Z_{dd}\right]^{-1} Z_{dq} \tag{4.32}$$

Once the impedance matrix for the entire plane shape is available, the matrix size can be reduced by eliminating the entries corresponding to the temporary ports.

4.3 Transmission-Line Models

Although transmission-line models are better suited for rectangular plane shapes, the approach can be extended to handle irregularly shaped planes using adaptive gridding or transmission matrix models.

4.3.1 Transmission-Line Grid Models for Rectangular Plane Shapes

As shown in Figure 4.13, rectangular plane shapes can be discretized by overlaying a square or rectangular grid, which divides the planes into unit cells [15–17]. The unit cells can be square; this will result in a different number of cells along the two axes for a rectangular plane. Alternately, the same number of cells can be used along both axes; this method retains the aspect ratio of the planes in the unit cell.

Each cell is then substituted with an equivalent circuit; this circuit represents the transmission-line behavior along the unit cell's sides or along their center lines. Figure 4.14 shows the two fundamental options in terms of assigning transmission lines to unit cells. When transmission lines are assigned to the borders of the unit cells, the resulting SPICE grid is closed, with no floating nodes. The transmission lines along the periphery, however, represent only half of the area compared to transmission lines inside the grid. It is easy to compensate for this by adjusting the parameters of the lines along the periphery.

When the transmission lines are assigned to the center lines of the unit cells, all of the transmission-line segments horizontally or vertically will have the same parameters, but now we encounter different problems: segments facing the plane edges will create open nodes, and portions of the original plane area along the edges will not be covered. These can be corrected for by applying dummy elements at the open ends to eliminate floating nodes and correcting for the uncovered plane area.

There are several possible options to model the transmission behavior in the unit cells. Some of the options are shown in Figure 4.15. Discrete RLGC circuits with fixed parameters can be used in either time or frequency-domain SPICE simulations; but, except for the lossless case, the model is not causal. Causal frequency-dependent RLGC parameters can be approximated with more complex subcircuits; each

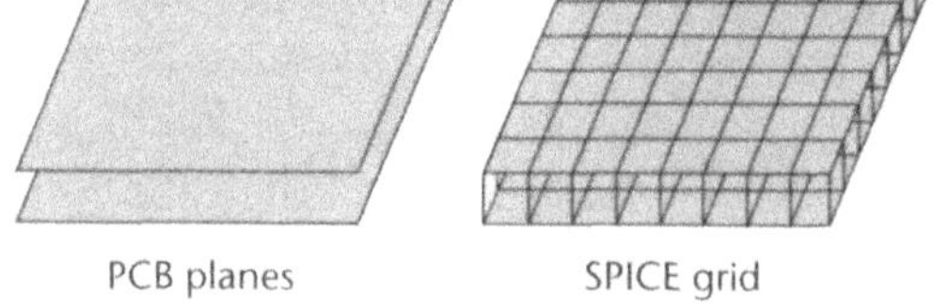

Figure 4.13 Discretization of a rectangular plane pair into unit cells.

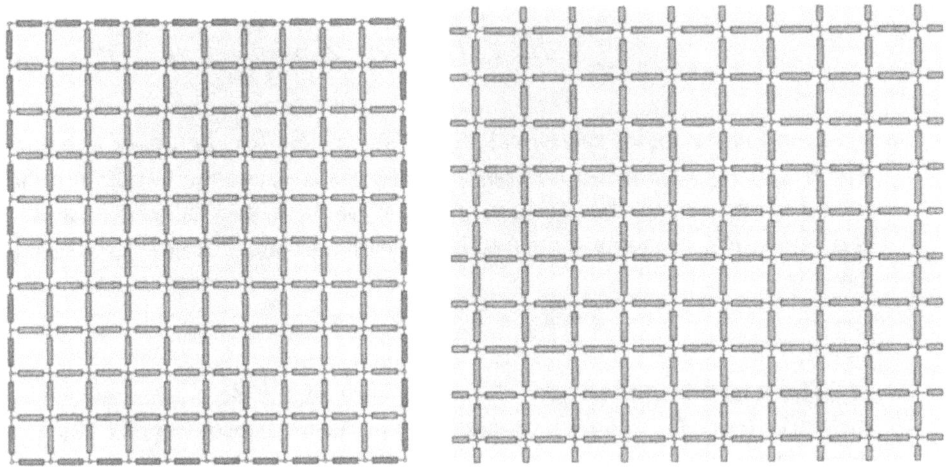

Figure 4.14 Mapping unit cells to transmission lines.

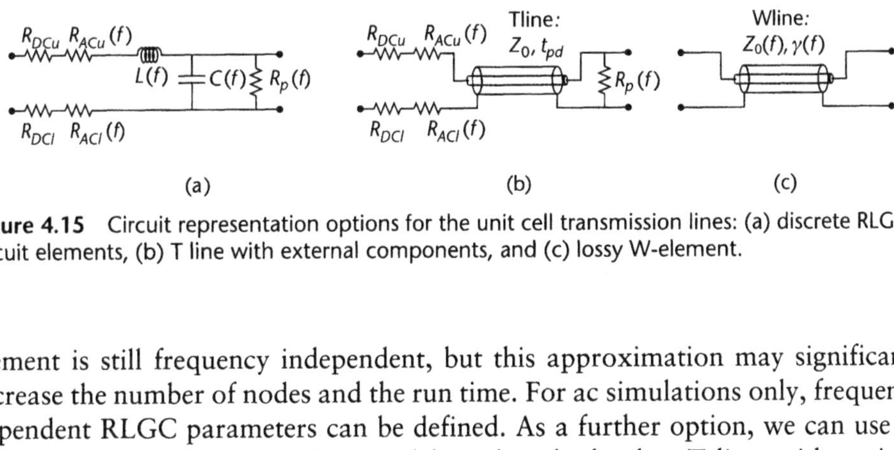

Figure 4.15 Circuit representation options for the unit cell transmission lines: (a) discrete RLGC circuit elements, (b) T line with external components, and (c) lossy W-element.

element is still frequency independent, but this approximation may significantly increase the number of nodes and the run time. For ac simulations only, frequency-dependent RLGC parameters can be defined. As a further option, we can use the built-in SPICE transmission line models: either the lossless T line, with optional external components to approximate losses, or the lossy W-line element. However, we have to keep in mind that by using transmission-line elements we loose the horizontal connectivity along the unit cells because the input and output potentials of the transmission-line elements are floating with respect to each other.

The transmission-line matrix model, also called the *bedspring model*, can capture arbitrary losses and different thicknesses and/or different conductivities in the upper and lower conductive planes.

To obtain the unit-cell parameters, we can start with the lossless characteristic impedance and propagation delay expressions and then add losses as perturbation. The quasi-static approximation of plane capacitance, C, calculates the plate capacitance for the cell area represented by the transmission line. The propagation delay, t_{pd}, along the length of the cell gives the second independent parameter. From the capacitance and propagation delay, the two dependent parameters, the characteristic impedance, Z_0, and the inductance, L, can be derived from the basic transmission-line equations:

$$t_{pd} = \sqrt{LC}, \quad L = \frac{t_{pd}^2}{C}, \quad Z_0 = \sqrt{\frac{L}{C}} \tag{4.33}$$

Conductive losses can be calculated separately for the upper and lower planes. The frequency-dependent resistance can be easily used in ac SPICE simulations. The model approximates the conductive loss, R, as the sum of dc resistance, R_{dc}, and skin resistance, R_{skin}, assuming a $\sqrt{f}$ frequency dependence for the skin resistance:

$$R = R_{dc} + R_{skin}, \quad R_{skin} = R_s\sqrt{f} \tag{4.34}$$

The parallel conductance, G, is the sum of the dc conductance, G_{dc}, and the dielectric loss, G_{diel}, which is assumed to have approximately linear frequency dependence.

$$G = G_{dc} + G_{diel}, \quad G_{diel} = G_d f \tag{4.35}$$

Note that (4.33) through (4.35) still do not result in causal solutions, because the frequency dependencies of L and C are not included. If necessary, causal solutions can also be included by using the causal RLGC solution described in Section 4.5.5.2.

4.3.2 Transmission-Line Grid Models for Arbitrary Plane Shapes

There are several limitations related to uniform rectangular SPICE grids applied to arbitrary power plane shapes [18]. As an example, consider the power-ground plane shape of Figure 4.16.

Figure 4.17 shows the outline of the inner plane shape with a 6.35-mm (0.25-inch) geometrically uniform square grid fitted over its envelope. The uniform

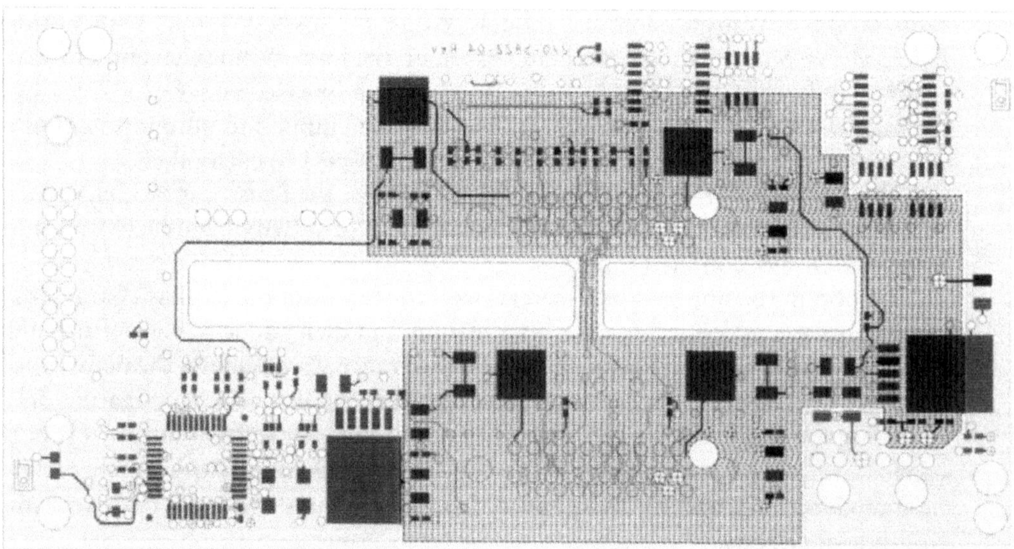

Figure 4.16 Within the outer rectangular board outline, there is an inner odd-outline plane shape with a varying degree of perforations, due to smaller and larger holes, as well as with a large cutout.

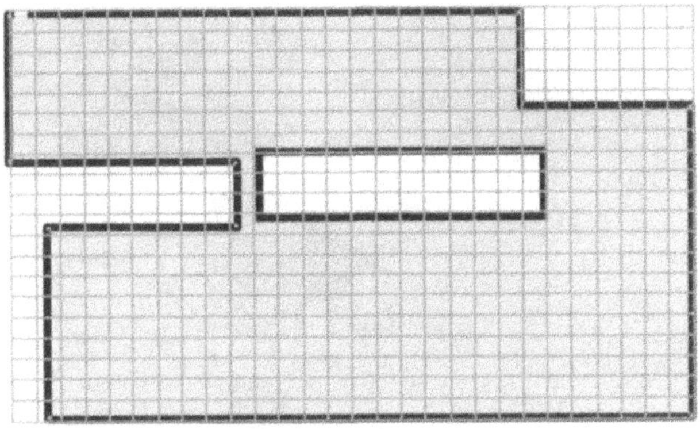

Figure 4.17 Outline of the highlighted inner plane pair shape from Figure 4.16, with a uniform square-unit-cell SPICE grid laid over it. Each side of the overlaid grid cells represents one piece of transmission line in the equivalent SPICE circuit.

grid has 28 cells horizontally and 20 cells vertically, totaling $28 \times 20 = 560$ cells and $(28 + 1) \times (20 + 1) = 609$ nodes.

Figure 4.17 shows that dependent on the actual outline and cutouts, there may be unnecessary cells and nodes in the uniform grid. In this example, out of the total of 560 cells (and 609 nodes), altogether there are 91 cells outside of the actual plane shape we want to simulate. Another problem with using uniform rectangular grids for irregular shapes is that in SPICE, run time grows sharply as the number of nodes increases. Any unnecessary nodes increase the run time without the benefit of higher resolution/accuracy. Furthermore, there may be areas where smaller grid cells may be necessary (e.g., around odd-shaped outline contours or in perforated areas); if the entire plane is meshed with the smallest grid size, the total grid number may again increase unnecessarily. Finally, modal resonances may not be captured correctly with uniform grids. One of the major roles of SPICE models of planes is to capture modal resonances so that bypass capacitors can be applied properly to smooth out the impedance profile. Modal resonance frequencies depend on the possible standing-wave pattern; that pattern is determined by the actual boundary shapes and cutouts. If it is not captured accurately, the simulated resonance frequencies are in error.

To handle complex outline shapes, results are shown below using an adaptive, variable-size cell SPICE grid. The shape is more coarsely gridded in solidly filled areas and gradually converges into a finer mesh around the shape's outline and (possible) inner cutout contours by using square unit cells. The resulting SPICE grid preserves the actual static plane-capacitance by dropping cells completely that are not at least partly on the plane shapes and adjusting the electrical parameters of unit cells that are either not entirely on the plane shape or are not solidly filled (e.g., due to antipads).

In each unit cell, the amount of metal within the unit cell's area is calculated separately for the two conductive planes; the conductive loss values are adjusted according to those fill ratios. The common set of the two planes' metal contents is also calculated (as shown in Figure 4.18), and this is used to adjust the transmission characteristics of the SPICE grid elements.

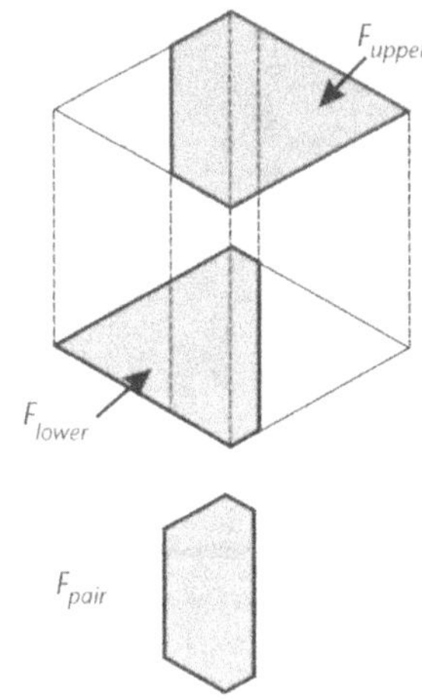

Figure 4.18 Illustration of metal fill ratios used for the adaptive grid.

To further preserve the static capacitance value of planes, as indicated in Figure 4.19, compensating capacitors are introduced in the SPICE grid to account for the missing coverage on the boundary of different-size unit cells.

Finally, following the same procedure, multiple plane pairs connected in parallel by vias can be handled as separate pairs first, then the SPICE grids of individual pairs can be linked. Figure 4.20 shows the adaptive grid for the example plane shape. Figure 4.21 shows the measured impedance profile compared to the simulated responses with uniform and adaptive variable-size cell grid.

Note that the adaptive grid captures the static capacitance and the modal resonances accurately. Since the uniform rectangular grid follows the outer envelope of the shape, it overestimates the static capacitance (as it does not account for the cutouts and missing portions along the jagged outline); it also overestimates the first modal resonance frequency. However, with the rectangular uniform grid, both conditions cannot be met at the same time by adjusting the envelope of the rectangular uniform grid: any attempt to decrease the outline to match the static capacitance more closely would increase the predicted first modal resonance frequency further, and vice versa.

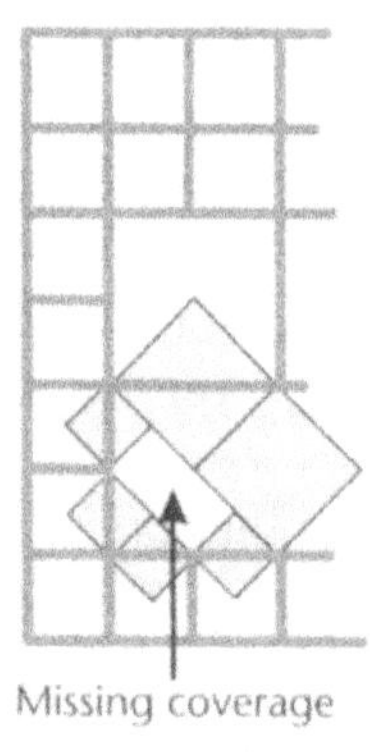

Figure 4.19 Compensation for missing unit-cell coverage.

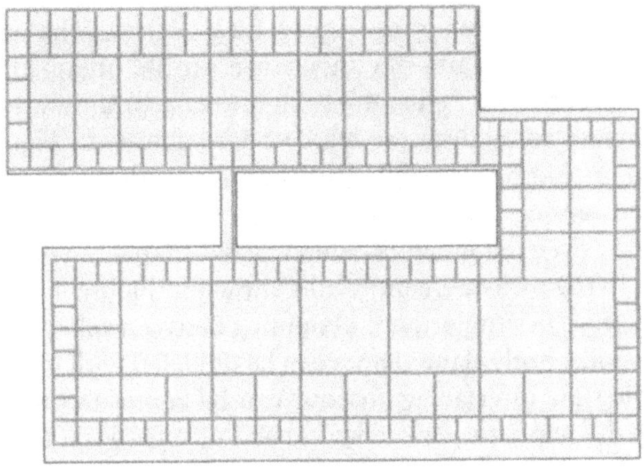

Figure 4.20 Result of adaptive subgridding on the inner plane shape shown in Figure 4.16.

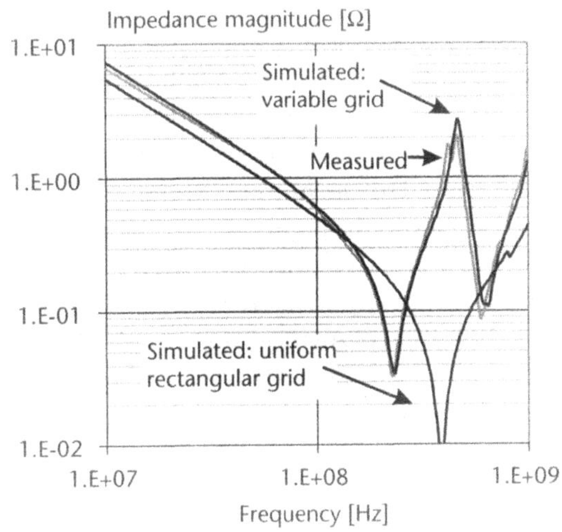

Figure 4.21 Impedance profiles of the inner plane shape shown in Figure 4.16, obtained from measurement and simulation using a rectangular uniform grid and adaptive grid.

The first modal resonance from the propagation delay along the longer side of the rectangular envelope can be calculated from the length and dielectric constant: it is 800 MHz for the first peak. Note that there is a small glitch at around 800 MHz in the impedance simulated with a rectangular uniform grid. It is not pronounced because of the location of the test point. If the plane had no cutout and were to follow the rectangular outline of the envelope, the modal resonance would be highly suppressed at this same location. However, due to the odd outline and cutout of the inner plane shape, the actual plane cut has a much lower first modal resonance frequency, about half of the frequency obtained from the uniform grid.

The dual peak at the first modal resonance in the measured impedance profile of Figure 4.21 is the result of a trace passing over both the inner and outer plane shapes shown in Figure 4.16. This was proven in measurement by cutting the inner plane shape along its periphery, thus cutting the trace while leaving intact the inner plane shape, and then remeasuring. Figure 4.22(a) shows the new measured impedance profile compared against the impedance profile obtained with the same adaptive grid that was used for Figure 4.21. Figure 4.22(b) shows the simulated results using Ansoft SIwave to analyze the whole board shown in Figure 4.16. The board was simulated, including and excluding coupling from the trace to the inner plane shape (all other coupling was enabled). The simulation results demonstrate that the trace that crosses over the inner and outer plane shapes is responsible for the dual peak observed in the measurement results shown in Figure 4.21.

To capture in simulation the coupling between split plane shapes using the adaptive grid mode, both plane shapes can be modeled with its own adaptive grid and the nodes along the interfacing contour can be connected with coupling capacitors to represent the plane-to-plane edge capacitance between the two shapes [19].

4.3.3 Transmission Matrix Model for Arbitrary Plane Shapes

The transmission matrix model [20] can be used to model both rectangular and arbitrary plane shapes. The method relies on the fact that power planes are linear networks, and as such the plane can be subdivided into smaller networks and the networks cascaded. The overall network response is simply the product of the individual transfer matrices. (Note that only transfer matrices are multiplicative, others, like $[S]$ or $[Z]$ are not.) This technique can be applied to irregular plane geometries, like the L-shaped structure in Figure 4.23. Although the size of the matrices associated with sections 1 and 2 of Figure 4.23 do not match, the matrix array for section 2 can be padded with zero matrix elements to match the matrix size of section 1.

The transmission matrix method starts with dividing the plane into unit cells, as shown in Figure 4.23. Each cell is represented by an equivalent circuit representation using either T or Π models. Starting with plane section 1 in Figure 4.23, the larger

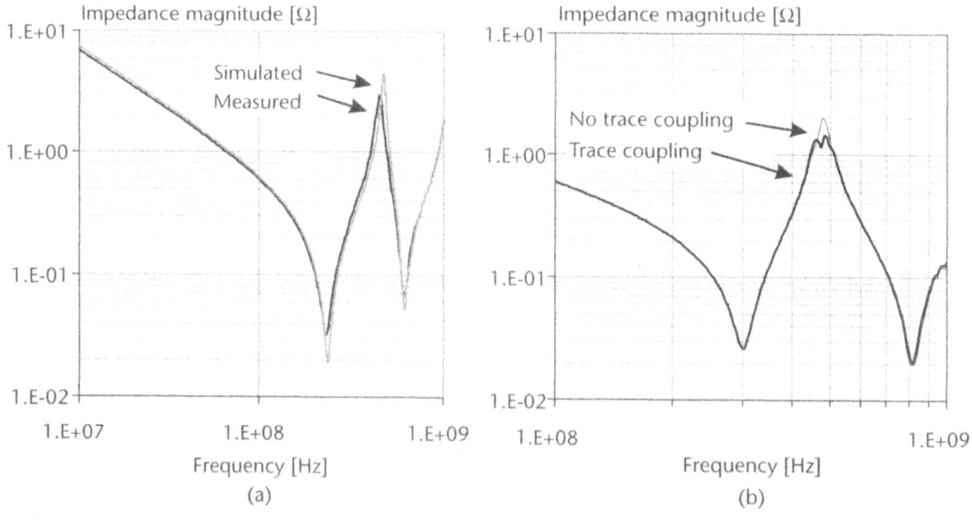

Figure 4.22 (a) Impedance profiles of the inner plane shape shown in Figures 4.16 obtained from measurement and simulation using an adaptive grid. Measured after cut around plane periphery. Part (b) shows field solver results.

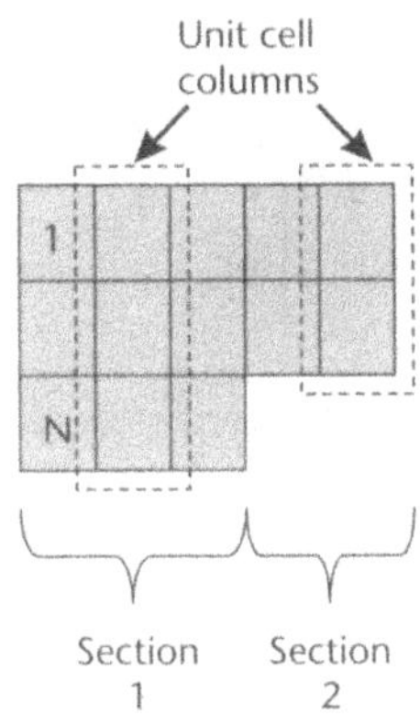

Figure 4.23 Top view of an L-shaped plane.

rectangular shape can be represented by a number of unit cell columns, in this case 3. Each $N \times 1$ unit cell can be represented as a $2N \times 2N$ matrix formed by N input ports and N output ports. Thus the transmission matrix for section 1 is a 6×6 matrix

$$[T_1] = \begin{bmatrix} T_{11} & T_{12} & T_{13} & T_{14} & T_{15} & T_{16} \\ T_{21} & T_{22} & T_{23} & T_{24} & T_{25} & T_{26} \\ T_{31} & T_{32} & T_{33} & T_{34} & T_{35} & T_{36} \\ T_{41} & T_{42} & T_{43} & T_{44} & T_{45} & T_{46} \\ T_{51} & T_{52} & T_{53} & T_{54} & T_{55} & T_{56} \\ T_{61} & T_{62} & T_{63} & T_{64} & T_{65} & T_{66} \end{bmatrix} \tag{4.36}$$

Equation (4.36) can be rewritten in a simpler form:

$$[T_1] = \begin{bmatrix} T_{1A} & T_{1B} \\ T_{1C} & T_{1D} \end{bmatrix} \tag{4.37}$$

where $[T_{1A}]$, $[T_{1B}]$, $[T_{1C}]$, and $[T_{1D}]$ are 3×3 matrices. Then the overall network for section 1 (of Figure 4.23) can be obtained by multiplying the individual matrices for each column. Since all the matrices for each of the three columns are the same, the response for the entire geometry can be obtained from a single 6×6 matrix as follows

$$[T_m] = [T_1]^3 \tag{4.38}$$

If the input and output ports are open circuited, then the $[T_m]$ would be multiplied by the identity matrix as follows

$$[T'] = [I_L][T_m][I_R] \tag{4.39}$$

Then section 2 of Figure 4.23 can be included by modifying the identity matrix on the right side of (4.39), $[I_R]$, as follows. The transmission matrix for section 2 is a 4×4 matrix that can be written as:

$$[T_2]=\begin{bmatrix} T_{11} & T_{12} & T_{13} & T_{14} \\ T_{21} & T_{22} & T_{23} & T_{24} \\ T_{31} & T_{32} & T_{33} & T_{34} \\ T_{41} & T_{42} & T_{43} & T_{44} \end{bmatrix} \tag{4.40}$$

which can then be incorporated into the identity matrix, $[I_R]$, as follows:

$$[I'_R]=\begin{bmatrix} T_{11} & T_{12} & 0 & T_{13} & T_{14} & 0 \\ T_{21} & T_{22} & 0 & T_{23} & T_{24} & 0 \\ 0 & 0 & I & 0 & 0 & 0 \\ T_{31} & T_{32} & 0 & T_{33} & T_{34} & 0 \\ T_{41} & T_{42} & 0 & T_{43} & T_{44} & 0 \\ 0 & 0 & 0 & 0 & 0 & I \end{bmatrix} \tag{4.41}$$

The transmission matrix for the entire structure is then

$$[T']=[I_L][T_m][I'_R] \tag{4.42}$$

Knowing the transmission matrix for the network, one can now determine the impedance matrix [Z], including the impedance at specific points on the plane.

4.4 Effect of Plane Parameters on Self- and Transfer Impedances

In this section, we examine how the plane parameters influence the plane impedance and resonances. Specifically, the impact of the dielectric thickness, plane thickness, number of power ground plane pairs, and dielectric constant and dielectric loss on the plane impedance is discussed [21]. The rectangular plane structures were simulated using the transmission-line grid method with lossless transmission lines and external components to approximate the losses. Even though the model is noncausal, it is sufficiently accurate to show how the impedance profile changes as a function of the plane parameters.

4.4.1 Impact of Dielectric Thickness with Regular Conductors

Resonances of bare planes can contribute to and increase simultaneous switching noise, ground bounce, or V_{cc} bounce. Thin dielectric materials by themselves can effectively help to suppress plane resonances of bare boards. The mechanism responsible for this is best understood by looking at the real part of the propagation constant of the transmission line segments in the equivalent model circuit. The attenuation of a matched interconnect is:

$$A(f)^{dB}=4.35\left(\frac{R_s(f)}{Z_o}+G_d(f)Z_o\right) \tag{4.43}$$

where Z_0 is the characteristic impedance of the transmission line, and $R_s(f)$ and $G_d(f)$ are the series conductive and parallel dielectric loss values versus frequency. As the dielectric thickness decreases, skin effect losses remain constant, but the characteristic impedance, that is, $Z_0 = \sqrt{L/C}$, decreases proportionally with the dielectric thickness. This happens because inductance and capacitance are proportional and inversely proportional to the dielectric thickness, respectively. With decreasing dielectric thickness, the dielectric loss term eventually decreases, thus leaving the skin loss responsible for the suppression of plane resonances. This simple approximation shows how thin dielectrics between power and ground planes have tremendous advantages for power distribution systems at high frequencies.

A uniform rectangular lossy SPICE grid model was applied to simulate a pair of 25 × 25 cm (10 × 10 inch) parallel planes with 35-μm (1-oz) copper on either side, but with variable thickness of dielectric separation. The dielectric constant was assumed to be 4. The grid size was 20 × 20, providing at least 1 GHz of useful upper limit for the model.

Figure 4.24 shows the magnitude and phase of simulated self-impedance measured between the upper and lower planes at the center. As the dielectric thickness is reduced, the impedance profile becomes smoother at high frequencies. There is another obvious advantage: due to the increase of static capacitance, the low-frequency impedance is reduced. In turn, this reduction helps to reduce the need for low-frequency bulk capacitors. Furthermore, this results in a complete suppression of plane resonances for dielectric thicknesses below 8 μm (0.3 mil). Phase figures manifest this trait as well [see Figure 4.24(b)]; with thin dielectrics, the phase of the self-impedance becomes more resistive. Also note that the slant of the self-impedance magnitude at high frequencies is due to the increase of skin resistance with the square root of frequency. Figure 4.25 proves the assumption that increasing series losses create a low-pass transfer function. While dielectric thicknesses above 25 μm (1 mil) yield a transfer function with noticeable peaks at high frequencies. A thickness of 2.5 μm (0.1 mil) creates a flat response. Even a thinner dielectric separation provides a monotonic low-pass function. The series losses also increase the upper

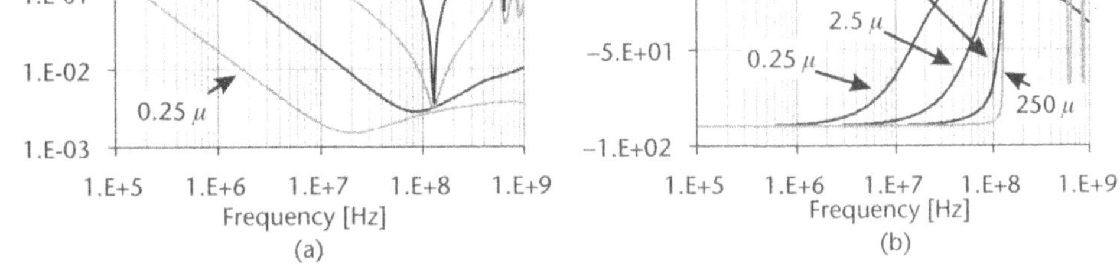

Figure 4.24 Effect of dielectric thickness on the self impedance of a pair of 25 × 25 cm planes, with 35-μm copper on either side. (a) The simulated self-impedance magnitude and (b) the phase are probed at the center of planes.

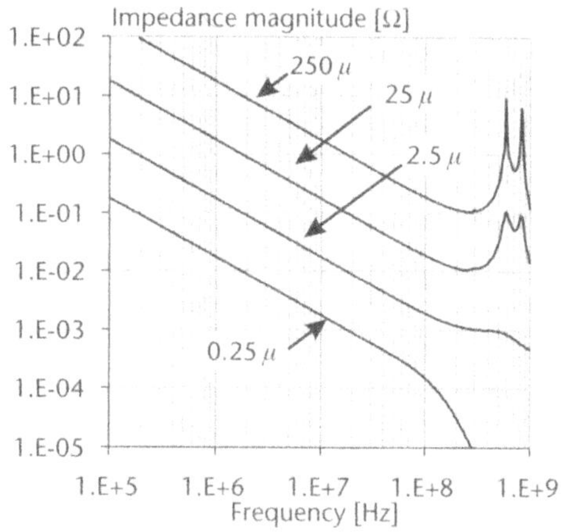

Figure 4.25 Effect of dielectric thickness on the transfer impedance of a pair of 25 × 25 cm planes, with 35-μm copper on either side. The simulated magnitude of the transfer impedance is probed between the center and one of the corners of the planes.

frequency limit of the simulation model by reducing the reflections, thereby effectively creating a resistor rather then transmission line grid at high frequencies. The illustrations in this section are shown for 10 × 10-inch plane sizes, smaller plane shapes result in higher characteristic impedance. Hence, from (4.43) the damping of resonances due to conductive losses will be less.

4.4.2 Impact of Plane Thickness

While the advantages of thin dielectrics are clear from the simulations results of Figures 4.24 and 4.25, it is not easy to manufacture and process a very thin dielectric layer of a few micrometer or less thickness with the usual several micrometer or more copper layers. With a 0.25-μm (0.01-mil) dielectric layer, the conductor layers may be about 100 times thicker.

To look at the other possible extreme, Figures 4.26 and 4.27 show the same structure under the same assumptions as Figures 4.24 and 4.25, except the conductive layer on both sides is assumed to be 0.25-μm (0.01-mil) copper. Note that the skin depth in copper at 1 GHz is approximately 2 μm (0.08 mil); that depth is about eight times higher than the selected copper thickness. Hence the series loss resistance is less dependent on frequency. By comparing Figure 4.24 to Figure 4.26 and Figure 4.25 to Figure 4.27, we can see that the series losses of the 0.25-μm conductive layers still leave considerable peaking in the impedance profile with thick (> 25-μm) dielectric layers.

In case of thin conductive layers, the high-frequency impedance does not drop inversely proportionally to the plane separation (as one would expect based on the equivalent inductance between the planes). Because the impedance now becomes limited by the series ac loss resistance. With a 0.025-μm (0.01-mil) copper conductors, the self-impedance profile is almost totally flat, because near 10 MHz the impedance of the static capacitance intercepts the series resistance. The higher series resistance also creates stronger low-pass filtering.

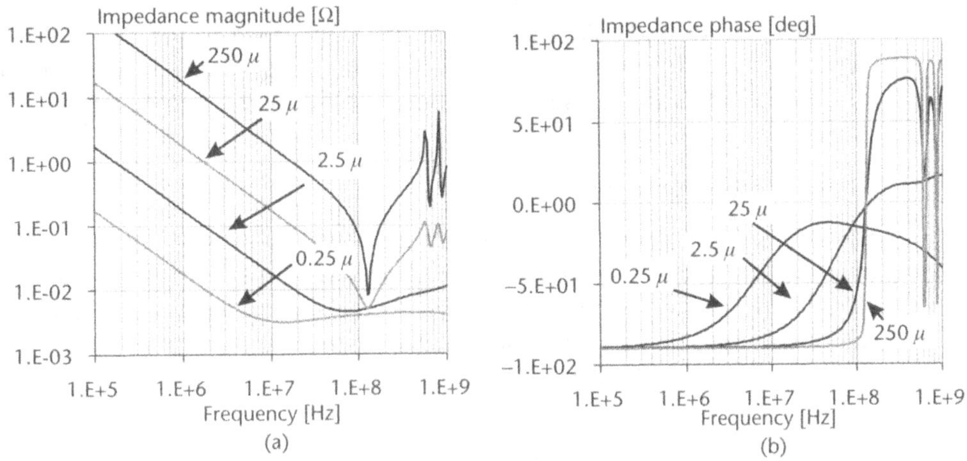

Figure 4.26 Effect of dielectric thickness on the self impedance of a pair of 25 × 25 cm planes, with 0.25-μm copper on either side. The simulated self-impedance (a) magnitude and (b) the phase, probed at the center of planes.

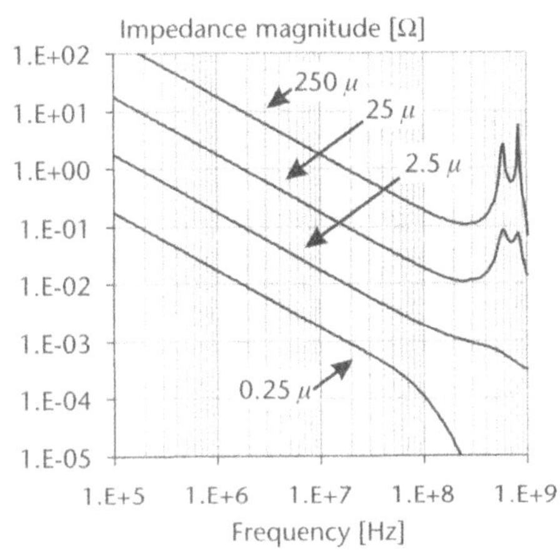

Figure 4.27 Effect of dielectric thickness on the transfer impedance of a pair of 25 × 25-cm planes, with 0.25-μm copper on either side. The simulated transfer impedance magnitude is probed between the two planes from the center to the corner.

On the other hand, using very thin conductive layers alone is not practical either because the copper weight may be needed to handle the large dc currents in the systems. However, by using both thin dielectric and thin conductive layers, we can provide the necessary copper weight by stacking up several of these thin layers. At the same time, this will reduce the impedance further because of the parallel connection of individual plane pairs. A typical 50-μm dielectric separation with 35-μm (1 oz) copper has a total thickness of 120 μm. If we used the 120 μm total thickness and we stacked up 240 pairs of 0.025-μm dielectrics with 0.025-μm conductive layers, we would end up with the same total thickness, same amount of total conductor weight on either side, and (neglecting the connecting impedance between the stacked layers) an approximately 25-$\mu\Omega$ flat resistive impedance in the 10–1,000-MHz frequency range. However, to use multiple thin conductive and dielectric layers in large-size rigid PCBs, several technological problems have to be addressed first.

4.4.3 Parallel Plane Pairs

Stacking power-ground plane pairs in parallel has most of its advantages if we reached the resistive bottom of the impedance profile with the thin dielectrics already. As illustrated in Figure 4.28, in the unsaturated range of the curves, a smoother impedance profile is obtained using one plane pair with thinner dielectrics as opposed to a stack of several thicker laminates.

4.4.4 Impact of Dielectric Constant and Dielectric Losses

The granularity of the power-ground plane models is important: each transmission line segment in the model should represent a small fraction of the wavelength of the highest frequency of interest. With a 15.24-cm (6-inch) square plane with 8 × 8 grid and $\varepsilon_r = 4$ dielectric constant, the accuracy of the model significantly deteriorates above 2 GHz. Since the propagation delay goes linearly with $\sqrt{\varepsilon_r}$, the same grid model is limited to about 1 GHz and 0.5 GHz, as the dielectric constant is increased to 16 and 64.

The typical PCB materials have been optimized for low-loss signal transmission; as a result, they do not provide sufficient suppression of plane resonances. If used only between the power/ground planes, intentionally high dielectric losses may be utilized. Figure 4.29 shows the effect of dielectric losses on the real part and magnitude of the self-impedance. The impedance magnitude curve show that a dielectric loss tangent of 0.3 or higher is sufficient to suppress almost completely the plane resonances.

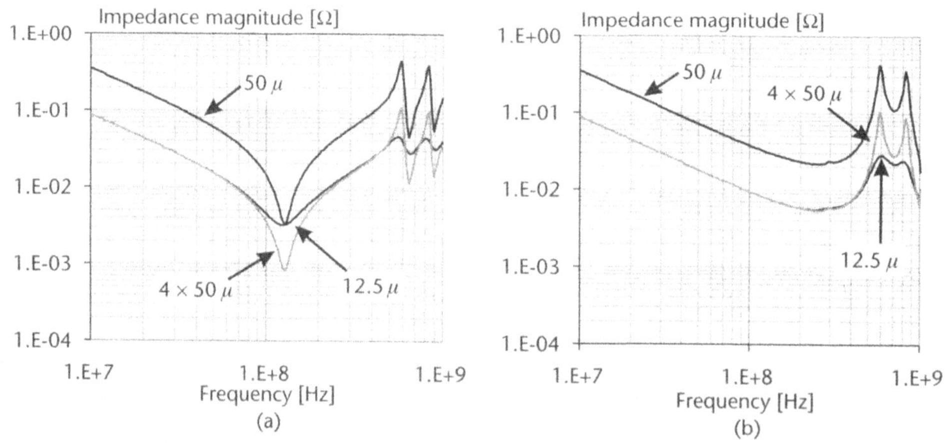

Figure 4.28 (a) Self- and (b) transfer impedance magnitude of a pair of 25 × 25 cm parallel planes with three different configurations: stand-alone pair with a 50-μm dielectric separation, stand-alone pair with a 12.5-μm dielectric separation and four pairs of planes with a 50-μm dielectric separation. The simulated self-impedance is probed between the planes at the center; transfer impedance is probed between the two planes from center to corner.

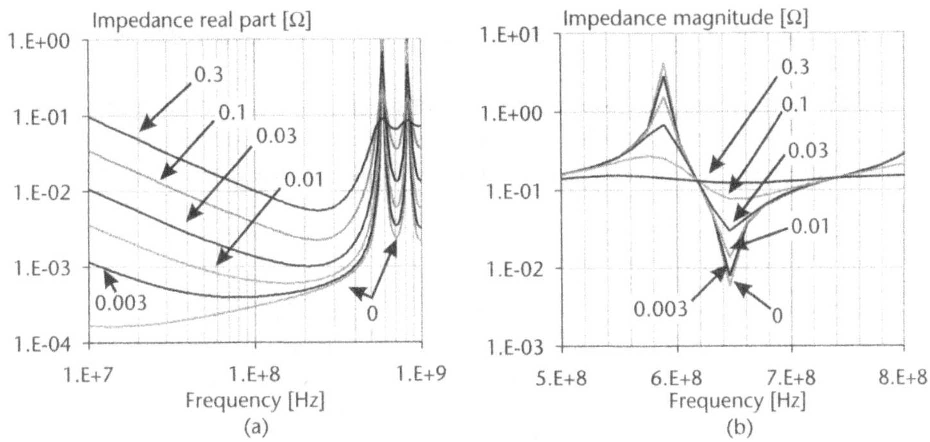

Figure 4.29 Effect of dielectric losses on the (a) real part and (b) magnitude of self-impedance measured at the center of a pair of 25 × 25-cm planes with a dielectric separation of 50-μm (2-mil), dielectric constant 4, and 35-μm (1 oz) copper.

4.4.5 Run Time Versus Number of Cells

The upper end of the valid frequency range of SPICE grid models depends linearly on the number of cells along each side. As the number of cells increases linearly along one side, the total number of cells and nodes in the SPICE equivalent circuit increases quadratically. The run time of SPICE is a nonlinear function of the number of nodes. Eventually, the total run time increases very sharply as we try to increase the upper frequency end by increasing the number of cells. Faster computers help, but the trend stays unchanged. Figure 4.30 shows the runtime as a function of number of cells along one side in a square grid. The run time is normalized to the value obtained with 10 cells along a side. Note that increasing the number of cells along a side from 5 to 30 (by a factor of 6) increases the number of cells and SPICE nodes by a factor of 36; this increases the run time by a factor of 91.

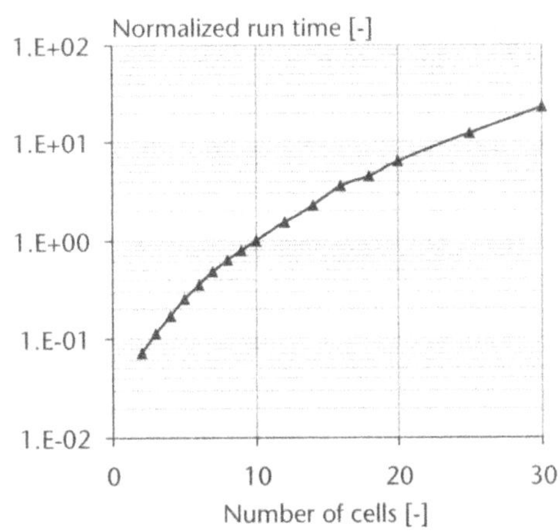

Figure 4.30 Normalized run time versus number of cells along one side of a square grid.

4.5 Characterization of Plane and Laminate Parameters

In this section, we discuss approaches to calculate the dc resistance of planes with and without perforations. We use these approaches to simulate the dc voltage drop on planes and characterize the mid- and high-frequency plane parameters.

4.5.1 DC Resistance of Planes

At high frequencies the plane thickness matters little, because the skin depth limits the current flow to a thin layer on the plane surface. At low frequencies (significantly below the frequency at which skin depth equals the plane thickness) the entire volume of plane carries the current. Especially in high-power applications, the dc resistance may become a major limiting factor. Therefore, it is important to understand how to characterize the plane resistance.

With the definitions of Figure 4.31, the dc resistance, R_{dc}, of a rectangular conductor shape composed of a homogeneous material of conductivity σ, can be calculated as:

$$R_{dc} = \frac{1}{\sigma}\frac{l}{wt} \tag{4.44}$$

where σ, l, w, and t are the conductivity of the plane in S/m, length, width, and thickness of plane in meters. Equation (4.44) assumes that not only the conductive material, but also the current flow is homogeneous in the material; in other words, the current distribution is assumed to be uniform through the entry and exit surfaces at the shaded sides. Equation (4.44) can be simplified by introducing the R_s sheet resistance:

$$R_s = \frac{1}{\sigma t}, \quad R_{dc} = R_s \frac{l}{w} \tag{4.45}$$

In printed circuit boards the typical conductor material is copper, either electrodeposited or rolled-annealed. The conductivity of raw bulk copper at room temperature is $\sigma = 5.8E^7$ S/m. The plane thickness is defined implicitly by the weight of the copper plane. The copper weight is given for 1 ft^2 (or 0.0929 m^2) of material. The density of copper is 8,920 kg/m^3. The weight of the copper is usually given in ounces where 1 oz is 28.35g. From the above numbers, a one-ounce copper plane corresponds to a nominal 34.2-μm plane thickness. Eventually, the porosity of material

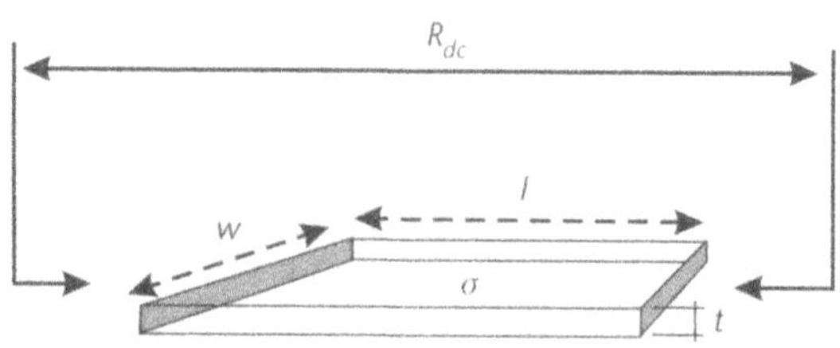

Figure 4.31 Parameters defining the dc resistance of a homogeneous rectangular plane between its opposing parallel sides.

and the surface roughness influence the sheet resistance also. The plane thickness may vary, too, with the processing steps during lamination. Treatment steps, especially when multiple repairs are used, will tend to reduce the plane thickness.

When the tooth structure of the copper has significant peaks and valleys with respect to the overall thickness of the plane, it is reasonable to assume that the dc resistance comes primarily from a plane thickness between the baselines of the tooth profiles, because we can expect little current to penetrate the individual bumps. The definition of equivalent plane thickness is illustrated in Figure 4.32.

Organic packages have copper planes; therefore the above numbers apply. For high-temperature cofired ceramic packages, the conductive layers are made of tungsten with a typical bulk resistivity of $7.5E^{-7}$ Ωm; this translates to $1.33E^{6}$ S/m, as opposed to $5.8E^{7}$ S/m for copper. The manufacturing process may require vent holes on solid planes; this perforation increases the resistance. The resistance increase can be calculated similar to the effect of the antipads.

Due to practical design constraints, using rectangular plane shapes with uniform current distribution is rather rare. A somewhat more practical scenario is when current spreads out in a radial fashion, for instance at the entry point around a copper slug. The spreading resistance is shown by the resistance of planes between two concentric circles.

With the definitions of Figure 4.33, the dc resistance, R_{dc}, of the homogeneous conducting plane between the concentric circles of radii r_1 and r_2 is given by:

$$R_{dc} = \frac{R_s}{2\pi} \ln\left(\frac{r_2}{r_1}\right) \tag{4.46}$$

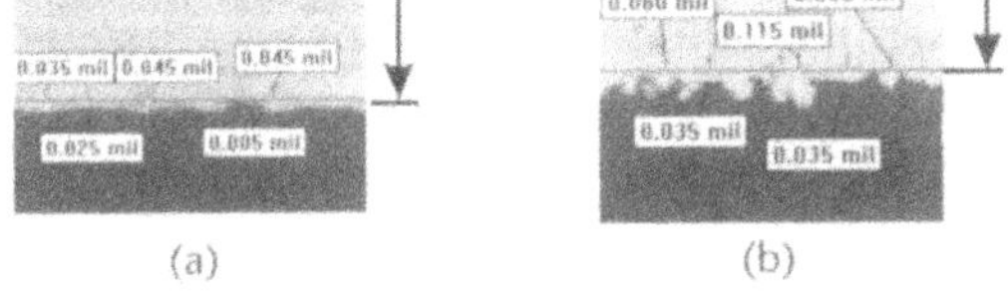

Figure 4.32 Cross sections of (a) rolled-annealed copper foil and (b) electrodeposited foil. Rough surface leaves less copper for current flow not only at high frequencies, but also at dc. (Courtesy of Sanmina-SCI.)

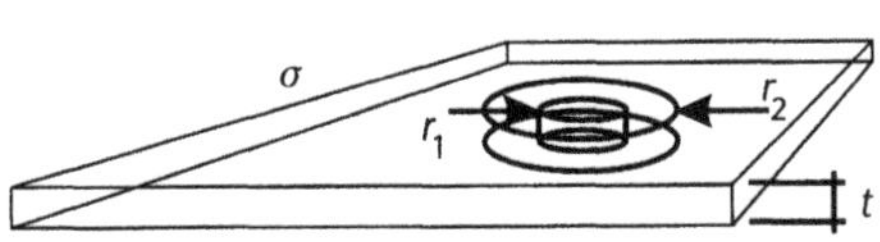

Figure 4.33 Parameters defining the dc resistance of a homogeneous solid rectangular plane between concentric circles.

where R_s is the sheet resistance in Ω-square and r_1 and r_2 are the radii of concentric circles in arbitrary but identical units. Here, too, the assumption is uniform current flow: the current density within each concentric circle is independent of direction.

Another realistic (but still simple) scenario is to calculate the resistance between two circular connections on a large plane. Reference [22] provides an approximate closed-form expression for the resistance. With the notions of Figure 4.34, and assuming a plane thickness of t, radii of connecting hemispherical electrodes of a, and electrode center-to-center spacing of d, the resistance is:

$$R_{ab} \approx \frac{1}{\pi\sigma a}\left[1+\frac{a}{t}\ln\left[\frac{\sqrt{1+\left(\frac{d}{2t}\right)^2}+1}{\sqrt{1+\left(\frac{a}{2t}\right)^2}+1}\right]\right] \tag{4.47}$$

4.5.2 Measuring DC Resistance of Planes

To measure the sheet resistance of a conductive foil, we can use the definition from Figure 4.31. A known number of squares can be measured by sending through a uniform dc current and then measuring the voltage drop. If we use a large number of squares (long, skinny plane shape), a single entry and exit point may be sufficient; all that needs to be measured is the voltage drop across one square of plane where the current is sufficiently uniform. Alternatively, to ensure uniform current distribution, the entry and exit connections can be formed of multiple connections. Figure 4.35 shows this measuring arrangement on a 25.4 × 2.54 cm (10 × 1 inch) strip of two-sided PCB laminate with 1-oz copper on either side. The copper foils are shorted at the end to create a loop. Eight power resistors are soldered to the copper foil, four to each end.

By measuring the voltage drop along a known number of squares generated by a uniform current, we obtain the sheet resistance directly. The accuracy of measurement depends on how accurately we can measure the voltage and current across the plane as well as how accurately we can count the number of squares. Larger plane sizes make it easier to maintain good accuracy when we cut and trim the conductive sheet.

In a finished package or PCB, measuring the dc resistance of the planes is not easy, usually because we do not have the number of connection points necessary for accurate measurements. If we measure around vias, the spreading resistance at the entry and exit points will distort the data. However, we can measure the dc drop on the planes with respect to a selected reference point. When we measure dc voltage,

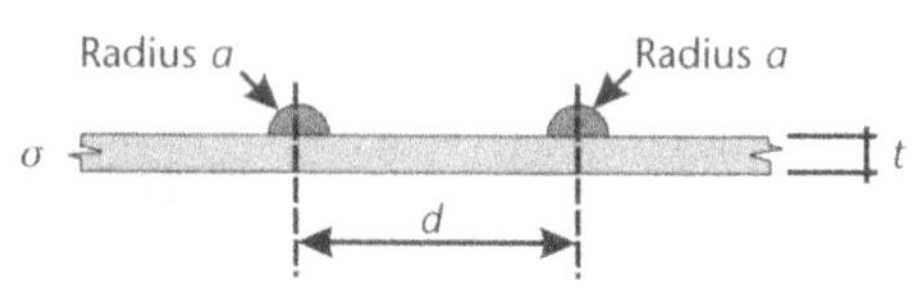

Figure 4.34 Two identical hemispherical electrodes of radii a on a plate of finite conductance and thickness of t.

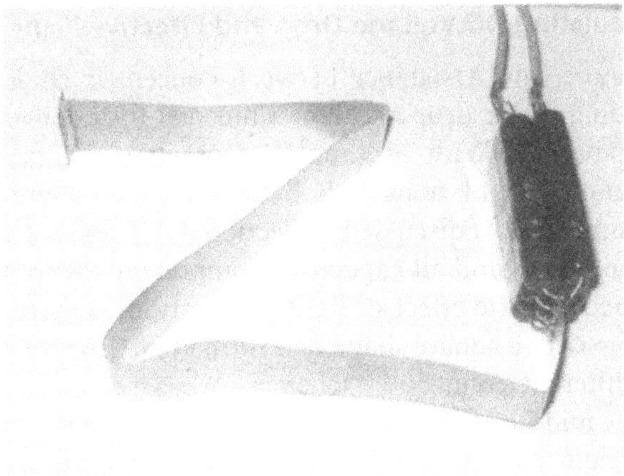

Figure 4.35 Setup showing the dc resistance measurement.

connecting to bypass capacitors is acceptable; as long as we use a voltmeter with high input resistance, the presence of capacitors will not alter the result.

4.5.3 Effect of Perforations on DC Plane Resistance

In PCBs and packages, the vertical signal connections require antipad holes around via barrels. The antipads reduce the amount of the conductive material and increase the plane resistance. With a uniform array of antipads, and as long as the remaining webbing is not very small (see Figure 4.36), we can approximate the sheet resistance by applying a correction factor:

$$R_s' = R_s \frac{total\ area}{total\ area - \sum cutout\ area} \tag{4.48}$$

With heavy perforations, where the remaining webbing is small, detailed meshing is necessary to take into account the current crowding at the narrow sections.

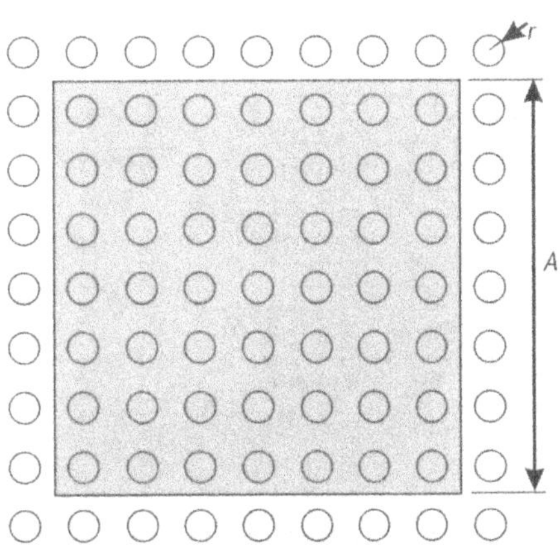

Figure 4.36 Illustration of plane perforations.

4.5.4 Simulating DC Voltage Drop and Effective Plane Resistance

Sheet resistance and resistance between concentric circles are useful concepts for understanding the dc drop on planes, but real-life connections seldom follow these simple geometries. With a complex geometry, the calculation of voltage drop requires detailed simulations. SPICE can also be used with a resistive grid. It is basically the same grid as discussed in Section 4.3.1, except we can omit the inductive part for the planes and all capacitive components. Also, the resistive grid can also take into account the effect of light perforations. Figures 4.37 and 4.38 show the voltage drop over a square shape of 35-μm (1-oz) copper plane, with 1-A dc current and two different connection schemes. We assume that the load connects at 25 nodes in the middle of the plane, whereas the source is connected to 7 nodes around the lower right corner. The simulation deck assumes a 1-A dc current total uniformly distributed among the 25 entry nodes at the center. The seven exit nodes at the lower right are all tied to SPICE node 0. The voltage surface exhibits a local peak where the 1-A current enters the plane. There is a large gradient of voltage in the lower right quadrant of the plane where the conductive plane is effectively utilized. The voltage gradient is low in the opposite direction at the upper left of the plane; this low gradient indicates that this portion of the plane, does not contribute effectively to carrying the current between the entry and exit points.

4.5.5 Characterization of Mid- and High-Frequency Plane Parameters

The procedure outlined below assumes rectangular planes, and a uniform and homogeneous cross section and materials. We also assume that the w_x and w_y dimensions of the planes (see Figure 4.1) are known: they can be obtained with sufficient accuracy either from the board CAD file or from mechanical measurements on the finished board. Also, as long as the dielectric material in all pairs is the same, the procedure below can be applied to multiple plane pairs connected in parallel by

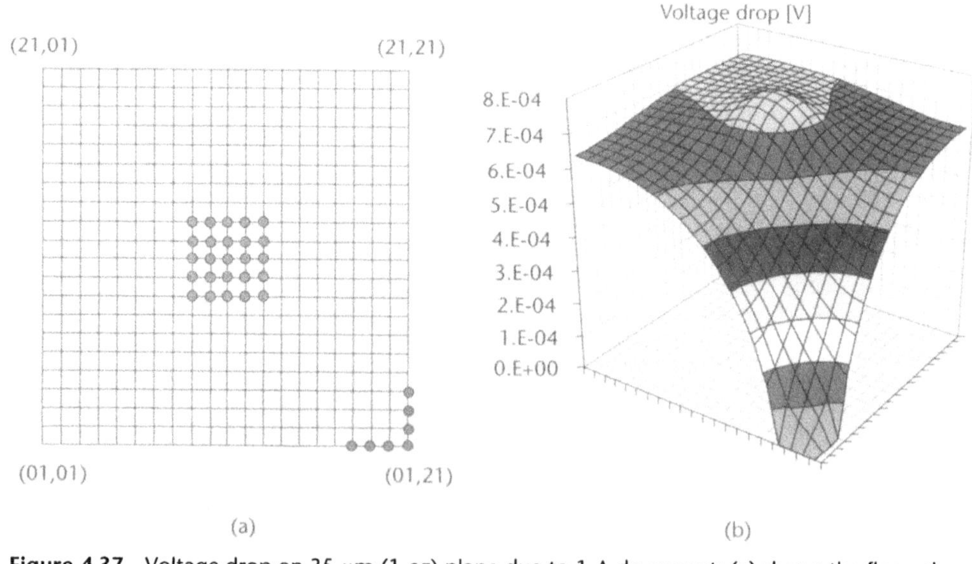

Figure 4.37 Voltage drop on 35-μm (1-oz) plane due to 1-A dc current: (a) shows the floor plan and connections and (b) shows the voltage drop surface.

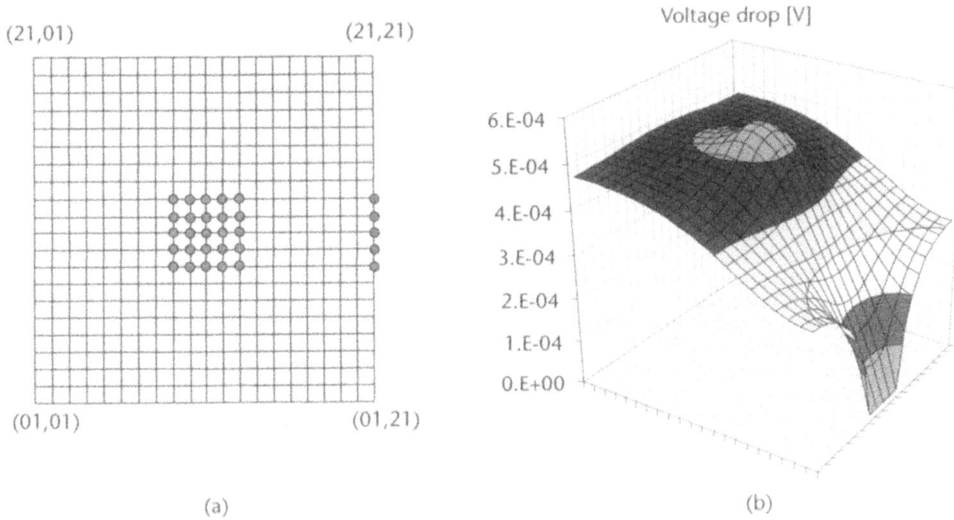

Figure 4.38 Voltage drop on 35-μm (1 oz) plane due to 1-A dc current: (a) shows the floor plan and connections and (b) shows the voltage drop surface.

many vias distributed evenly over the planes. In this case, the h separation will be the parallel equivalent of the individual laminate thickness values. For two parallel plane pairs with h_1 and h_2 plane separations, the equivalent plane separation is:

$$h = \frac{1}{\frac{1}{h_1} + \frac{1}{h_2}} \tag{4.49}$$

There are parameters, however, that on a finished board cannot be measured directly without destructive probing: the plane separation, thicknesses of planes, and surface roughness of planes. Furthermore, the electrical properties of the dielectrics and metals are not usually measurable on finished boards, due to cutouts and the stackup.

To obtain an estimate for the dielectric constant and the plane separation, we can use the formulas for the static capacitance and modal resonance frequencies, as illustrated in Figure 4.39. The measured static plate capacitance can be equated to its calculated value. This value has two unknowns: ε_r and h (since we assumed that the horizontal dimensions of the rectangular planes, w_x and w_y are known). The parallel-plate modal resonance frequencies are the integer multiples of where the half wavelength equals the w_x or w_y dimensions. The second formula in (4.50) takes the lowest resonance frequency along the w_x dimension.

$$C_p = \varepsilon_0 \varepsilon_r \frac{w_x w_y}{h} \tag{4.50a}$$

$$f_{res} = \frac{1}{2w_x\sqrt{\varepsilon_0 \varepsilon_r \mu_0}} \tag{4.50b}$$

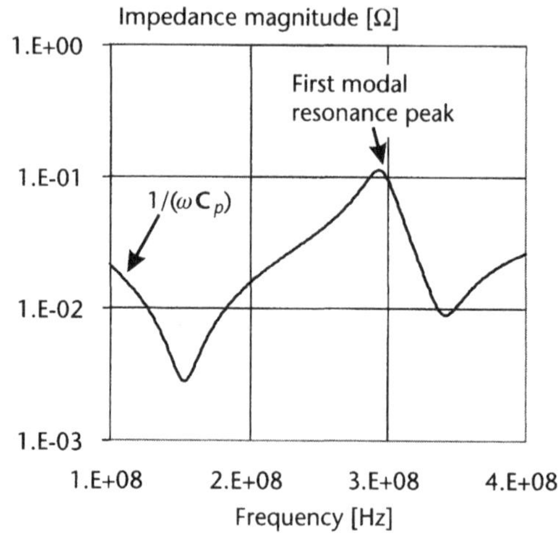

Figure 4.39 Impedance magnitude versus frequency plot, identifying the two parameters in (4.50).

In the formulas, ε_0 and μ_0, as well as w_x and w_y, are frequency-independent constants. Therefore, as long as ε_r is approximated as frequency independent, we can solve for ε_r and h. By rearranging (4.50), we get:

$$\varepsilon_r = \left(\frac{1}{4w_x^2 f_{res}^2 \varepsilon_0 \mu_0} \right), \quad h = \varepsilon_0 \varepsilon_r \frac{w_x w_y}{C_p} \tag{4.51}$$

If the capacitance is frequency independent, its value can be obtained from a low-frequency measurement point with (5.27). The frequency has to be high enough to ensure a reliable and low-noise measurement but low enough to ensure that the modal impedance minimum will not distort the result.

4.5.5.1 Determining the Modal Resonance Frequency of Lossy Dielectrics

If conductors and dielectrics were all ideally lossless, the first modal resonance frequency, f_{res}, could be obtained by probing the plane pair almost anywhere. However, with conductive and dielectric losses, the impedance profile and the frequencies of the modal resonance peaks do depend on the location over the planes. Furthermore, the peaks are not unique anymore: the frequency of the peak depends on how we define the peak itself. The peak implies maximum magnitude of impedance. Resonance, however, is usually understood to happen at frequencies where the imaginary part is zero; hence, the phase of impedance is zero. As shown in Figure 4.40, the values extracted according to these two definitions will be close, but not exactly the same. Figure 4.40 shows the extracted first modal resonance frequency from the simulated impedance profile of a lossy pair of FR4 planes with 35-μm (1-oz) copper on either side and with plane dimensions of w_x = 25.4 cm (10 inches), and w_y = 12.7 cm (5 inches). Note that the frequency extraction is not unique; the values depend on the location on the planes. If, instead of impedance magnitude peak or phase zero crossing, we define the resonance frequency where the phase derivative has its

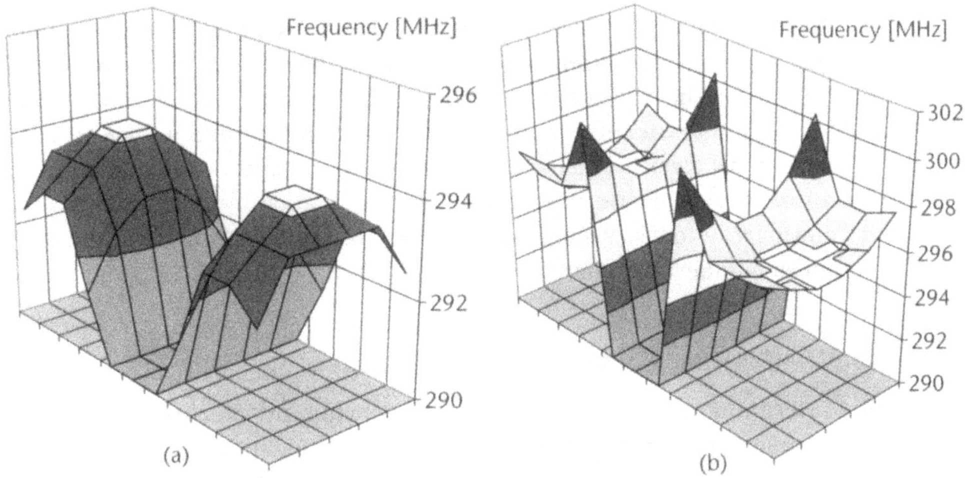

Figure 4.40 Extracted first modal resonance frequency from the simulated impedance of a lossy pair of planes with lossy dielectric. Part (a) shows the frequencies where the impedance magnitude is the highest. Part (b) shows the frequencies where the phase of impedance crosses 0. The gap in the middle reflects locations where the modal resonance is suppressed by the 2:1 aspect ratio. (The sharp slope in the middle is due to plotting artifacts.) The floors of the charts represent the surface of the planes.

extreme value, the extracted frequency becomes the same, regardless of the location on the planes.

Figure 4.41(a) shows the same pair of planes with the first derivative of phase plotted at one given location on the planes. Figure 4.41(b) shows the first modal resonance extracted from the extreme value of phase derivative (in case of first modal resonance: minimum) over the surface of the planes.

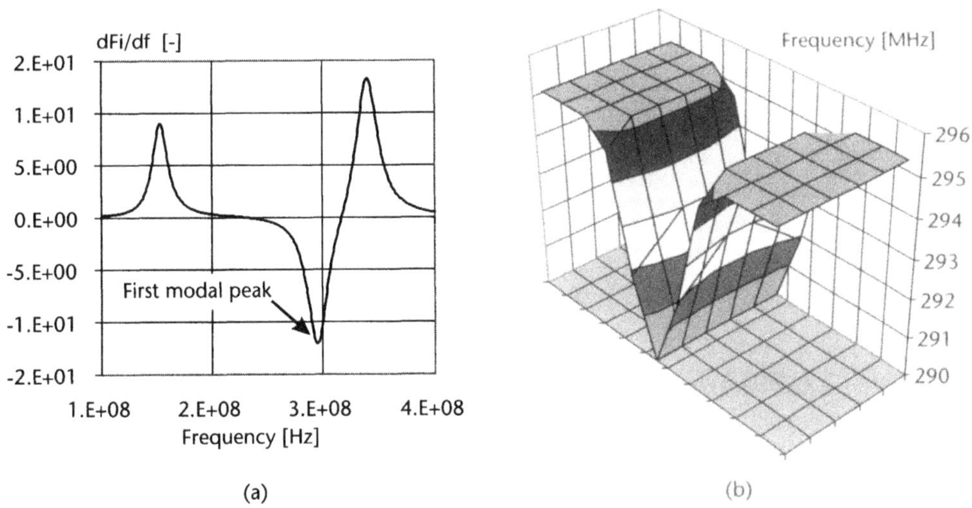

Figure 4.41 Extracted first modal resonance frequency from the simulated impedance of a lossy pair of planes with a lossy dielectric: (a) shows the first derivative of frequency at the corner of the planes and (b) shows frequencies where the first derivative of the phase has its extreme. The floor of the graph represents the surface of the planes.

As a summary, for plane pairs where the losses are not negligible (increasingly the case with thinner laminates), the first modal resonance frequency should be extracted from the extreme value of the phase derivative. Note, however, that phase itself is already more noisy in measurements than magnitude. The derivative of the phase becomes even noisier due to the high-pass nature of the derivative process. Therefore, measuring the plane parameters based on this procedure requires averaging and narrow measurement bandwidth to sufficiently suppress noise. Measured examples will be shown later in Section 4.5.5.5.

4.5.5.2 Dielectric Constant and Dielectric Loss

Now we can return to (4.50) and continue the parameter extraction using the static plane capacitance and the frequency of lowest modal resonance. The relative dielectric constant appears in both constraints, but with (5.28) we cannot extract the static plane capacitance at the first modal resonance frequency; there will be at least one, possibly two or more decades of frequency separation between the frequencies of the static plane capacitance and first modal resonance. We can uniquely solve for the two unknowns from (4.49) only if ε_r were frequency independent. If ε_r is not frequency independent, we have to know and model its frequency dependency. We have to obtain a scaling factor for ε_r to the frequency of first modal resonance before we solve for the two unknowns.

By measuring the complex impedance of the DUT, we could obtain the capacitance from the imaginary part and the loss tangent from the phase of the impedance. Unfortunately, with low-loss materials, the phase angle is very close to −90°; the finite accuracy of measuring instruments will not allow us to accurately measure the loss tangent. Knowing how the frequency-dependent capacitance and frequency dependent loss tangent relate to each other can help us in characterization. From the causality constraint of impedances, we know that the real and imaginary parts of the impedance are interrelated; their values and their frequency dependencies are linked to each other [23]. Dielectric constant and dielectric loss and inductance and resistive loss share tightly coupled frequency dependency [24]:

$$H(\omega) = \alpha(\omega) + j\beta(\omega) \tag{4.52}$$

$$\alpha(\omega) = \alpha(0) - \frac{\omega^2}{\pi}\int_{-\infty}^{\infty} \frac{\beta(u)}{u\left(u^2 - \omega^2\right)} du \tag{4.53}$$

$$\beta(\omega) = \frac{\omega}{\pi}\int_{-\infty}^{\infty} \frac{\alpha(u)}{u^2 - \omega^2} du \tag{4.54}$$

The dielectric constant and dielectric loss are expressed with the real and imaginary parts of the complex permittivity:

$$\varepsilon_r(\omega) = \varepsilon_r'(\omega) - j\varepsilon_r''(\omega) = \varepsilon_r'(\omega)(1 - j\tan\_\delta) \tag{4.55}$$

Equation (4.55) also defines the loss tangent, tan_δ. In its general form, (4.51) allows for a number of different frequency dependencies. The real and imaginary parts of the complex permittivity show very complex frequency dependencies as frequency varies from a few Hertz up to light wave frequencies. At lower frequencies, dipolar and ionic relaxation are typical; at higher frequencies atomic and electronic resonances are typical. Relaxation losses typically exhibit a changing real part of permittivity in a relatively narrow frequency range accompanied by a peak in the imaginary part. Relaxations are usually modeled by the Debye relaxation model [25], named after the chemist Peter Debye. The relaxation model for a single contributor can be expressed as:

$$\varepsilon(\omega) = \varepsilon_{\infty} + \frac{\Delta\varepsilon}{1 + j\omega\tau} \tag{4.56}$$

where the first term is the high-frequency asymptotical permittivity, $\Delta\varepsilon$ is the permittivity change between low and high frequencies, and τ is the relaxation time of the material.

Multiple relaxation processes over a broader frequency range can be modeled by a sum of N first-order relaxation terms:

$$\varepsilon(\omega) = \sum_{i=1}^{N} \frac{a_i}{1 + j\omega\tau_i} \tag{4.57}$$

which is suitable to capture any frequency dependency.

Experimental data shows that for a large number of commonly used PCB materials the dielectric constant drops, whereas the dielectric loss tangent increases with the same percentage value over a logarithmic frequency scale. In other words, on a linear-logarithmic scale, both the dielectric constant and the loss tangent are straight lines. Reference [26] shows that the measured capacitance of FR-4 laminates is constant up to about 1 kHz and drops linearly on a logarithmic frequency scale from 1 kHz to 10 GHz. To maintain causality and at the same time to meet common-sense expectations, the responses should deviate from this linear shape at very low and very high frequencies. If the curves maintained their slope over all conceivable frequencies, it would result in infinite capacitance at dc and infinite loss tangent at infinite frequency.

To obtain a simpler formula to approximate the empirical data, (4.57) can be modified to replace the finite sum either with an integral [26] or infinite sum [27] so that the multipole Debye model yields a linear change on the logarithmic frequency scale between a user-selected ω_1 lower frequency and ω_2 upper frequency:

$$\sum_{i=1}^{N} \frac{\Delta\varepsilon'_i}{1 + j\dfrac{\omega}{\omega_i}} \rightarrow \frac{\Delta\varepsilon'}{m_2 - m_1} \int_{x=m_1}^{m_2} \frac{dx}{1 + j\dfrac{\omega}{10^x}} = \frac{\Delta\varepsilon'}{m_2 - m_1} \frac{\ln\dfrac{\omega_2 + j\omega}{\omega_1 + j\omega}}{\ln 10} \tag{4.58}$$

where $m_1 = \log_{10}(\omega_1)$ and $m_2 = \log_{10}(\omega_2)$.

The causal frequency dependence of the dielectric constant and loss tangent from (4.58) are illustrated in Figure 4.42.

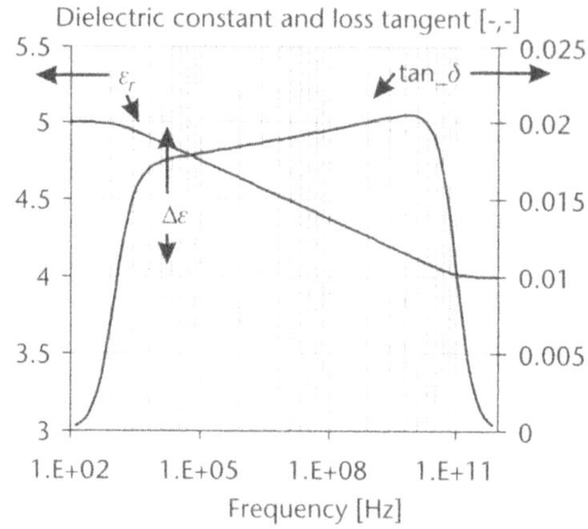

Figure 4.42 Simulated relative dielectric constant (ε_r) and dielectric loss tangent (tan_δ) following the modified Debye model of (4.58), with the following parameters: $m_1 = 3$, $m_2 = 11$, $\Delta\varepsilon = 1$, $\varepsilon_\infty = 4$.

Note that (even though we may want to maintain causality in our models) if we need only the frequency range where the data is a straight line on a linear-logarithmic scale, an even simpler approximation can be used. From a dielectric constant and slope at a working frequency, f_w, we can calculate the approximate lines for the dielectric constant and loss tangent. Note also that we turn to the causality constraint primarily to help in the measurement procedure; this constraint allows us to deduce the loss tangent from the change of capacitance versus frequency. As it was stated in Chapter 1, strictly speaking causality is less of a concern in PDN characterization, as we usually do not need the wave shape of the noise to be very accurate.

Though (4.58) provides a causal solution, we very rarely need laminate characterizations over such a wide frequency range that we would actually enter the frequency range below ω_1 and above ω_2. We can further simplify (4.58) by realizing that in the frequency range of interest the slopes of the ε' and tan_δ curves have the same magnitude but opposite signs: based on the above model, the capacitance always decreases with increasing frequency while the loss tangent always increases. By selecting an arbitrary working frequency of f_w somewhere conveniently within the range of linear slope (see Figure 4.43), the real part of permittivity and the loss tangent can be expressed by the value and first derivative of ε':

$$\tan\_\delta(f_w) = -\frac{m(\varepsilon_r)}{\varepsilon_r(f_w)} \frac{\pi/2}{\ln(10)} \tag{4.59}$$

where

$$m(\varepsilon_r) = \lim_{\Delta\to 0} \frac{\varepsilon_r(f_{w+\Delta}) - \varepsilon_r(f_{w-\Delta})}{\log_{10}\left(\dfrac{f_{w+\Delta}}{f_{w-\Delta}}\right)} \tag{4.60}$$

Equations (4.59) and (4.60) help us to characterize the laminates primarily by measuring the capacitance versus frequency curve in a frequency range where the data has the lowest noise, permitting a straightforward fit to the Debye model.

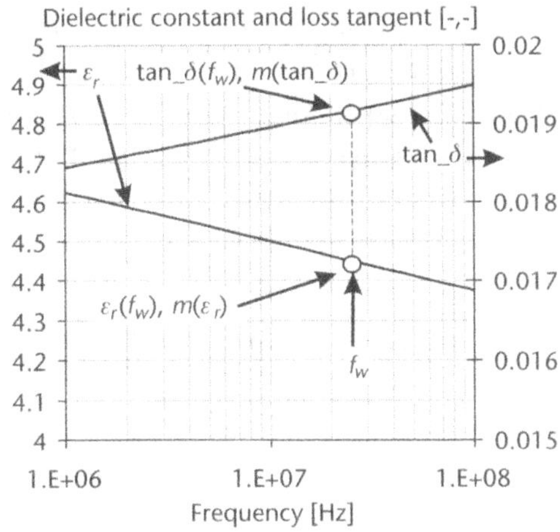

Figure 4.43 Illustration of (4.59) through (4.60) with the dataset of Figure 4.42 enlarged in the 1–100-MHz range.

4.5.5.3 AC Plane Resistance and Inductance

As dielectric constant and loss tangent are linked by the causality constraint, the same causality constraint also applies to the series branch of the transmission-line equivalent circuit: resistance and inductance. The resistance starts with the R_{dc} value at dc, then it gradually moves toward the resistance dictated by the high-frequency sheet resistance determined by the skin depth (4.8). Inductance is easier to follow from high frequencies going toward low frequencies. At very high frequencies, the current flows in thin layers on the facing surfaces of the planes; this results in the *external inductance*. In case of plane pairs, it is also called interplane inductance (in H/square):

$$L_{\infty} = \mu_0 \mu_r h \tag{4.61}$$

As frequency decreases, the current penetrates the planes more deeply; also as the skin depth increases, so does the inductance. The portion of inductance that comes from the current loop inside the planes is called *internal inductance*. The finite current penetration in the planes results in a complex surface impedance with frequency independent 45° of phase angle [27]. Using the definitions shown in Figure 4.1, the complex surface impedance is the sum of the surface impedances of the upper and lower planes:

$$Z_{surface}(\omega) = \frac{1-j}{\delta_u \sigma_u} \cot\left((1-j)\frac{t_u}{\delta_u}\right) + \frac{1-j}{\delta_l \sigma_l} \cot\left((1-j)\frac{t_l}{\delta_l}\right) \tag{4.62}$$

If the upper and lower planes are identical in thickness and conductivity, the two terms of (4.62) can be combined.

The series plane impedance can be expressed as the sum of the dc resistance, complex surface impedance, and the reactance of external interplane inductance:

$$Z_s(\omega) = R_{dc} + Z_{surface}(\omega) + j\omega L_{\infty} \tag{4.63}$$

The complex surface impedance of planes given by (4.63) assumes smooth plane surfaces. Realistic plane surfaces are never ideally smooth. As the skin depth becomes comparable to and smaller than the surface roughness with increasing frequency, the surface impedance will increase. There is a widely quoted empirical formula (originally from microwave trace measurements and simulations) describing the approximate rise of surface impedance. The scalar multiplier, K_{sr}, is:

$$K_{sr}(\omega) = 1 + \frac{2}{\pi}\arctan\left(1.4\left(\frac{\Delta}{\delta(\omega)}\right)^2\right) \tag{4.64}$$

where Δ is the rms surface roughness and $\delta(\omega)$ is the skin depth in the same, but arbitrary units.

With (4.64), (4.63) becomes:

$$Z_s(\omega) = R_{dc} + K_{sr}(\omega)Z_{surface}(\omega) + j\omega L_{\infty} \tag{4.65}$$

4.5.5.4 Resonator Quality Factor

One of the challenges in accurately characterizing plane pairs is that there are several interrelated unknowns. Unless we use trusted and proven models to fit the measured data, there is no direct way to measure the parameters one-by-one. As it was shown in Section 4.5.5.2, we can measure the static plane capacitance at frequencies much lower than the first modal resonance, but to obtain the plane separation from (4.51), we need the dielectric constant at the modal resonance frequency. We can measure the dc resistance of a plane; however, its high-frequency losses eventually depend not only on the plane thickness, but also on the conductivity and surface roughness. Another means of cross-correlating data is to measure the quality factor, Q, of the modal resonances and then equate it to the quality factor calculated from the assumed loss contributors. For microwave resonators, Q is expressed in terms of the quality factors resulting from conductive losses, Q_c, dielectric losses, Q_d, and radiation losses, Q_r [28]:

$$\frac{1}{Q} = \frac{1}{Q_c} + \frac{1}{Q_d} + \frac{1}{Q_r} \tag{4.66}$$

The quality factors due to conductive and dielectric losses are directly related to the complex surface impedance and loss tangent, respectively:

$$Q_c = f(Z_{surface}) \tag{4.67}$$

$$Q_d = \frac{1}{\tan\_\delta} \tag{4.68}$$

For plane pairs, where the plane separation is much smaller than the w_x and w_y horizontal dimensions, the Q_r radiation loss can be neglected.

4.5.5.5 Measurement-Model Correlations

Dedicated methods and setups exist to measure and characterize conductive and dielectric materials for PCBs as accurately as possible. Many of these methods rely on the measurement of a combination of conductive and dielectric layers, such that the static capacitance and the resonance pattern of the cavity can be measured [29]. There are several measurement methods recommended by IPC; two widely used options are the stripline method [30] and the full-sheet resonance method [31]. A good practical overview of available options can be found in [32].

In simple and unsophisticated setups, our best means of achieving the highest accuracy is to eliminate unnecessary variables as much as possible. When we measure dielectric materials, the uncertainty due to conductor parameters can be greatly reduced by preparing a dielectric sample and adding the necessary conductive layers as parts of the measurement fixture as electrodes. In this way the electrodes can be prepared and characterized more accurately and carefully. Similarly, when measuring the parameters of conductive layers, eliminating the insulating material helps to reduce uncertainties. While these characterizations are possible, they do not represent the final construction of the PCB. In a real PCB, the building block is a plane-dielectric-plane sandwich, as shown in Figure 4.1. This comes from a core or is the result of a prepreg facing the planes on the cores above and below or is the result of a layer-by-layer buildup process.

To characterize a basic two-sided laminate, we can start with a rectangular shape either as a stand-alone three-layer structure or as part of a finished PCB or package. By measuring the dc resistance and the impedance profiles with open boundaries and with shorts, we can fit the model of our choice to the measured data. The geometric parameters can be measured either by simple visual inspection or after cross-sectioning to yield the dielectric and plane thicknesses, as well as surface-roughness numbers.

As we show in the first characterization and correlation example, in a limited frequency range, sufficient correlation can be achieved, even with mostly frequency-independent models. Figure 4.44 shows the photo of a 20-layer test board with w_x = 25.4-cm (10-inch) and w_y = 12.7-cm (5-inch) dimensions. The test board had five plane pairs with 50-μm (2-mil) laminate separation (see Figure 4.45), arranged into two groups: the upper two thin laminate pairs are connected by the test vias; the bottom three laminate pairs were unconnected. The first plane pair was 75 μm (3

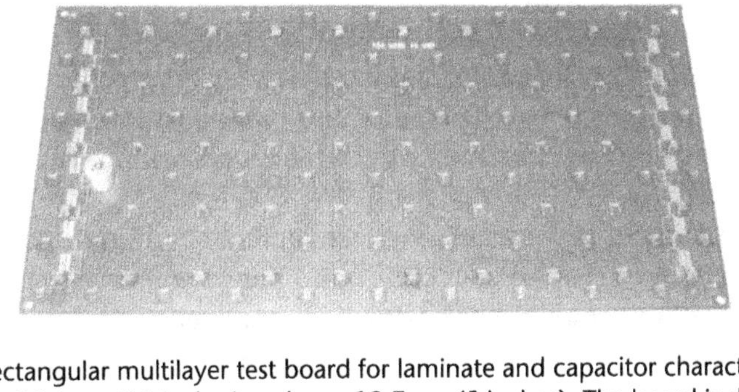

Figure 4.44 Rectangular multilayer test board for laminate and capacitor characterizations. The board size is w_x = 25.4 cm (10 inches) and w_y = 12.7 cm (5 inches). The board is shown with one bulk capacitor attached.

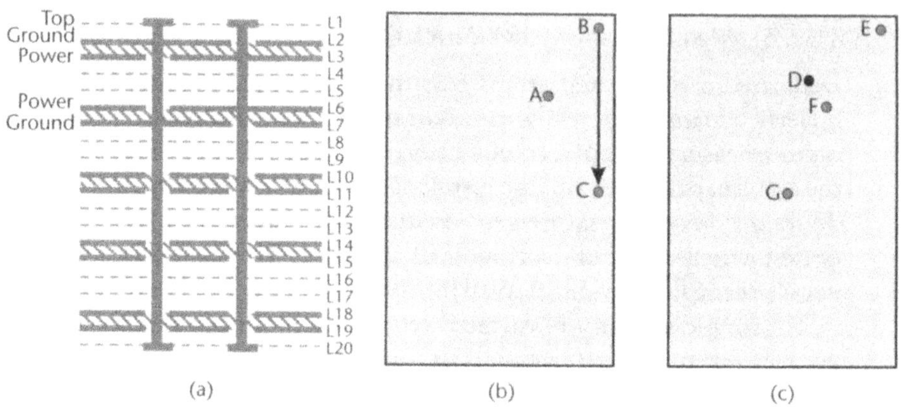

Figure 4.45 (a) Stackup and connections in the test board of Figure 4.44. (b) Connection points for the bare-board correlations shown in Figure 4.46. (c) Connection points for the shorted correlations shown in Figure 4.47.

mils) below the surface. There were two sets of vias, both on a 2.54-cm (1-inch) grid, with a 1.27-cm (0.5-inch) offset with respect to each other. One set of vias were through holes for probing purposes, connecting to the planes on layers 2 and 3. The second grid of vias was groups of blind vias connecting the same planes to surface pads for eight-terminal capacitors.

The board was measured at various test points with nothing attached to it (bare), as well as with shorts attached in different combinations to the capacitor sites. First, the dielectric constant and plane separations were calculated from the average of static capacitances and first modal resonances measured at three different locations. Based on (4.51) this resulted in $h_1 = h_2 = 52.3\,\mu$m (2.06 mils) and $\varepsilon_r = 4.22$. A uniform SPICE grid with a 6.35-mm (0.25-inch) cell size was created to simulate the planes where the dielectric constant, loss tangent, and plane inductance were assumed to be frequency independent. Second, the resistance and inductance of the individual shorts (vias, pads, and shorting bar) were obtained by fitting the unknown parameters to measured data. It was found that 0.5-mΩ dc resistance, $1.6\text{E}^{-7}\,\Omega/\sqrt{f}$ skin resistance and 47-pH inductance gave good agreement for all measured locations. The attached impedance of the shorted structure thus was (in ohms):

$$Z_{attached} = 5\text{E}^{-4} + 1.6\text{E}^{-7} * \sqrt{f} + j2\pi f * 47\text{E}^{-12} \tag{4.69}$$

Finally, the measured and simulated impedance profiles were compared for various self- and transfer-impedance combinations for the bare and shorted cases. Figures 4.46 and 4.47 show the degree of correlation for the bare and shorted cases, respectively.

The process outlined above is convenient to obtain the approximate laminate characteristics in real-life stackups and board constructions. The test via pairs dispersed evenly over the board provide a convenient way to measure the laminate characteristics at various locations. However, the test-via pairs require small cutouts (i.e., antipads) around via barrels. These cutouts change the plane parameters: by reducing the static capacitance and slightly distorting the resonance pattern.

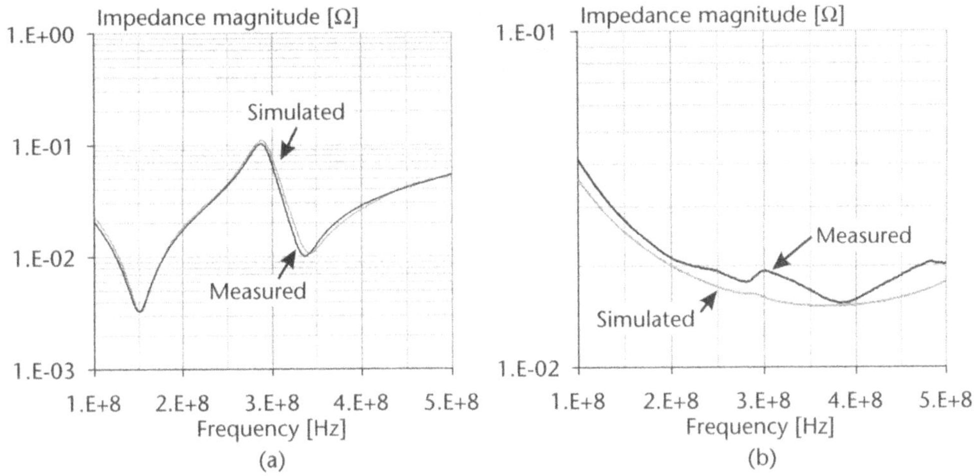

Figure 4.46 Simulated and measured impedance magnitudes at two different locations on the test board. (a) Self-impedance at location A. (b) Transfer impedance between B and C.

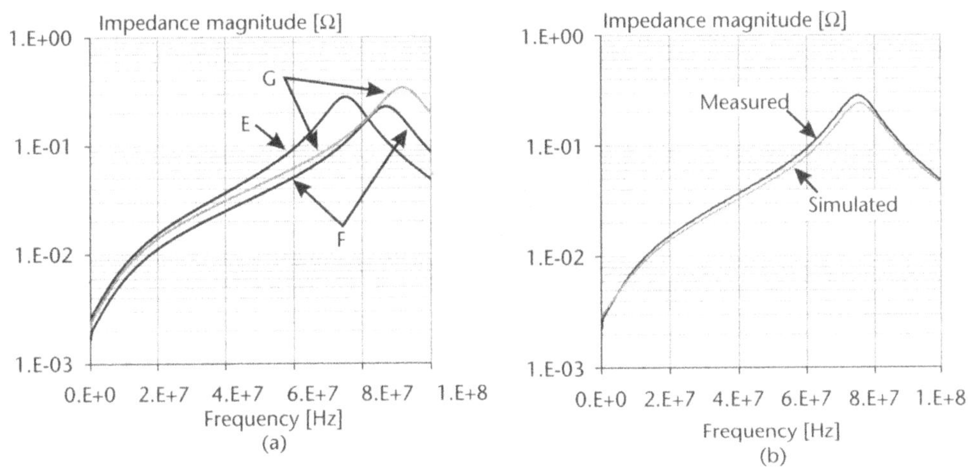

Figure 4.47 (a) Measured self-impedance plots at various locations of *shorts* as identified on Figure 4.45(c). Measurement point is D for all three traces. (b) Correlation between measurement and simulation for trace E.

The same test-board construction was used to compare various laminates, as reported in [33]. Figures 4.48 and 4.49 show the frequency-dependent capacitance and inductance of various laminates, all of which were tested in the same nominal board construction. Table 4.1 summarizes the major parameters of the laminates and decodes the labels. The dielectric thickness and copper thickness values are nominal; they were not checked by cross-sectioning. The column of relative dielectric constant, ε_r, shows the measured values on the test boards at 1 MHz. The dielectric loss, tan_δ column gives the measured value calculated from the rate of capacitance change with (4.59) and (4.60).

The data in Figure 4.48 is normalized to the low-frequency capacitance value of each board, and thus the impact of dielectric thickness and dielectric constant differ-

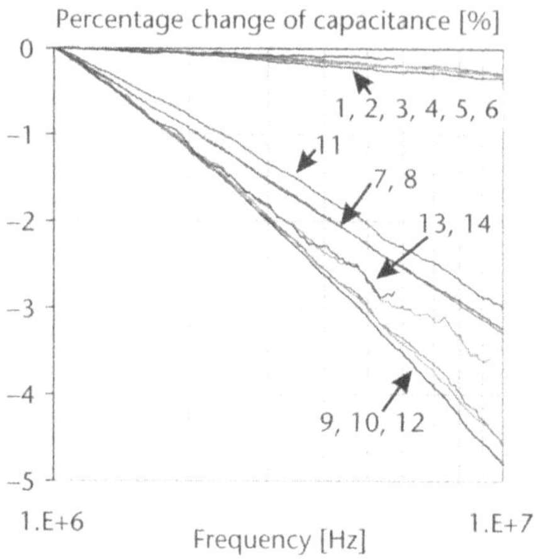

Figure 4.48 Percentage change of capacitance of boards with laminates from Table 4.1, measured in bare boards of construction, as shown in Figures 4.44 and 4.45.

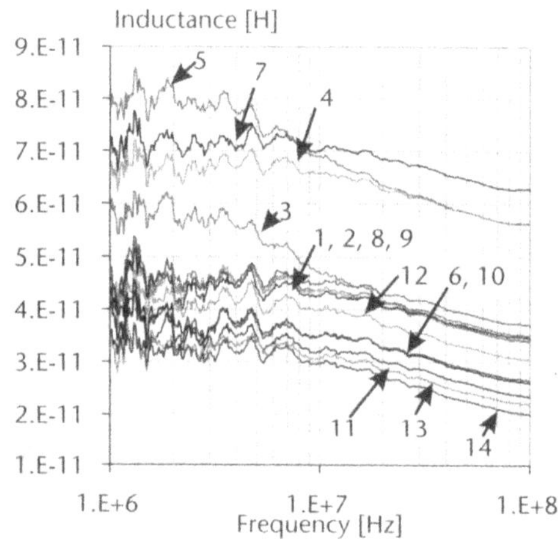

Figure 4.49 Inductance of boards with laminates from Table 4.1, measured in shorted boards of construction, as shown in Figures 4.44 and 4.45.

ences is removed. All six unfilled polyimide laminates (1–6) exhibit approximately, –0.3%/decade capacitance drop. The two glass-reinforced laminates (7 and 8) show –3.2%/decade capacitance change. The highest rate of capacitance drop comes from laminates 9, 10, and 12.

The inductance values in Figure 4.49 were measured on the same test boards when all capacitor sites around the board periphery were shorted. There are three major phenomena we can observe on the inductance data. The first obvious trend is that the high-frequency inductance is proportional to the nominal dielectric thickness. The second trend is that the saturated inductance at low frequencies is proportional to the average thickness of the dielectric *and* copper. The third trend is the drop of inductance with increasing frequency as the current penetration inside the planes is reduced by the skin effect.

Table 4.1 Summary of Parameters of Laminates Shown in Figures 4.48 and 4.49

	Dielectric Thickness [mm]	*ε_r at 1 MHz [-]*	*tan_δ at 1 MHz [-]*	*Dielectric Type*	*Copper Thickness [mm]*	*Copper Type*	*Reference*
1	25	3.15	0.002	unreinforced, unfilled	36	ED	HK042536E
2	25	3.17	0.0024	unreinforced, unfilled	36	RA	HK042536R
3	25	3.13	0.0019	unreinforced, unfilled	72	RA	HK042572R
4	50	3.14	0.0021	unreinforced, unfilled	36	RA	HK045036R
5	50	3.15	0.0021	unreinforced, unfilled	72	RA	HK045072R
6	50	14.7	0.0032	unreinforced, filled	36	RA	HK1014
7	50	3.4	0.022	reinforced, unfilled	36	ED	ZBC2000
8	25	3.7	0.022	reinforced, unfilled	36	ED	ZBC1000
9	24	4.4	0.033	unreinforced, unfilled	36	RA	BC24
10	16	4.0	0.031	unreinforced, unfilled	36	RA	BC16
11	12	3.8	0.02	unreinforced, unfilled	36	RA	BC12
12	24	16.8	0.031	unreinforced, filled	36	ED	C-Ply24
13	12	14.2	0.024	unreinforced, filled	36	ED	C-Ply12
14	8	13.1	0.023	unreinforced, filled	36	ED	C-Ply8

The second characterization example uses the same structure that was shown in Chapter 2, Figure 2.1. The sample piece of a nominal size of $w_x = 5.08$ cm (2 inches) and $w_y = 2.54$ cm (1 inch) was manually cut out from a multilayer board; this cut is similar to the one shown in the previous example, except for the fact that the outermost plane pair was connected to the test-via pair. The laminate thickness was nominally 50 μm (2 mils). The cut was made in such a way that the test vias were located in the middle; this was the only point on the planes where self-impedance was measured. The measuring instrument was an Agilent N5230, with a logarithmic sweep in the 1-MHz–10-GHz frequency range. We use this example to show in more detail how to extract the frequency-dependent ε_r and tan_δ. Figure 4.50 shows the measured self-impedance magnitude and phase. The same data is shown on both charts: (a) shows the full-span impedance and (b) shows the zoomed impedance profile around the first modal resonance.

The size of the sample was remeasured more accurately; it was found that the longer dimension was $w_x = 5.22$ cm (2.05 inches). From the impedance data shown in Figure 4.50(b), the first modal resonance frequency was extracted as $f_{res} = 2.9768$ GHz. From this resonance frequency, (4.50b) yields a relative dielectric constant of 3.7254. To get the plane separation from (4.50a) and to allow for frequency-dependent parameters, we need the measured capacitance and the extracted dielectric constant at the same frequency. To do this, we extract the capacitance versus frequency curve at frequencies below the first impedance minimum, fit a causal multipole Debye model on the extracted capacitance, and then, from the model, we extrapolate the dielectric constant at f_{res}. The extracted capacitance versus frequency and the dielectric constant, as well as the loss tangent from the Debye model, are shown in Figure 4.51.

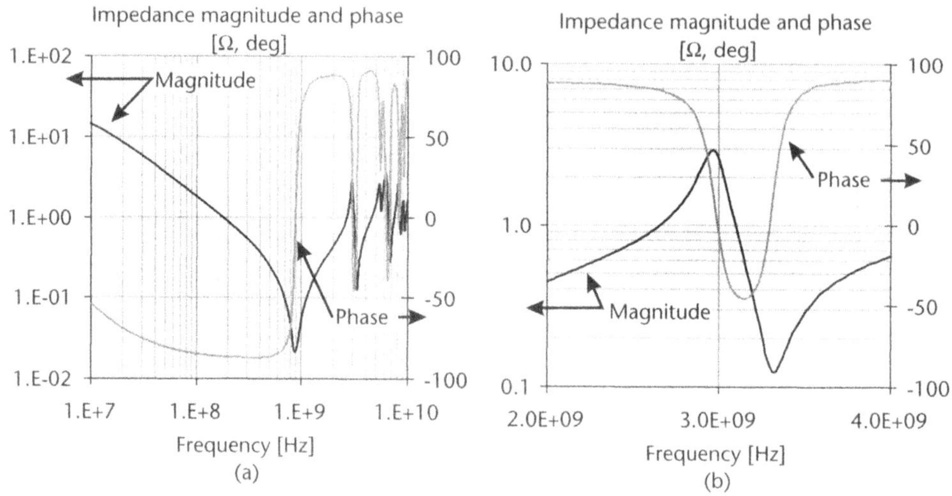

Figure 4.50 Self-impedance (a) magnitude and (b) phase of the 50-μm laminate sample.

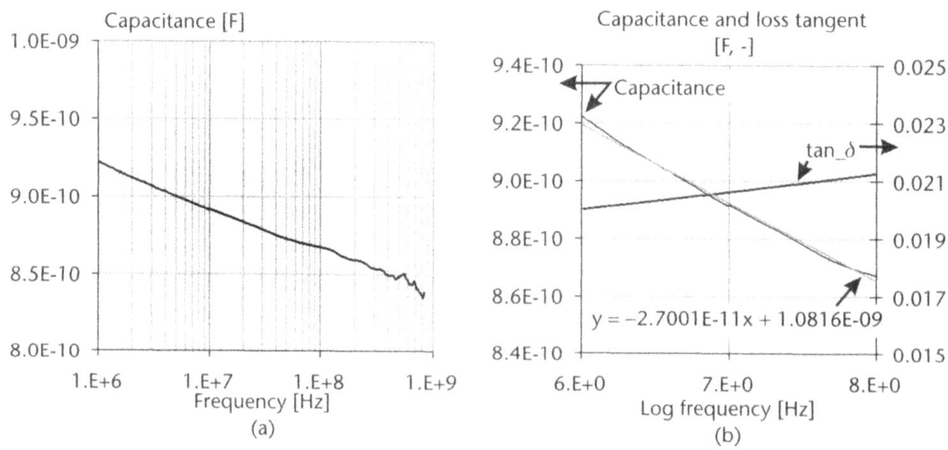

Figure 4.51 (a) Capacitance versus frequency and (b) capacitance and loss tangent from the Debye model for the 50-μm laminate.

The capacitance versus frequency plot on Figure 4.51(a) uses (5.28) and accounts for the plane inductance. Figure 4.51(b) shows both the measured capacitance and the straight line best-fit approximation of the Debye model. With the straight-line approximation of capacitance on the logarithmic frequency scale, we can extrapolate the capacitance out to f_{res}: the result is $C(f_{res}) = 825.8$ pF. Using the above $C(f_{res})$ estimate, (4.50a) yields $h = 59.8\ \mu$m (2.35 mils) for the laminate thickness. Finally, the sample was cross sectioned, and the laminate thickness, upper and lower plane thicknesses and surface-roughness values were extracted, as illustrated in Figure 4.52.

The correlation between the measurement and simulation results was shown in Chapter 2.

The third and final characterization example uses a bare sheet of two-sided composite laminate, a DuPont Pyralux LF911R, which was probed at the edges with

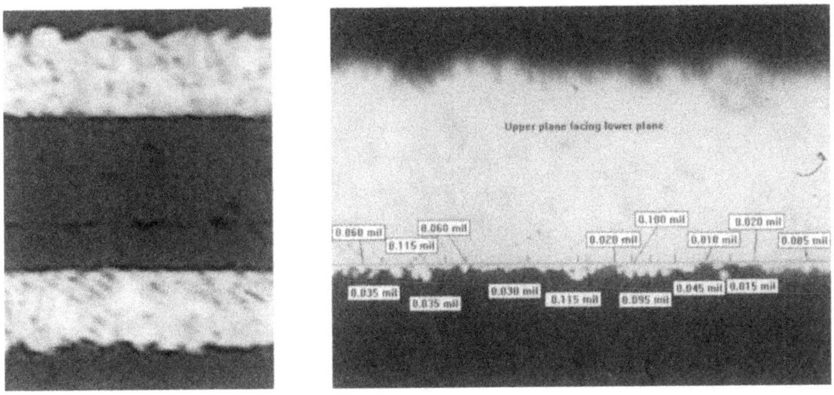

(a) (b)

Figure 4.52 Cross-section (a) and surface-roughness measurement (b) on the 50-μm laminate. Cross-section and photo courtesy of Sanmina-SCI.

250-μm wafer probes. The nominal dimensions and probe connections are shown in Figure 4.53.

Measurements were carried out with two separate VNAs. The 100-kHz–1,800-MHz range was measured with a logarithmic sweep with an Agilent 4396A VNA with the 85046 *S*-parameter test kit. The 1–10-GHz frequency range was measured with linear sweep with an Agilent N5230 VNA. For both VNAs, calibration was done to the tips of the probes with a GGB Industries CS-14 calibration substrate.

The square sample was measured in three configurations: self-impedance at the corner (A), self-impedance in the middle of a side (B), and transfer impedance between the corner and middle of a side (see Figure 4.53). Figures 4.54 and 4.55 plot the measured self- and transfer impedances. Note that in the frequency range of 1.0–1.8 GHz there are two measured traces originated from two different instruments. The independent datasets run on top of each other; this indicates that the data is of good quality.

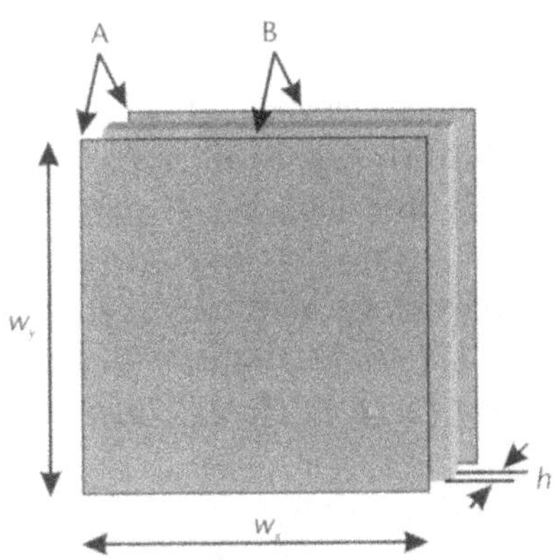

Figure 4.53 Bare laminate characterization. The sample size was $w_x = w_y = 6.35$ cm (2.5 inches) with a nominal dielectric thickness of $h = 75$ μm (3 mils).

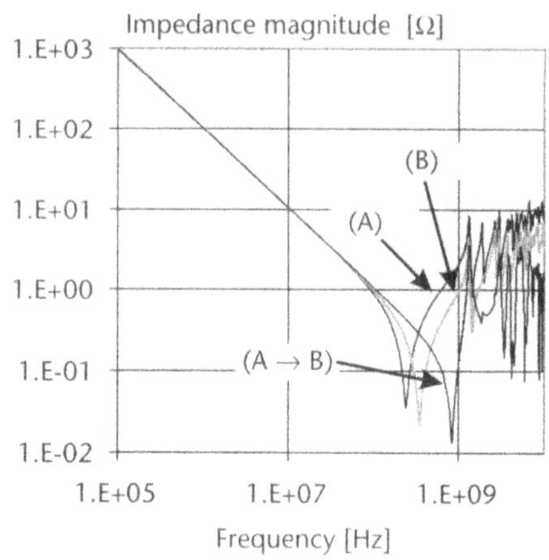

Figure 4.54 Self- and transfer impedance magnitudes measured on the sample shown in Figure 4.53.

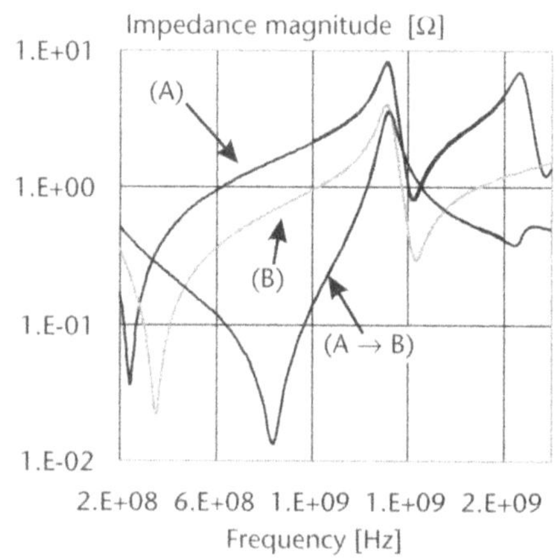

Figure 4.55 Self- and transfer impedance magnitudes measured on the sample shown in Figure 4.53. Zoomed horizontal scale.

The capacitance versus frequency curves were extracted from the measured impedance for all three cases. The capacitance measured at (B) is shown on Figure 4.56(a). Capacitance is shown as a function of the logarithm of frequency. The linear best-fit approximation over the highlighted 300 kHz–15 MHz is also shown on the chart. Figure 4.56(b) shows the dielectric constant and loss tangent independently extracted from the three datasets. The relative dielectric constant curves run on top of each other. The loss tangent curves are also close, but have a slight separation.

The first modal resonance frequency is f_{res} = 1.3144 GHz, the calculated laminate thickness is h = 78.8 μm (3.1 mils). The quality factor at the first modal resonance at location (B) is Q = 26.8. The calculated Q from the loss tangent and surface impedance is Q = 21.5.

The cross section of the laminate is shown in Figure 4.57. Figure 4.57(a) shows the entire two-sided laminate: the light strips on the top and bottom are the rolled anneal copper foils; the dark strip in the middle is the dielectric layer. By graphical postprocessing the photo can be enhanced to show the three layers of dielectrics: the

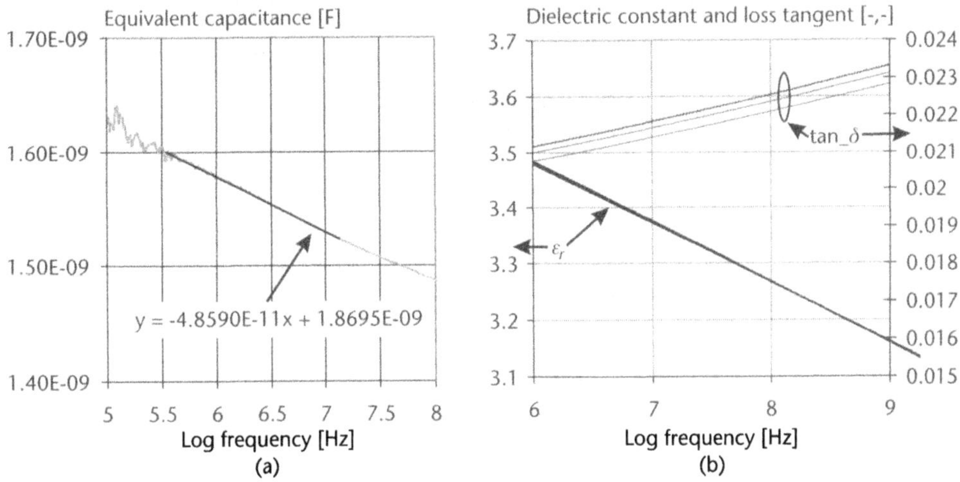

Figure 4.56 (a) Extracted and modeled capacitance from data at location (B). (b) Multipole Debye dielectric constant and loss tangent models from all three measured locations. Horizontal scale is the logarithm of frequency. There is no difference among the three dielectric constant (ε_r) curves; the top loss tangent (tan_δ) curve is from location (B); the middle trace is extracted from the transfer impedance; the bottom trace was extracted from data measured at location (A).

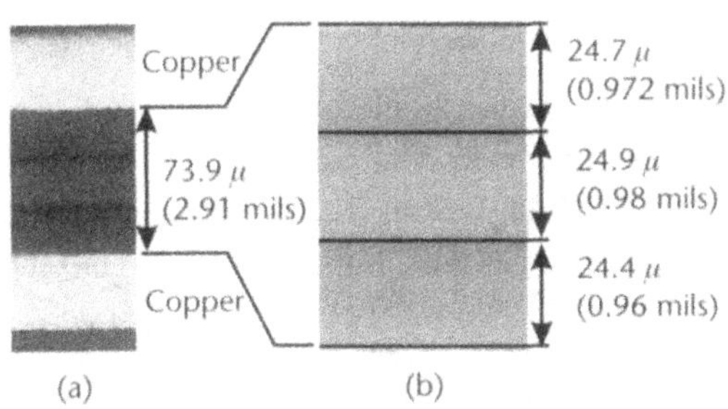

Figure 4.57 (a, b) Cross-section of the 75-μm laminate. (Laminate is courtesy of DuPont, cross section photo courtesy of Sanmina-SCI.)

polyimide layer in the middle and the two C-staged modified acrylic layers above and below. This is shown in Figure 4.57(b). The measured dielectric thickness is 73.9 μm, and from (4.50a) we get 77.2 μm, which agree within less than 5%. The surface roughness was measured at four different locations on the inner surfaces of the upper and lower planes; the root-mean-square surface roughness varied between 0.56 μm and 1.19 μm. Reference [34] is useful for analyzing multilayer structures.

Note that measuring the laminate parameters as shown above yields the average electrical properties of the three stacked dielectric layers.

To close the correlation loop, full-wave field-solver simulations were performed with the plane separation and dielectric parameters extracted from the measured impedance data. The frequency-dependent dielectric constant and loss tangent were entered into the tool. A surface impedance approximation was used to capture the frequency dependence of the inductance and resistance. The probes were modeled

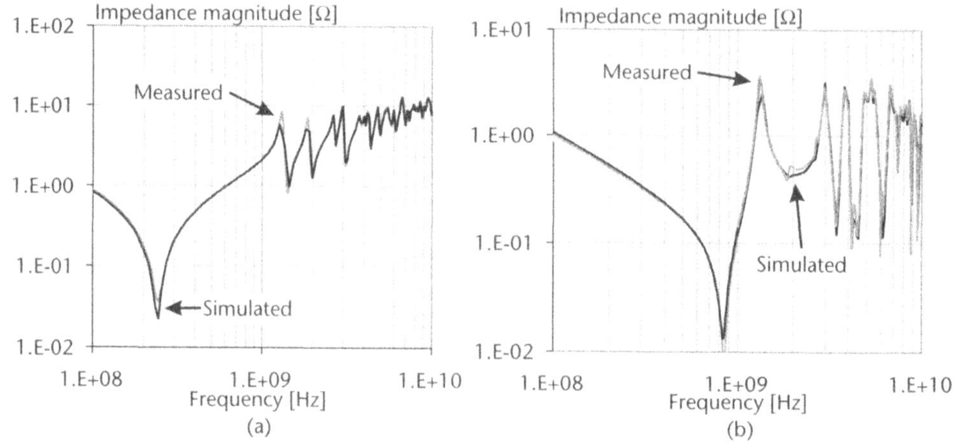

Figure 4.58 (a) Correlation between measured and simulated impedance magnitude at location (A) and (b) transfer impedance between locations (A) and (B).

with 75 × 75 μm lumped ports, and the self-impedance at location (A) and the transfer impedance between locations (A) and (B) were simulated in the 0.1–10-GHz frequency range. The correlation is shown in Figure 4.58.

References

[1] SpeedXP Suite, http://www.sigrity.com.

[2] EZ-PowerPlane, http://www.ems-plus.com

[3] Cadence Allegro PCB PI Option XL, http://www.cadence.com

[4] Carver, K. R., and J. W. Mink, "Microstrip Antenna Technology," *IEEE Trans. on Antennas and Propagation*, Vol. AP-29, 1981, pp. 2–24.

[5] Okoshi, T., and T. Miyoshi, "The Planar Circuit—An Approach to Microwave Integrated Circuitry," *IEEE Trans. Microwave Theory and Technique*, Vol. MTT-20, 1972, No. 4, pp. 245–252.

[6] Wang, Z. L., et al., "Convergence Acceleration and Accuracy Improvement in Power Bus Impedance Calculation with a Fast Algorithm Using Cavity Modes," *IEEE Trans. on Advanced Packaging*, Vol. 47, No. 1, February 2005.

[7] Novak, I., "Accuracy Considerations of Power-Ground Plane Models," *Proceedings of the 8th Topical Meeting on Electrical Performance of Electronic Packaging*, San Diego, CA, October 25–27, 1999, pp. 153–156.

[8] Wang, D. -D., and Z. -F. Li, "Fast Analysis of Bounces on Power/Ground Planes Using Even-Odd Partition," *IEEE Transactions on Advanced Packaging*, Vol. 26, No. 1, February 2003, pp. 65–72.

[9] Park, M. -J., et al., "A New, Efficient Analytic Expression for the Impedance of a Rectangular Power Plane," *Proceedings of Electrical Performance of Electronic Packaging*, October 21–23, 2002, Monterey, CA, pp.163–166.

[10] Shlepnev, Y., "Transmission Plane Models for Parallel-Plane Power Distribution System And Signal Integrity Analysis," *22nd Annual Review of Progress in Applied Computational Electromagnetics*, March 12–16, 2006, Miami, FL, pp. 382–389.

[11] Xu, M., H. Wang, and T. H. Hubing, "Application of the Cavity Model to Lossy Power-Return Plane Structures in Printed Circuit Boards," *IEEE Trans. on Advanced Packaging*, Vol. 26, No. 1, February 2003.

[12] Na, N., et al., "Modeling and Transient Simulation of Planes in Electronic Packages," *IEEE Trans. on Advanced Packaging,* Vol. 23, No. 3, August 2000, pp. 340–352.

[13] Okoshi, T., Y. Uehara, and T. Takeuchi, "The Segmentation Method—An Approach to the Analysis of Microwave Planar Circuits," *IEEE Trans. on Microwave Theory and Techniques*, Vol. 24, No. 10, October 1976, pp. 662–668.

[14] Wang, C., et al., "An Efficient Approach for Power Delivery Network Design with Closed-Form Expressions for Parasitic Interconnect Inductances," *IEEE Trans. on Advance Packaging,* Vol. 29, No. 2, May 2006, pp. 320–334.

[15] HDT Application Note 16, "SPRINT and SIGHTS Simulation of Power and Ground Distribution Planes," *High Design Technology,* Italy, 1993.

[16] Lee, K., and A. Barber, "Modeling and Analysis of Multichip Module Power Supply Planes," *IEEE Trans. on Components, Packaging, and Manufacturing Technology—Part B*, Vol. 18, No. 4, November 1995, pp. 628–639.

[17] Smith, L., R. Anderson, and T. Roy, "Power Plane SPICE Models and Simulated Performance for Materials and Geometries," *IEEE Trans. on Advanced Packaging,* Vol. 24, No. 3, August 2001, pp. 277–287.

[18] Novak, I., J. R. Miller, and E. Blomberg, "Simulating Complex Power-Ground Plane Shapes with Variable-Size Cell SPICE Grids," *Proceedings of Electrical Performance of Electronic Packaging*, Monterey, CA, October 21–23, 2002.

[19] Zhang, L., et al., "A Circuit Approach to Model Narrow Slot Structures in a Power Bus," *Proceeding of IEEE Symposium 2004*, Santa Clara, CA, August 9–13, 2004.

[20] Kim, J. -H., and M. Swaminathan, "Modeling of Irregular Shaped Power Distribution Planes Using Transmission Matrix Method," *IEEE Trans. on Advanced Packaging*, Vol. 24, No. 3, August 2001.

[21] Novak, I., et al., "Lossy Power Distribution Networks with Thin Dielectric Layers and/or Thin Conductive Layers," *IEEE Trans. on Advanced Packaging*, Vol. 23, No. 3, August 2000, pp. 353–360.

[22] Loyka, S. L., "A Simple Formula for the Ground Resistance Calculation," *IEEE Trans. on Electromagnetic Compatibility*, Vol. 41, No. 2, May 1999, pp. 152–154.

[23] Deutsch, A., et al., "Extraction of $\varepsilon_r(f)$ and tan_$\delta(f)$ for Printed Circuit Board Insulators Up to 30GHz Using the Short-Pulse Propagation Technique," *IEEE Trans. on Advanced Packaging*, Vol. 28, No. 1, February 2005, pp. 4–12.

[24] Johnson, H., and M. Graham, *High-Speed Signal Propagation: Advanced Black Magic*, Upper Saddle River, NJ: Prentice Hall, 2003, Chapter 2.15.5.

[25] Bur, J., "Dielectric Properties of Polymers at Microwave Frequencies: A Review," *Polymer*, Vol. 26, July 1985, pp. 963–977.

[26] Djordjevic, A. R., et al., "Wideband Frequency-Domain Characterization of FR-4 and Time-Domain Causality," *IEEE Trans. on Electromagnetic Compatibility*, Vol. 43, No. 4, November 2001, pp. 662–667.

[27] Svensson, C., and G. H. Dermer, "Time Domain Modeling of Lossy Interconnects," *IEEE Trans. on Advanced Packaging*, Vol. 24, No. 2, May 2001, pp. 191–196.

[28] Bunno, N., and I. Novak, "Frequency Domain Analysis and Electrical Properties Test Method for PCB Dielectric Core Materials," *Proceedings of DesignCon East*, Boston, MA, June 23–25, 2003.

[29] Baker-Jarvis, J., "Dielectric and Conductor-Loss Characterization and Measurements on Electronic Packaging Materials," *National Stand Technology Technical Note 1,520*, 2001.

[30] IPC-TM-650 Test Methods Manual, No. 2.5.5.5.1, "Stripline Test for Complex Relative Permittivity of Circuit Board Materials to 14 GHz," *The Institute for Interconnecting and Packaging Electronic Circuits*, March 1998.

[31] IPC-TM-650 Test Methods Manual, No. 2.5.5.6, "Non-Destructive Full Sheet Resonance Test for Permittivity of Clad Laminates," *The Institute for Interconnecting and Packaging Electronic Circuits*, May 1989.

[32] Bogatin, E., S. Begley, and M. Resso, "The Role of Dielectric Constant and Dissipation Factor Measurements in Multi Gigabit Systems," *Proceedings of DesignCon 2007*, Santa Clara, CA, January 29–February 1, 2007.

[33] Novak, I., "Part IX. Frequency Dependent Capacitance and Inductance of Thin and Very Thin Laminates," in TecForum TF9, "Thin and Very Thin Laminates for Power Distribution Applications: What Is New in 2004?" *DesignCon 2004, High-Performance Systems Design Conference*, Santa Clara, CA, February 2–5, 2004.

[34] Bhattacharyya, A. K., *Electromagnetic Fields in Multilayered Structures*, Norwood, MA: Artech House, 1994.

CHAPTER 5

Impedance Measurements Basics

In this chapter we cover the generic measurement considerations around PDNs. Some of the specific measurement setups for PDN structures will be covered in later chapters where we describe the modeling and characterization details.

5.1 Selecting the Measurement Concept for PDN Impedance

With increasing power levels and constantly dropping supply voltages, the PDN impedance of high-power systems must be low, sometimes in the milliohm or submilliohm range. Even if we do not necessarily need very low PDN impedance (i.e., in the case of lower-power systems and low-current power rails) dc-dc converters with high dc loop gain will create submilliohm output impedances in the Hertz and low kilohertz frequency range. Moreover, low-loss test fixtures and low-ESR bypass capacitors may have effective impedances in the milliohm range also. These low impedance values themselves create unique challenges in the selection of instruments, setups, and connections. As we will see, a big part of the challenge is practical: the constraints of high-density product designs are in conflict with the size and cost of connections needed for repeatable and consistent measurement of low PDN impedance.

There are several basic methods for the purposes of impedance measurements. Reference [1] lists six commonly used options:

- Bridge method;
- Resonant method;
- I-V method;
- RF I-V method;
- Network analyzer method;
- Autobalancing method.

Figure 5.1 shows the core concept behind each method. According to [1], the bridge method has high accuracy, low cost, but narrow frequency coverage; so, the balancing may need to be done manually. The resonant method inherently assumes a certain type (R-L on the simplified schematics) of network impedance to be measured, so it is not well suited for frequencies above 100 MHz. The autobalancing bridge method can be highly accurate, but it is also limited to about 100 MHz or lower frequencies. The I-V method can take DUTs with one terminal grounded, suitable to connections with probes, but when implemented with a transformer in

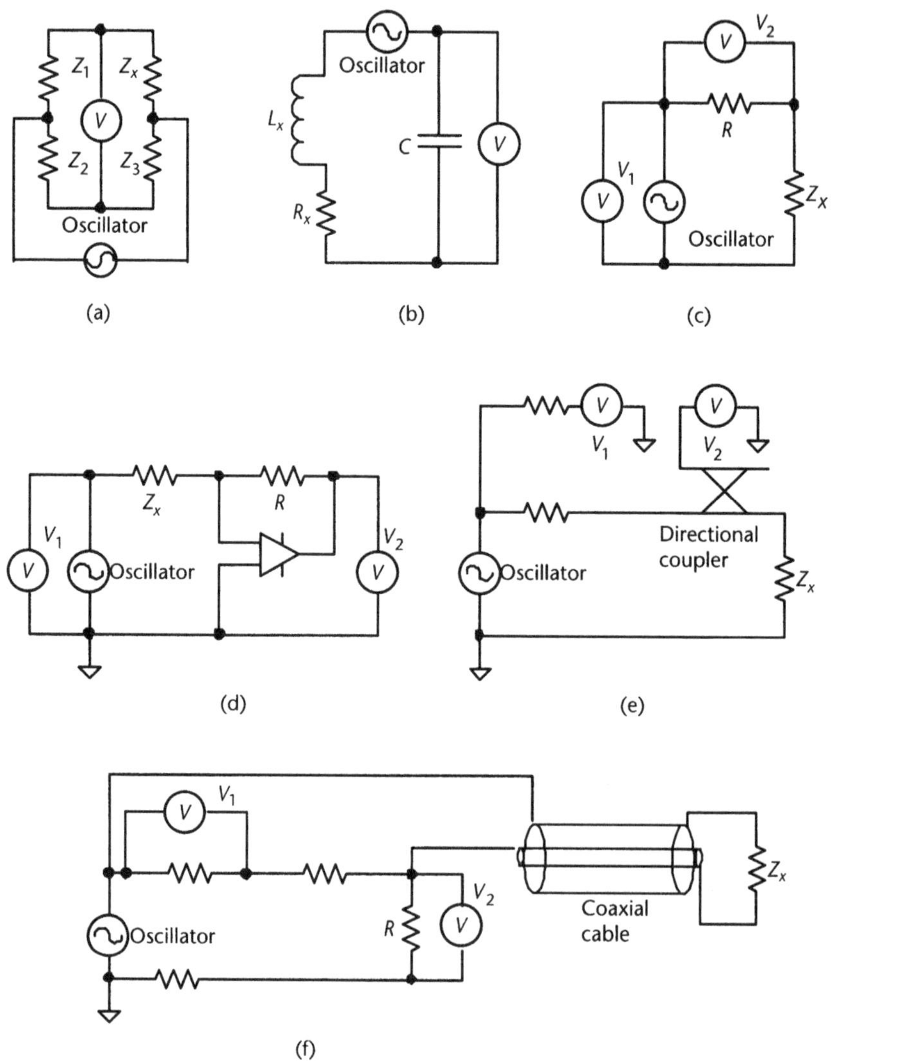

Figure 5.1 Simplified schematics of commonly used impedance measuring methods: (a) bridge method, (b) resonant method, (c) I-V method, (d) autobalancing bridge method, (e) network analysis method, and (f) low-impedance RF I-V method. (*After:* [1].)

the probe, it limits the frequency range to about 10 kHz–100 MHz. The RF I-V method exhibits high accuracy and covers a wide impedance range; 1-MHz–3-GHz instruments are available. The network analysis method based on S_{11} data has good high-frequency characteristics, but it is not particularly accurate for impedances much different from the 50Ω impedance of the instrument.

Overall, the available methods shown in Figure 5.1 all share a common challenge when we apply them to low-impedance PDN measurements. If only two connections are made to the DUT, the uncertainty of the connection geometry will become the greatest limitation eventually because PDN components, subsystems and systems usually can afford connectors only in a few and specialized cases. When measuring individual small components, custom test fixtures shaped to match the component outline may help to create a repeatable and high-performance connection. For complex circuit boards and systems, however, building custom connection features into the design becomes increasingly prohibitive.

For instance, consider the Network Analysis method. From the measured voltage reflection coefficient (Γ, or S_{11}), we can calculate the unknown Z_x impedance of the DUT as:

$$Z_x = Z_o \frac{1+\Gamma}{1-\Gamma} \quad \Gamma = \frac{Z_x - Z_0}{Z_x + Z_0} \tag{5.1}$$

where Z_0 is the reference impedance of the instrument. The value Z_0 is typically 50Ω and real, but Z_x and Γ are both complex in general. If we measure all combinations of the **S** matrix on a multiport PDN, we could translate the results into Z-parameters by using the generalized definition of voltage reflection coefficient [2]:

$$[S] = \{[Z]-[Z_0]\}\{[Z]+[Z_0]\}^{-1} \quad [Z] = [Z_0]\{[U]+[S]\}\{[U]-[S]\}^{-1} \tag{5.2}$$

where [$\mathbf{Z}_0$] and [**U**] are the characteristic impedance and unity matrix, respectively. Both are diagonal matrices where all elements in the main diagonals are Z_0 and 1, respectively, and all off-diagonal elements are zero. Z_0 is the normalization impedance of the scattering parameters, usually 50Ω. The implementation options will be detailed in Section 5.4.

Figure 5.2 shows the network analysis method from Figure 5.1, but here we also highlight the $Z_{connection}$ parasitic impedance in series to the unknown Z_x impedance.

For instance, assume that we want to validate the PDN on a populated board where the impedance should be 10 mΩ or less from dc to 100 MHz. If the board is large, we will probably need a cable to reach connection points in the middle of the board.

Good calibration standards have coaxial connectors, but if we need to measure the impedance only up to about 100 MHz, with some experience we can make decent approximations of OPEN, SHORT, and LOAD for reflection calibration at the open end of a cable by connecting nothing (OPEN), a solid conductive plane (SHORT), and a matching-size 50Ω (or 49.9Ω) termination resistor (LOAD) by hand soldering the standards to the end of the cable. Once the calibration is done, we solder the same cable to the DUT. Unfortunately, we would need to connect the DUT exactly at the same point where we connected the calibration standards; we cannot do this with absolute precision. The shorting metal sheet, the resistor piece

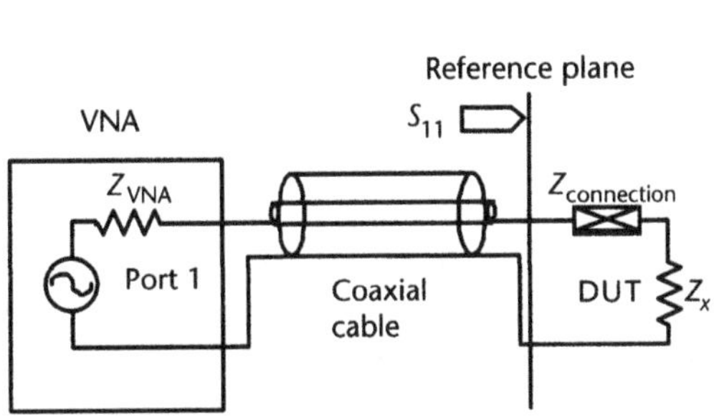

Figure 5.2 Setup scheme for the network analysis method, showing the extra impedance in series to the DUT.

providing the matched load both have finite thickness; so, we may not be able to solder them precisely to the same point along the open center wire of the cable.

Let us assume that the uncertainty of the connection point is 1 mm (40 mils). Assuming a standard 50Ω coaxial cable with a nominal velocity of 66% and an un-stranded center wire of AWG 25 size, each millimeter length of the cable accounts for approximately 250-pH inductance and 0.1-mΩ dc resistance. The 0.1-mΩ resistance uncertainty becomes a problem with very-low-resistance DUTs, such as when measuring active power sources, dc-dc converters, or shorted structures. The quarter of a nanohenry inductance, however, proves to be a much more serious limitation: this represents a 0.16Ω impedance error at 100 MHz. Clearly, when our PDN is supposed to provide 10 mΩ of impedance at 100 MHz, we cannot afford a measurement scheme that has an added 0.16Ω error of measured impedance.

When we have one port (two wires) connecting to the DUT, one major limitation is that we will eventually measure $Z_x + Z_{connection}$, instead of Z_x. This limitation, in fact, is not related to the measurement scheme or measuring procedure itself; any kind of instrument, based on any of the measuring schemes would be limited by this error if only one port is used to connect to the DUT.

As we will see in Section 6.2.3, we could use wafer probes with a VNA and calibrate the tips of the probes, to remove most of the error caused by the uncertainty of connection points. This would remove the error related to $Z_{connection}$, but still we would face a second major limitation of the one-port network analyzer method: the finite error in measuring the voltage reflection coefficient of a highly reflective DUT. The assumed low Z_x impedance creates an almost full reflection, therefore the incident and reflected wave magnitudes are similar. The low impedance value must be resolved from the difference of the incident and reflected waves with almost equal magnitude, therefore the relative measurement error increases sharply as impedance values get lower. The measurement uncertainty of typical VNAs in one-port S_{11} measurements depends on the frequency and measured value of voltage reflection coefficients. Approaching full reflection, the uncertainty typically rises to 0.01–0.02. The 0.01–0.02 error near full reflection means an impedance uncertainty of $50^*(0.01 - 0.02)/2 = 0.25 - 0.5\ \Omega$. This error makes it impossible to measure impedances in the milliohm range or lower; only impedances in the ohm range and above can be measured with reasonable accuracy with one-port VNA connections.

We will show in the next section that using two ports of a VNA instead only one port largely solves both of the above two limitations.

5.2 The Importance of Two-Port Connections

VNAs in a two-port shunt-through connection offer a convenient way of measuring low impedances, similar to the four-wire dc resistance measurement setup.

As shown in [3, 4], two-port VNA connections use Port 1 to launch a known test signal into the unknown impedance, and Port 2 is used to measure the voltage drop. Figure 5.3 explains the connection scheme. Figure 5.3(a) shows the four-wire

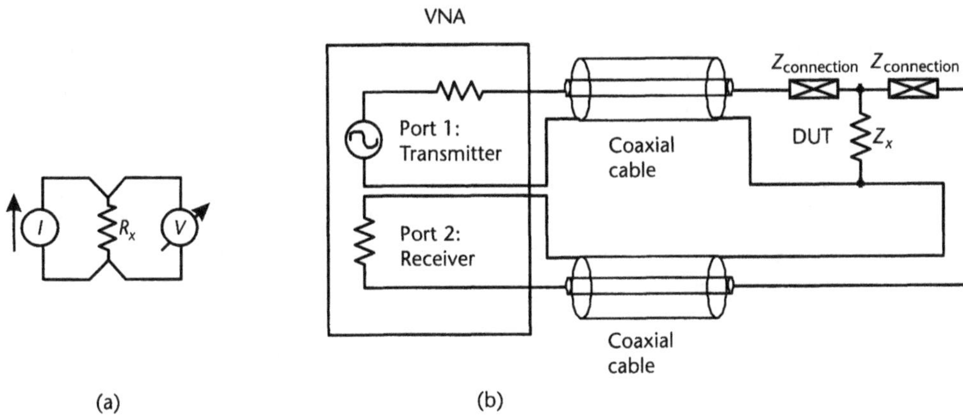

Figure 5.3 (a) Four-wire Kelvin connection and (b) setup with VNA for two-port shunt-through impedance measurements. This scheme greatly reduces the error due to the series connection impedance, by transforming them into the 50Ω loops of VNA ports.

Kelvin connection; this is the preferred scheme in measuring low dc resistances. Figure 5.3(b) shows its high-frequency equivalent. The extra $Z_{connection}$ impedance that appears directly in series to the unknown impedance in one-port VNA measurements is now transformed into the two loops of VNA ports, each having a nominally 50Ω impedance. Though we now have two connections to care for instead of one, the through calibration establishes the reference plane at the center point of the two $Z_{connection}$ impedances; this effectively removes them from the circuit. Therefore, they introduce very little error up to several gigahertz frequencies.

Assuming that the DUT has $Z_x << 50\Omega$ impedance, the error in one-port VNA measurements is proportional to $Z_{connection}/Z_x$, whereas the error in two-port measurements is proportional to $Z_{connection}/Z_{VNA}$. The improvement ratio between measurements in one-port versus two-port connections is Z_{VNA}/Z_x.

At the same time, the error floor of the receive-only input is also much lower; we usually get up to 100–140-dB dynamic range, making it possible to make repeatable measurements in the submilliohm range.

From Figure 5.3 we can estimate the error caused by $Z_{connection}$. The connection impedance can be approximated by a series $L_{connection}$ inductance because in series to the low-impedance Z_{DUT}, the connection discontinuity will become mostly inductive. Assuming the same inductance in both connections, we get:

$$Z_{connection} = j\omega L_{connection} \tag{5.3}$$

As was shown in [4], as a result of the $j\omega L$ inductive reactance, the measured Z_x' will be different from Z_x as:

$$\frac{Z_x'}{Z_x} \approx \frac{1}{1 + j\dfrac{\omega}{\omega_c}} \tag{5.4}$$

where $\omega_c = \dfrac{1}{\dfrac{L_{connection}}{Z_{VNA}}}$ and we assume identical Z_{VNA} impedance and $Z_{connection}$ on both VNA ports.

If we assume the same 1-mm connection uncertainty as before, the associated 250-pH inductance comes with an ω_c = 31-GHz corner frequency. This results in a negligible error up to several gigahertz frequencies. For frequencies far below the ω_c corner frequency, the magnitude error can be approximated by:

$$\frac{Z'_x}{Z_x} \approx \frac{1}{1+\frac{1}{2}\left(\frac{\omega}{\omega_c}\right)^2} \approx 1-\frac{1}{2}\left(\frac{\omega}{\omega_c}\right)^2 \qquad (5.5)$$

With 250-pH inductance, (5.5) yields –0.0005% error at 100 MHz, –0.05% error at 1 GHz, and –1.25% magnitude error at 5 GHz. This stands in contrast to the +160% magnitude error of the 10 mΩ Z_{DUT} when measured in a one-port VNA connection.

The two-port VNA measuring methods are elaborated in more detail in [5]. Figure 5.4 shows the two subcategories: two-port series-through and two-port shunt-through.

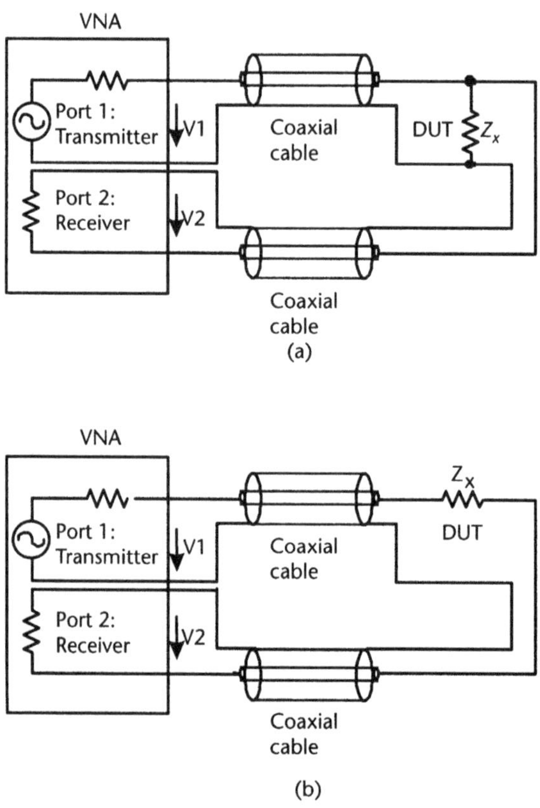

Figure 5.4 Equivalent schematics of two-port VNA connections for impedance measurements in (a) shunt-through and (b) series-through connections.

The two-port series-through connection requires floating DUT when none of the DUT terminals has to be connected to ground. This connection may be preferred for physically small-size and/or high-impedance DUTs, such as small surface-mount capacitors or inductors. The two-port shunt-through connection is preferred for low-impedance and/or large-size DUTs. This connection is convenient if we have to connect one terminal of the DUT to ground also.

In the rest of the book, the measurement setup of choice for low-impedance PDN components and systems (unless otherwise stated) is the two-port shunt-through scheme from Figure 5.4.

5.3 Self- and Transfer Impedance

By using VNAs in either a series-through or shunt-through connection, we literally always measure the transfer response between the connections from port 1 to port 2. If, however, the two ports are connected to the same (or almost the same) physical location on the PDN, the transfer impedance asymptotically approaches the self-impedance at the point of connection. This creates a single and unified way to measure both self- and transfer-impedances with the same instrumentation and setup.

This is illustrated with the self- and transfer impedances of bare planes. The simulated surface graph in Figure 5.5 shows the gradual change of transfer impedances into self-impedance over a bare pair of 25.4 × 25.4-cm (10 × 10-inch) planes with 50-μm (2-mils) plane separation, and dielectric constant of 4. To show the

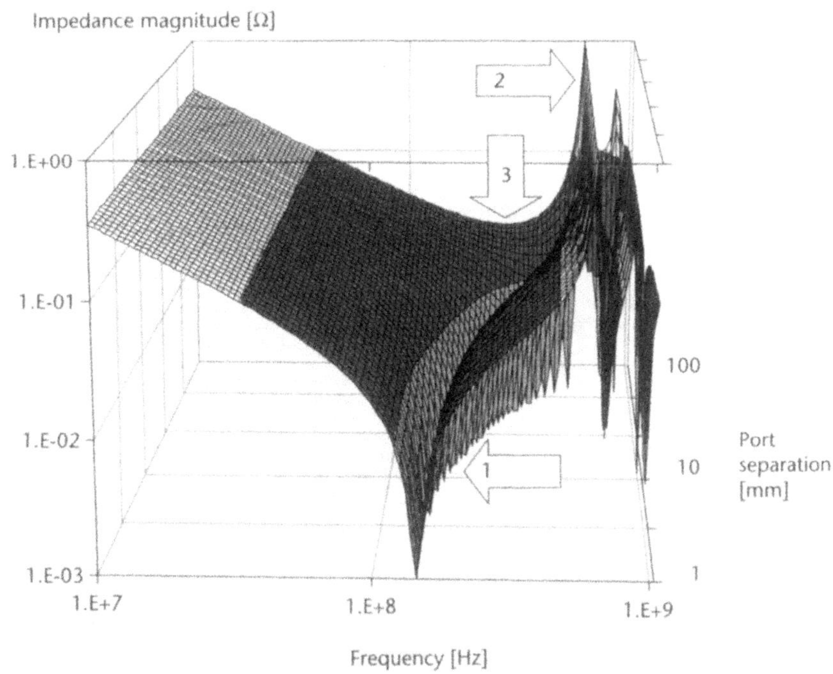

Figure 5.5 Illustration of simulated transfer-impedance magnitudes on a large pair of bare power planes, as a function of frequency and port separation.

effects clearly, dielectric and copper losses were removed to produce sharp modal-resonance peaks and dips.

The vertical scale on the graph shows impedance magnitude, the front horizontal scale is logarithmic frequency from 10 MHz to 1,000 MHz, the side horizontal scale is logarithmic distance between the two ports on the planes ranging from 1 mm to 100 mm. Note that the left side of the surface shows a steady slope coming from the static capacitance of the planes. Since the static capacitance does not depend on the port location and port distance, this part of the graph is independent of port separation. The numbered block arrows show three important features on the graph. Arrow 1 points towards the bent contour of the series resonance frequency footprint. The frequency minimum shows almost no change with port separations up to about 3 mm, but with a 10-mm port separation there is already a small visible shift in the resonance frequency. With larger separations the frequency minimum gradually shifts to higher values. Finally, with port separations above 30 mm, the sharp minimum disappears, and it goes over into a curve with a smooth and shallow minimum. This part of the surface is identified by arrow 3. While the series resonance changes with port separation, the parallel resonances do not change. The straight line of the first modal resonance peak is identified on the graph by arrow 2.

The port separation where transfer impedance noticeably departs from self-impedance on planes scales with frequency. Physically smaller structures or laminates with lower dielectric constants have higher resonance frequencies; so, the allowable port separation required to approximate self-impedance gets smaller. Figure 5.6 illustrates this with the simulated impedance surface on a 6.35 × 6.35-cm (2.5 × 2.5-inch) square bare laminate with 75-μm dielectric spacing. The surface plot shows the impedance magnitude in the 100-MHz to 1-GHz frequency range as one of the ports moves from the corner towards the center of the side. The second port stays at the corner. This structure was measured also and the correlation details

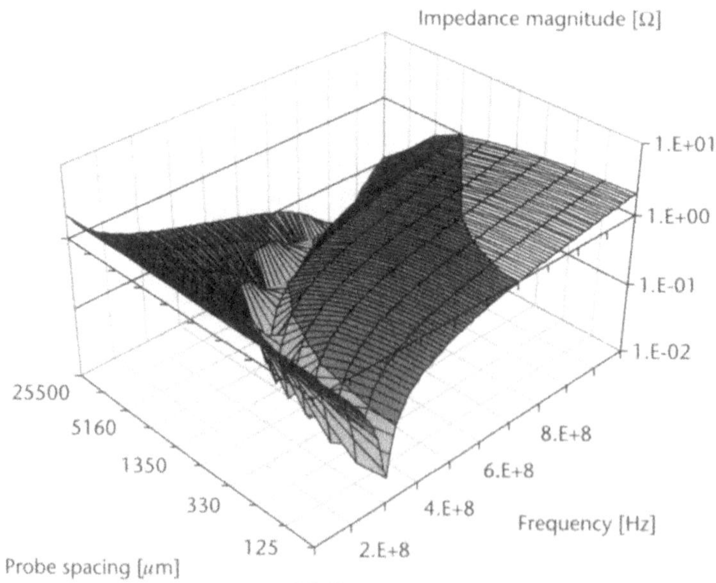

Figure 5.6 Illustration of simulated transfer impedance magnitudes on a small pair of bare power planes as a function of frequency and port separation.

were shown in Chapter 4, Figures 4.53 through 4.58. Note that the four times smaller structure results in a series resonance frequency that is about four times higher and a maximum port separation that is about four times smaller; within this separation the self-impedance and transfer impedance are almost the same.

At lower frequencies, as long as the spatial dimensions are much smaller than the shortest wavelength of interest, the usual assumption might be that self- and transfer impedances do not differ much. At very low impedance values, however, the series losses of the distribution network together with the parallel bypass capacitors may create different impedances at and between various points.

This is illustrated on the measured low-frequency plots in Figure 5.7. Both sets of curves were measured on the same supply rail designed for a maximum current rating of 50A. The PDN was measured with the dc-dc converter unpowered and powered. There were six large bulk capacitors on the planes; these created the shallow minima around 100 kHz (visible on both sets of curves). There were two power/ground plane pairs stitched together with multiple vias in the stack up. There were pairs of test vias where the VNA ports were connected. The curves labeled "self bottom" and "self top" were measured with both VNA ports connected to the bottom pads or to the top pads on one of the test via pairs, respectively, thus having the longer or shorter part of the via loop in series to the PDN impedance. The curves labeled "self opposite" were measured with the two VNA ports connected to the opposite sides of the same test via pair. The curves labeled "transfer, 20 mm" were measured with one VNA port connected to one test via pair, the other VNA port connected to another via pair 20 mm (0.8 inch) away.

With no input power applied to the dc-dc converter, all measured traces slope upwards below 30 kHz, and there is no appreciable difference in their magnitudes. At 100 kHz, there is a clear difference among the four curves: the lowest is the transfer 20-mm curve, followed by the self opposite and self top, and the highest impedance magnitude is from self bottom. This order is maintained throughout the higher end of the measured frequency range. With input power applied, the minimum at 10 kHz is due to the dc-dc converter. At 200 kHz and above, the order of the four traces is the same in both graphs. Below 200 kHz, however, the order of traces self

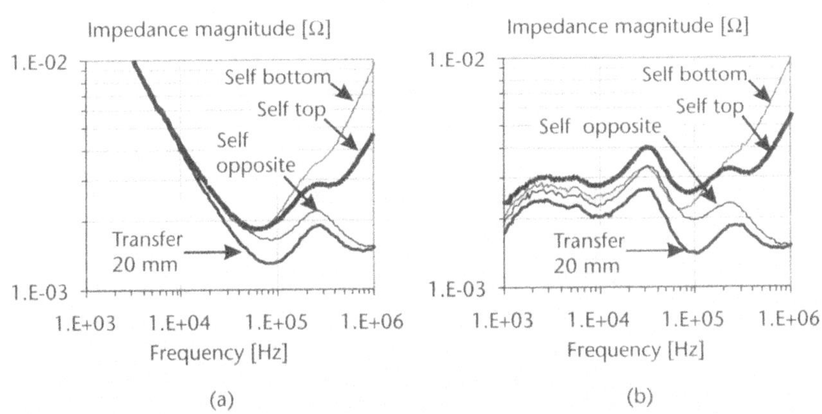

Figure 5.7 Measured low-frequency self- and transfer impedances on a populated board: (a) measured with no input power; and (b) measured with input power applied. The connection geometries are illustrated in Section 6.1.2.

bottom and self opposite changes several times over frequency, being dependent on the geometry of the current flow in the planes, vias, bulk capacitors, and output pins of the converter. Note that at 10 kHz where the wavelength in the PCB material is near 15,000 meters (almost 10 miles), a 20-mm distance results in an impedance spread of 2.0–2.8 mΩ. This spread occurs because the low-ESR bulk capacitors form an attenuator with the series resistance of the planes.

5.4 Transforming Measured *S*-Parameters

In the two-port shunt-through VNA measurement setup, we get *S*-parameters from the instrument.

In self-impedance measurements, to convert the S_{21} values to impedance, we can use the equivalent circuits in Figure 5.8.

In the two-port shunt-through connections, the Z_{VNA} port impedances from port 1 and port 2 create a 2:1 voltage divider to the source voltage and create an equivalent source impedance of $Z_{VNA}/2$. Eventually we end up with the equivalent circuit on the right. From this voltage, divider we can calculate the voltage across the unknown impedance as:

$$V = \frac{V_s}{2} \frac{Z_x}{\frac{Z_{VNA}}{2} + Z_x} \tag{5.6}$$

Note that many VNAs today have 50Ω port impedance, but other impedances (for instance, 75Ω) may also be used; (5.6) works for any port impedance as long as the impedances at the two ports are the same.

During the reflection calibration process, we compensate for impedance deviations from the reference impedance so that, up to the reference plane of calibration (including cables and probes if applicable), we can then assume Z_{VNA} impedance. During the through part of the calibration without the DUT being present, we establish the $S_{21} = 1.0$ reading to the voltage that appears after the 2:1 voltage divider, namely $V_s/2$. This gives us the link between the S_{21} reading and the impedances:

$$S_{21} = \frac{V}{\frac{V_s}{2}} \frac{Z_x}{\frac{Z_{VNA}}{2} + Z_x} \tag{5.7}$$

In (5.7) S_{21} and Z_x are complex numbers.

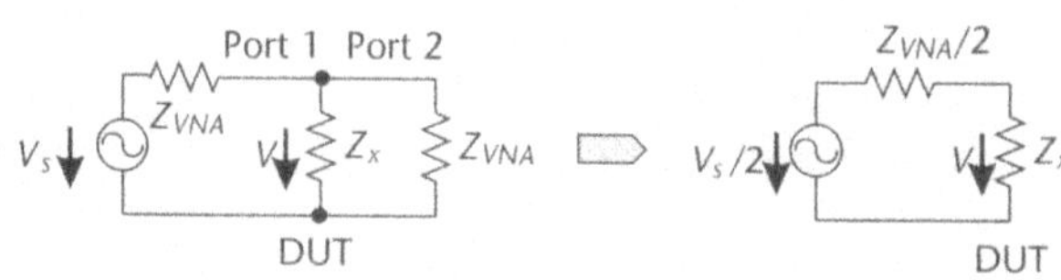

Figure 5.8 Equivalent circuits for calculating self-impedance from S_{21} obtained in two-port shunt-through connection.

5.4.1 Measuring Self-Impedance with Magnitude $|Z_x| << 25\Omega$

As long as the unknown impedance magnitude is much lower than 25Ω, (5.7) describing the complex voltage divider of Figure 5.8 can be simplified, by neglecting the unknown impedance next to the constant term of $Z_{VNA}/2$ in the denominator. This yields:

$$Z_x \approx \frac{Z_{VNA}}{2} S_{21}, \text{ for } Z_{VNA} = 50, \quad Z_x \approx 25\, S_{21} \tag{5.8}$$

If we use the Bode-plot style output from the VNA with magnitudes given in decibels, first the decibel numbers have to be converted to ratios:

$$S_{21} = 10^{\frac{S_{21}[dB]}{20}} \tag{5.9}$$

Note that (5.8) still has complex numbers, so even with the applied simplification we can obtain the complex impedance (real and imaginary or magnitude and phase) of the unknown impedance, as long as the complex S_{21} is known. For instance, if all we know is S_{21} magnitude, by using (5.9) but ignoring the phase information or by using a scalar network analyzer, we will get the magnitude of the unknown impedance without its phase.

If we do not need high accuracy or the phase of the unknown impedance, with certain VNA types at low frequencies there is little need for calibration (see Section 6.3). We simply have to establish the nominal full-scale impedance reading as 25Ω.

Figure 5.9 shows a low-cost setup for measuring impedance magnitudes of low values. The setup uses a tunable source and a receiver. Spectrum analyzers with the appropriate tracking generator provide both in one unit. Alternately, we could use a source that we tune either manually or automatically and use a receiver tuned to the same frequencies to which we tune the source. As the lowest-accuracy option, we can also use a spectrum analyzer in peak-hold mode to pick up the signal as the source tunes through the frequency range. This way, however, we lose the benefit of increased dynamic range due to the synchronously tuned transmitter/ receiver pair.

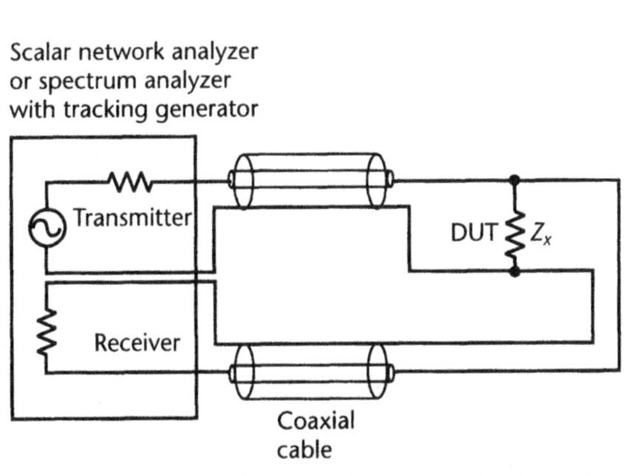

Figure 5.9 Low-cost measurement setup for impedance magnitudes of Z_x << 25Ω values.

5.4.2 Measuring Arbitrary Self-Impedance Values

The impedance of an inductor or capacitor varies linearly or inversely with frequency. Therefore, in some frequency ranges, the approximation of (5.8) may not be applicable. This is also to be expected when we have parallel resonating *L-C* circuits or power planes with large undampened modal resonance peaks. In all such cases, we have to solve the complex voltage divider expression of (5.5) for the unknown impedance.

In the two-port shunt connection scheme, we establish $S_{21} = 1.0$ (or 0 dB) during calibration; this corresponds to $V_s / 2$ voltage reading. By rearranging (5.6) and substituting S_{21}, we get:

$$Z_x = \frac{Z_{VNA}}{2} \frac{S_{21}}{1 - S_{21}} \tag{5.10}$$

In order to find the complex value of the unknown impedance, calibration must be performed, because we also need the correct phase information to solve the complex equation. For medium accuracy, a simple through calibration may be sufficient. To obtain the highest possible accuracy, and at high frequencies, a full two-port calibration has to be done.

Due to the robustness of the two-port shunt-through scheme, small discontinuities in the connections to the DUT may be neglected up to several gigahertz frequencies.

By rearranging (5.10), the real and imaginary parts of the unknown impedance can be expressed as:

$$\mathrm{Re}(Z_x) = \frac{Z_{VNA}}{2} \frac{\mathrm{Re}(S_{21})\{1 - \mathrm{Re}(S_{21})\} - \mathrm{Im}(S_{21})^2}{\{1 - \mathrm{Re}(S_{21})\}^2 + \mathrm{Im}(S_{21})^2} \tag{5.11}$$

$$\mathrm{Im}(Z_x) = \frac{Z_{VNA}}{2} \frac{\mathrm{Im}(S_{21})}{\{1 - \mathrm{Re}(S_{21})\}^2 + \mathrm{Im}(S_{21})^2} \tag{5.12}$$

Note that the expressions of (5.11) and (5.12) can be easily programmed in a spreadsheet that takes the measured VNA data.

To illustrate how this works, we show the extracted impedance magnitude and phase measured on a small-size PCB with multiple paralleled plane pairs. The horizontal and vertical dimensions of the small test site are shown in Figure 5.10. The test board had multiple test sites with various plane allocations. The test site shown here had the first plane pair from the top connected to the capacitor site and test vias; all other planes were left unconnected. The graph of Figure 5.11(a) was extracted with (5.8). With this approximation, we get an impedance curve that saturates at 25Ω. Moreover, instead of the –90° phase angle that we expect from a capacitance, at low frequencies the uncorrected phase angle approaches 0°. The plot of Figure 5.11(b) uses (5.11) and (5.12) to translate between S_{21} and the complex impedance; the impedance magnitude line becomes a straight line on the Bode plot with a constant phase of about –90°.

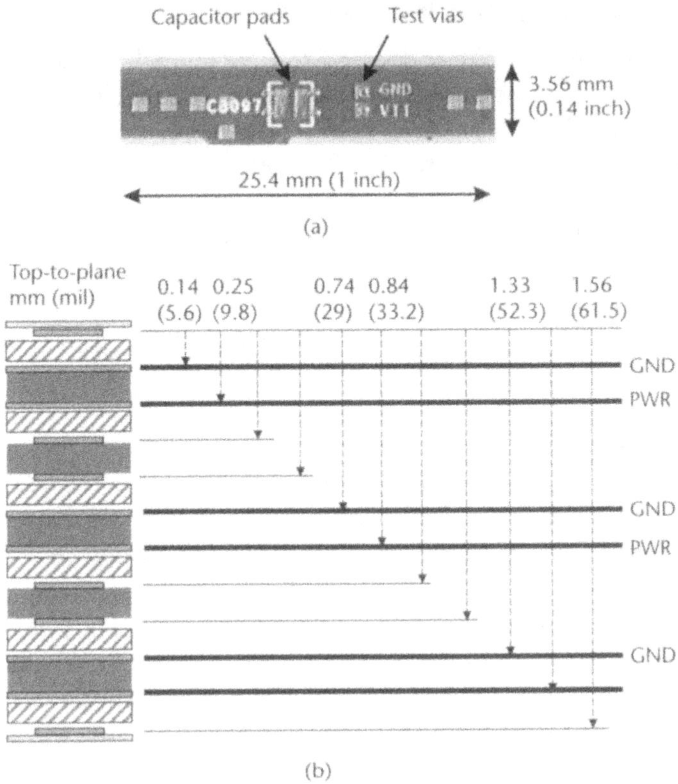

Figure 5.10 Test board used for Figure 5.11: (a) photo with horizontal dimensions and (b) vertical board construction.

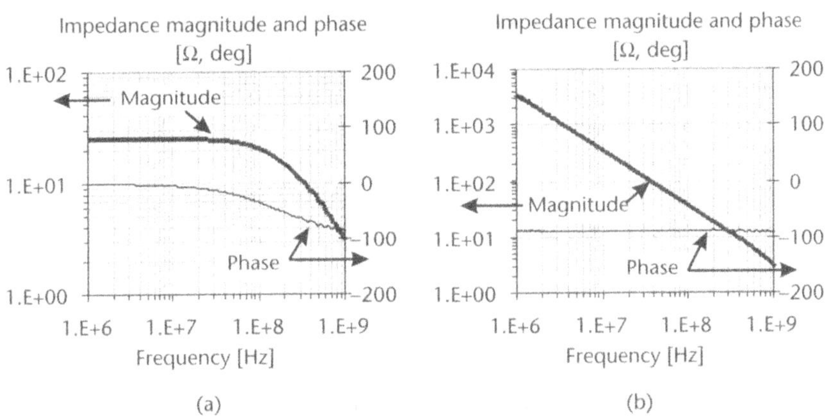

Figure 5.11 Measured impedance magnitude and phase of multiple bare plane pairs of 3.56 × 25.4 mm (0.14 × 1 inch) size in the test board shown in Figure 5.10. (a) Impedance obtained without complex inversion of voltage divider formula, using (5.8). (b) Same measured data using complex inversion with (5.11) and (5.12). On both graphs, left axes show impedance magnitude on logarithmic scale, while right axes show phase on the linear scale.

5.4.3 Measuring Large Impedance Values

Though finished PDN circuits tend to have much less than 50Ω impedance values, individual components over a wide frequency range usually span a large range of impedances. This is because PDN components behave either like capacitors or inductors, but not necessarily like resistors. Neglecting parasitic elements, and assuming frequency-independent capacitance or inductance, the impedance magnitude of an ideal inductance or capacitance varies linearly or inversely with frequency, respectively:

$$|Z_L| = |j\omega L| = \omega L, \quad |Z_C| = \left|\frac{1}{j\omega C}\right| = \frac{1}{\omega C} \tag{5.13}$$

where $\omega = 2\pi f$.

A 10-μF capacitance at 10 MHz accounts for a 1.5-mΩ impedance, which is much smaller than the 50Ω connection impedance of instruments. However, the same capacitance at 100 Hz represents 150-kΩ reactive impedance, which is much higher than 50Ω. Similarly, a 100-nH inductance represents 0.628-mΩ impedance at 1 kHz, whereas its impedance rises to 628Ω at 1 GHz. Resonating structures, like undamped power planes or parallel-connected capacitors with different values, may also span large impedances creating one or more impedance maxima.

Measuring impedance values much bigger than 50Ω will create a problem similar to the one encountered when measuring impedances much smaller than 50Ω: one-port measurements will add too much error due to the high reflection coefficients and parasitics related to the connection. When we measure very low impedances, the limiting parasitics is the series inductance and/or resistance. In contrast, when measuring very large impedances, the limiting parasitics is the parallel stray capacitance and/or conductance.

By using the two-port shunt-through connection, we can eliminate the error related to the one-port voltage-reflection coefficient measurements; but, when we measure large impedances, the magnitude of S_{21} also approaches unity, increasing the sensitivity to errors in S_{21}. This can be seen in (5.10), where the denominator approaches zero as S_{21} approaches one. Eventually this error is determined largely by the scaling error and quantization granularity of the measured S_{21}.

In practice, well-behaved network analyzers will allow us to measure impedances up to about 1 kΩ in a two-port shunt-through configuration, saturating in the tens or hundreds of kilo-ohms range. This kind of saturation will limit us, for instance, when we measure small-value capacitors at low frequencies.

Figure 5.12 illustrates this with measured data on a pair of power planes in a small-size PCB. The magnitude and phase of S_{21} parameters were measured in two-port shunt-through connection. Note that the S_{21} magnitude curve saturates at a value of 1.0 below 100 kHz, where the capacitive impedance gets much larger than 25Ω. The phase saturates at 0°. The impedance plot shows three impedance magnitude curves, extracted in different ways from the S_{21} data on the left. The trace labeled "uncorrected" uses the approximation of (5.8). The extracted uncorrected impedance saturates at 25Ω because it is just a scaled version of S_{21}. We get the trace labeled "corrected" when we apply (5.11) and (5.12). Note that this curve also satu-

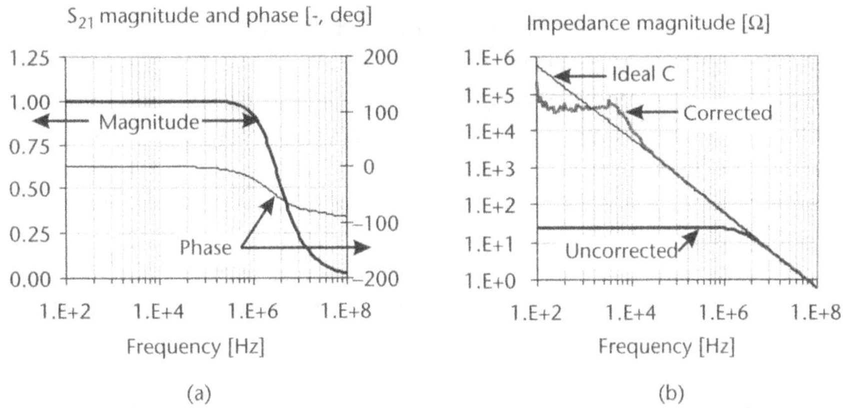

Figure 5.12 Illustration of the two different saturations when measuring large impedance values with two-port shunt-through connections. (a) Measured S_{21} data on bare plane pairs; and (b) extracted impedance values.

rates, but at a much higher value of around 30 kΩ. This saturation would not occur with error-free and perfectly accurate measuring instrumentation. The saturation value depends on the noise, quantization, granularity and scaling errors of the S_{21} reading. This saturation limit may quickly drift to lower values, as time passes after calibration.

Lastly, the trace labeled "ideal C" refers to the impedance magnitude of an ideal capacitor, which matches the capacitance of the DUT at 10 MHz. Note that this measured data set spans the 100-Hz to 100-MHz frequency range, or six decades of frequencies. In this wide frequency range, the DUT's impedance closely follows that of an ideal capacitor, spanning six decades in impedance magnitude. The six decades of impedance magnitude corresponds to a 120-dB range; this range stretches the dynamic range of most setups.

The dynamic-range limitation of the measurement setup is not necessarily imposed by the measuring instrument itself. Precision impedance analyzers (see, e.g., [6]) can achieve 200 dB or an even larger dynamic range. However, their frequency range may be limited, and even more limitation may come from the often-times inevitable external connections. VNAs provide much wider-band coverage and provide a means to calibrate to the end of user-configured connections, but their dynamic range is limited to about 40 dB in one-port measurements and 80–140 dB in two-port measurements.

If we want to measure impedances in the tens of kilo-ohms range or above with VNA, we could use the two-port series-through connection from Figure 5.4. This, however, will be limited by the stray parasitic components between the DUT and reference ground, and is not very well suited for DUTs, where one terminal naturally may require ground connections.

5.4.4 Measuring Arbitrary Transfer-Impedance Values

The VNA's two-port shunt-through connection is convenient to measure self-impedance values from a fraction of a milliohm to several hundred ohms. We make this measurement by connecting both ports of the VNA to the DUT points where we need the impedance reading and leaving all other connection points on the DUT open. We measure S-parameters with the VNA; this, by definition, would require matched terminations on all ports, not just the one port that we measure. Since we immediately convert the S_{21} reading to impedance with (5.11) and (5.12), and because the impedance matrix by definition calls for open termination on all ports, we do not have any contradiction when we measure only self-impedance port-by-port.

The situation, however, is different when we want to obtain transfer impedances on a multiport network. To show this, we will demonstrate the procedure for a two-port DUT. When we measure port 1 of the DUT with the two-port shunt-through VNA scheme, we connect both VNA ports to port 1 of the DUT and leave port 2 of the DUT open. Through (5.10) this correctly yields Z_{11}, which requires port 2 of the DUT to be left open. Repeating the same at port 2 of the DUT correctly yields Z_{22}. When we connect port 1 of the VNA to port 1 of the DUT, and port 2 of the VNA to port 2 of the DUT, we correctly measure S_{21} of the DUT. Assuming a reciprocal network, we at the same time also have S_{12}, because for a reciprocal DUT $S_{21} = S_{12}$. The problem is, now we have a mix of impedance and scattering parameters, Z_{11}, Z_{22} and S_{12}, S_{21}, so we cannot apply (5.2) to get Z_{12} and Z_{21}.

One possible solution is to continue to use the S-parameters. However, we cannot simply connect the two ports of the VNA to the two ports of the DUT, take the S-matrix readings, and then use (5.2), because we concluded that at low DUT impedances the S-parameters cannot be obtained accurately enough to calculate the impedance matrix.

However, we can measure the input impedance at port 1 and port 2 with a matched termination placed at the opposite port. Then, we can use (5.1) to determine S_{11} and S_{22}. The S_{12} and S_{21} parameters do not need any further preprocessing; so from that point we can use (5.2). Figure 5.13 outlines this measurement procedure for two ports. In the figure and in the corresponding equations, Z_{VNA} is the nominal port impedance of the VNA, typically 50Ω.

We complete the measurement in three steps. During these steps, we measure three S_{21} values and one S_{12} value. The superscript index of S_{21} (that appears in parentheses) refers to the step in which we obtain the S_{21} parameter. In Step 1, both ports of the VNA are connected to port 1 of the PDN; so, we measure $S_{21}^{(1)}$. The input impedance at port 1 with matched termination at port 2 can be obtained by substituting $S_{21}^{(1)}$ into (5.10). This input impedance is then substituted into (5.1) to get S_{11} of the PDN. The two steps can be combined, to yield the following formula:

$$S_{11} = \frac{3/2\, S_{21}^{(1)} - 1}{-1/2\, S_{21}^{(1)} + 1} \tag{5.14}$$

In Step 2, both ports of the VNA are connected to port 2 of the PDN. Following the same steps as above, we obtain S_{22}:

$$S_{22} = \frac{3/2\, S_{21}^{(2)} - 1}{-1/2\, S_{21}^{(2)} + 1} \tag{5.15}$$

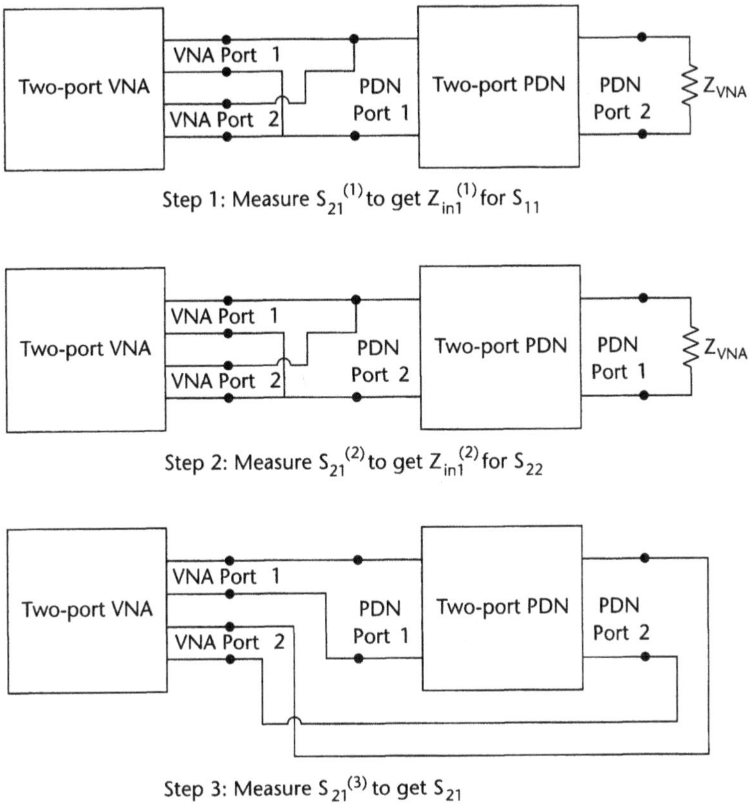

Figure 5.13 Scheme to measure arbitrary transfer impedance values and the full scattering matrix on a two-port network, based on S-parameters. Z_{VNA} is the nominal port impedance of the VNA, typically 50Ω.

In Step 3, we measure $S_{21}^{(3)}$, which directly equals S_{21} of the DUT and $S_{12}^{(3)}$, which equals S_{12} of the DUT. Finally, the four S-parameters obtained above are converted into the four Z-parameters through (5.2).

While this process yields the correct results, inconveniently, it requires us to connect a matched termination to the opposite port when we measure the single-port input impedances. When the DUT has more than two ports, we have to repeat this process for all port combinations of interest.

Fortunately, there is a more convenient and direct method: first, measure Z_{11} and Z_{22}, then measure S_{12} and S_{21}, and, finally, apply the necessary conversions to properly account for the mix of matrices. In this way, we work with a mix of impedance and scattering parameters. The process is outlined in Figure 5.14. The steps are the same as in Figure 5.13, except that we do not use the additional matched terminations on the opposite ports. From steps 1 and 2 now we get Z_{11} and Z_{22}:

$$Z_{11} = \frac{Z_{VNA}}{2} \frac{S_{21}^{(1)}}{1 - S_{21}^{(1)}} \tag{5.16}$$

$$Z_{22} = \frac{Z_{VNA}}{2} \frac{S_{21}^{(2)}}{1 - S_{21}^{(2)}} \tag{5.17}$$

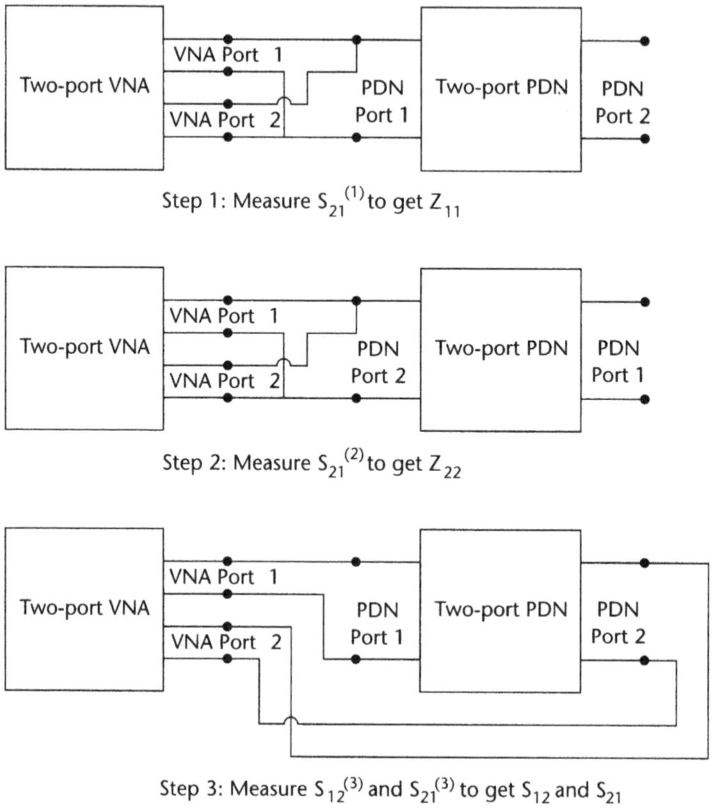

Figure 5.14 Scheme to measure arbitrary transfer impedance values on a two-port network based on mixed *Z*- and *S*-parameters.

To convert $S_{21}^{(3)}$ and $S_{12}^{(3)}$ into Z_{21} and Z_{12}, we start with the network equations for two ports.

$$\begin{aligned} v_1 &= Z_{11} i_1 + Z_{12} i_2 \\ v_2 &= Z_{21} i_1 + Z_{22} i_2 \end{aligned} \tag{5.18}$$

When we measure S_{21} or S_{12}, there are two additional constraints that need to be taken into account. For the case of measuring S_{21}, the matched source is at port 1 and the matched load is at port 2. Following the notation of Figure 5.15, we have the following constraints:

$$\frac{v_s - v_1}{Z_{VNA}} = i_1 \quad \frac{v_2}{Z_{VNA}} = -i_2 \tag{5.19}$$

For the case of measuring S_{12}, the indices have to be swapped:

$$\frac{v_s - v_2}{Z_{VNA}} = i_2 \quad \frac{v_1}{Z_{VNA}} = -i_1 \tag{5.20}$$

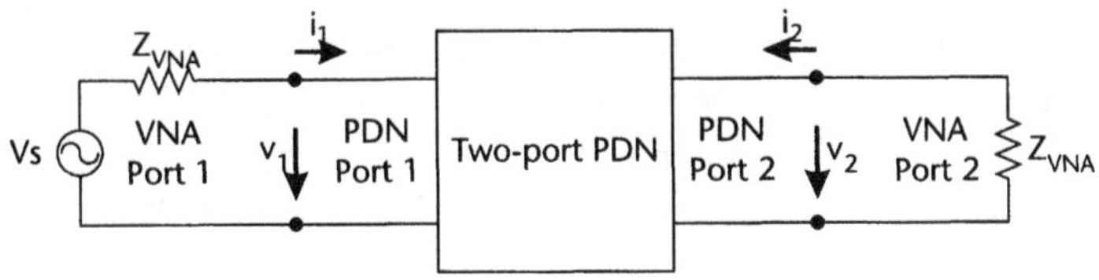

Figure 5.15 Termination constraints in Step 3 of Figure 5.12, measuring S_{21}. Z_{VNA} is the nominal port impedance of the VNA, typically 50Ω.

For sake of simplicity, for the remaining calculations we assume a reciprocal DUT so that $S_{21} = S_{12}$ and $Z_{21} = Z_{12}$. This means we can continue with only (5.19). Rearranging (5.18) and (5.19) yields:

$$Z_{21} = Z_{12} = Z_{tranf} = \frac{Z_{VNA}}{2} S_{21}^{(3)} \frac{1 + \frac{Z_{11}}{Z_{VNA}} + \frac{Z_{22}}{Z_{VNA}} + \frac{Z_{11}}{Z_{VNA}} \frac{Z_{22}}{Z_{VNA}}}{1 + S_{21}^{(3)} \frac{Z_{11}}{2Z_{VNA}}} \tag{5.21}$$

If the circuit is also electrically symmetrical, so that $Z_{11} = Z_{22} = Z_{self}$, (5.21) reduces to:

$$Z_{21} = Z_{12} = Z_{transf} = \frac{Z_{VNA}}{2} S_{21}^{(3)} \frac{1 + 2\frac{Z_{self}}{Z_{VNA}} + \left(\frac{Z_{self}}{Z_{VNA}}\right)^2}{1 + S_{21}^{(3)} \frac{Z_{self}}{2Z_{VNA}}} \tag{5.22}$$

We can see that when Z_{11}, Z_{22}, and S_{21} are small compared to the Z_{VNA} impedance, (5.21) and (5.22) reduce to (5.8). When the DUT has more than two ports, we have to repeat the process for all port combinations of interest.

5.4.5 Measuring Transfer Ratios

Self- and transfer impedances are convenient for PDN structures, where the PDN impedance is much smaller than the source and load impedances; for those structures the response of the PDN can be modeled as the ratio of unloaded output voltage and input current. There are situations, however, when this approximation is not convenient. This may be the case, for instance, when we have coupling between power rails through overlapping power planes or when a filter PDN circuit for a low-current analog pin has its input connected to a high-current low-impedance main PDN. If we want to characterize the coupling between the planes or the effectiveness of this filter, the noise appearing across the input of the filter can be considered as voltage imposed across its input. Furthermore, if the load to the filter circuit is high impedance, then the filter will produce its unloaded output voltage; in this case, a more convenient characterization for the filter is the voltage ratio of unloaded output voltage to input voltage. If the loading of the filter output is not negligible, we can either combine the load impedance with the filter circuitry for our simulations and measurements or we can take the loading into account separately by the voltage divider between the output impedance of the filter and the load impedance. Figure 5.16 shows a simple block schematic diagram of an example. For the purposes of characterizing the filter PDN, the main PDN can be considered as a one-port circuit.

With the assumption of infinite load impedance, $i_2 = 0$ and (5.18) reduces to:

$$\begin{aligned} v_1 &= Z_{11} i_1 \\ v_2 &= Z_{21} i_1 \end{aligned} \tag{5.23}$$

The voltage transfer ratio A thus becomes:

$$A = \frac{v_2}{v_1} = \frac{Z_{21}}{Z_{11}} \tag{5.24}$$

Note that as long as all PDN components are linear, it does not matter whether the main PDN is present at the filter input or not; both ways should give us the same voltage transfer ratio. If the main PDN is present, both Z_{21} and Z_{11} must be obtained this way. However, a very low impedance main PDN will result in small Z_{21} and Z_{11} values, making the division in (5.24) more sensitive to measurement and computational noise. Z_{21} and Z_{11} can be obtained by any of the methods described in previous sections.

Current transfer ratios may be useful to describe PDN blocks, which are driven by a low-impedance source and terminated in a low-impedance load.

With the assumption of zero load impedance, $v_2 = 0$ and (5.18) reduces to:

$$\begin{aligned} v_1 &= Z_{11} i_1 + Z_{12} i_2 \\ 0 &= Z_{21} i_1 + Z_{22} i_2 \end{aligned} \tag{5.25}$$

The current transfer ratio B thus becomes:

$$B = \frac{i_2}{i_1} = -\frac{Z_{21}}{Z_{22}} \tag{5.26}$$

Note the negative sign in (5.26) results from the convention of defining both current vectors as pointing inward (i.e., toward the block).

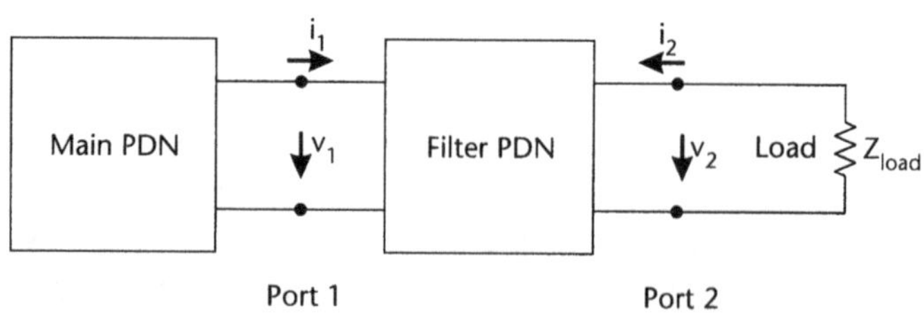

Figure 5.16 Block scheme of a low-current filter circuit after a high-current main PDN.

5.5 Extracting Component Parameters from Measured Data

Often a measured PDN component can be approximated with a simple equivalent circuit in a given frequency range. As shown in Figure 5.17, the most common equivalent circuits for PDN components and subsystems are series *R-L*, *C-R-L*, or parallel *C-R-L* circuits.

Series *C-R-L* equivalent circuit approximations can be applied to bare plane pairs (from very low frequencies up to close to the first modal parallel resonance) and to bypass capacitors. Parallel *C-R-L* equivalent circuits can be applied to shorted plane pairs and to capacitors mounted on plane pairs (around their parallel resonance). The parameter extractions can be built into spreadsheet calculations that may capture the measured data also. When a frequency independent *R-L* or *C-R* model is sufficient, the extraction is straightforward and accurate. When series or parallel *C-R-L* models have to be used or when the components show non-negligible frequency dependency, an iterative solution can help to obtain the values for all of the equivalent-circuit elements.

5.5.1 Extracting Capacitance

Assume we have to measure a single capacitor, or an open-terminated PCB trace or PCB parallel planes. Far below their first series resonance frequencies, the measured impedance will follow the $X_c = 1/(2\pi fC)$ trend. We can then reverse-calculate the capacitance from the imaginary part of DUT impedance. As we need the phase information from the measured impedance, a minimum of a through calibration is necessary before taking data (full two-port calibration is necessary for best accuracy). The capacitance estimate becomes:

$$C = -\frac{1}{2\pi f \operatorname{Im}\{Z_{DUT}\}} \tag{5.27}$$

where f is frequency and Z_{DUT} is the measured impedance of device.

Note that for low-loss capacitors, where the phase angle of the impedance is very close to –90°, $\operatorname{Im}\{Z_{DUT}\}$ can be replaced with magnitude$\{Z_{DUT}\}$.

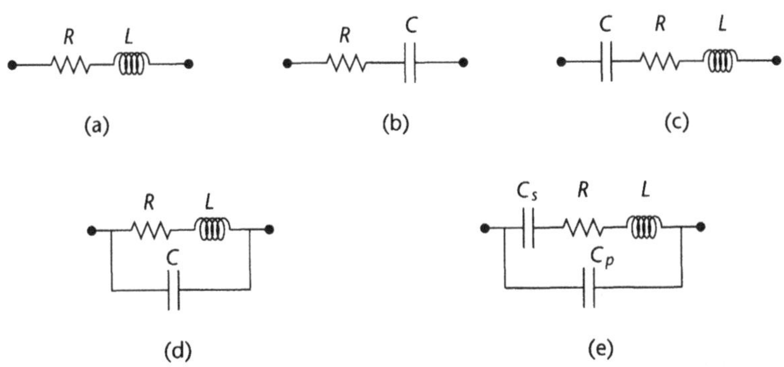

Figure 5.17 Simple equivalent circuit approximations of PDN components: (a) series *R-L*, (b) series *R-C*, (c) series *C-R-L*, (d) series *R-L* with parallel *C*, and (e) series *C-R-L* with parallel *C*.

Figure 5.18 shows the proper and improper ways to extract the equivalent capacitance of the 25.4 × 3.56-mm (1 × 0.14-inch) bare plane pairs from Figure 5.10, the impedance plot of which was shown in Figure 5.11. Figure 5.18(a) gives the equivalent capacitance calculated from the approximate impedance value using (5.8). Since this approximate impedance saturates at low frequencies at 25Ω, this translates to an exponentially increasing equivalent capacitance with decreasing frequency (note the logarithmic vertical scale). The graph in Figure 5.18(b) shows the capacitance extracted properly, with (5.27) from the correct impedance formula for the imaginary part using (5.12). Note that at low frequencies, as the impedance becomes much higher than the 25Ω normalization impedance, the extracted capacitance still becomes somewhat noisy. Apart from this increasing noise below 10 MHz, there is a small but definite negative slope on the data trace up to 300 MHz, which indicates the frequency dependence of the dielectric constant of the printed-circuit-board material. The 2–3-pF ripple that shows up above 100 MHz is the residual error from calibration. The steep upslope above 400 MHz is due to the approaching series resonance.

Figure 5.19 is an example on a low-frequency bulk capacitor that shows its impedance magnitude and phase as well as the extracted capacitance using (5.27). In the given frequency range, the impedance is always well below 25Ω; therefore, there would be no noticeable difference whether (5.8) or (5.10) is used to plot the impedance and to extract the capacitance. However, the extracted capacitance has a sharp increase as the frequency approaches the 100-kHz series resonance frequency of the mounted part. This is the same phenomenon that causes the capacitance upslope above 400 MHz of Figure 5.18(b). The correction for this phenomenon will be shown in the next section.

5.5.1.1 Compensating for Series Inductance

As shown in Figure 5.19(b), the capacitance value given by (5.27) will create a gradually increasing error as we approach the series resonance frequency of the DUT.

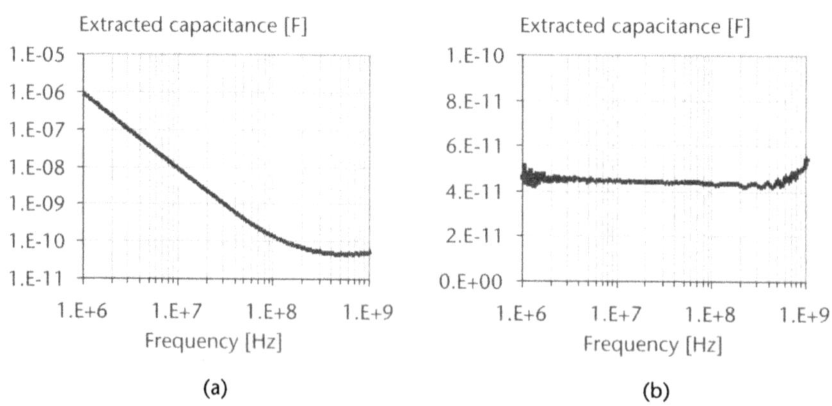

Figure 5.18 Extracted capacitance of the 3.56 × 25.4-mm (0.14 × 1-inch) bare plane pair, the impedance plot of which is shown in Figure 5.11 (a) gives the equivalent capacitance calculated from the uncorrected impedance value using (5.8) and (b) shows the capacitance extracted from the corrected impedance using (5.12). Note the logarithmic vertical scale on the left and the zoomed linear vertical scale on the right.

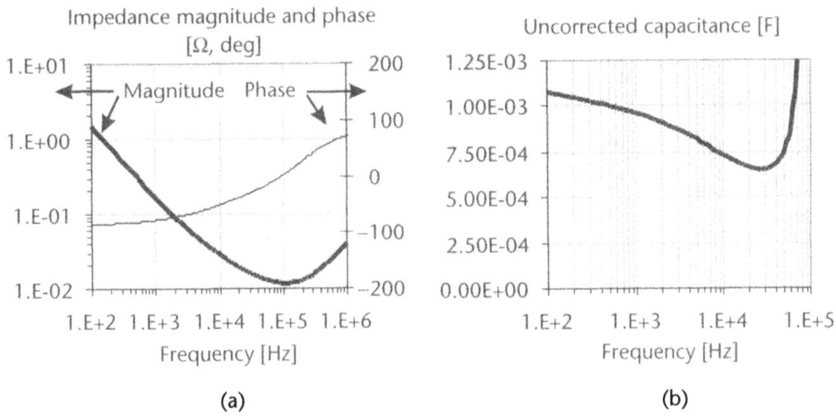

Figure 5.19 (a) Measured impedance magnitude and phase of a 1,200-μF polymer electrolytic capacitor. Left axis: impedance magnitude. Right axis: phase. (b) The extracted capacitance versus frequency using (5.12) to convert S_{21} to complex impedance, followed by (5.27) to calculate capacitance.

Note the sharp rise of capacitance immediately before 100 kHz. This rise occurs because the imaginary part of the impedance contains the capacitive reactance in series to the inductive reactance. The positive inductive reactance is added to the negative capacitive reactance, making the absolute value of reactance smaller and effectively increasing the capacitance. While this effect is negligible very far below the series resonance frequency, it becomes the source of an increasing error as we approach resonance. At the resonance frequency, the imaginary part of impedance is zero and (5.27) yields infinite capacitance.

As long as the inductance in the equivalent circuit can be estimated with reasonable accuracy, a correction can be applied to (5.27). We assume that the equivalent circuit of Figure 5.17(c) is applicable. With an inductance estimate of L, we subtract its inductive reactance from the measured imaginary part of the impedance. The corrected capacitance becomes:

$$C = -\frac{1}{2\pi f\left(\text{Im}\left\{Z_{DUT}\right\} - 2\pi f L\right)} \tag{5.28}$$

The estimation of series inductance is covered later in Section 5.5.4.

Figure 5.20 shows the frequency dependence of the extracted capacitance and the large error between the measured impedance plot and simulated impedance, assuming frequency-independent capacitance value. To account for the frequency-dependent capacitance, various frequency-dependent capacitor models will be shown in Chapter 8.

5.5.2 Extracting Equivalent Series Resistance

Any complex one-port linear time-invariant circuit can be approximated with a single equivalent impedance or admittance at a given frequency. The equivalent series resistance (ESR) is the real part of the equivalent impedance. In general, the equivalent impedance and its elements are frequency dependent. For capacitors, ESR is

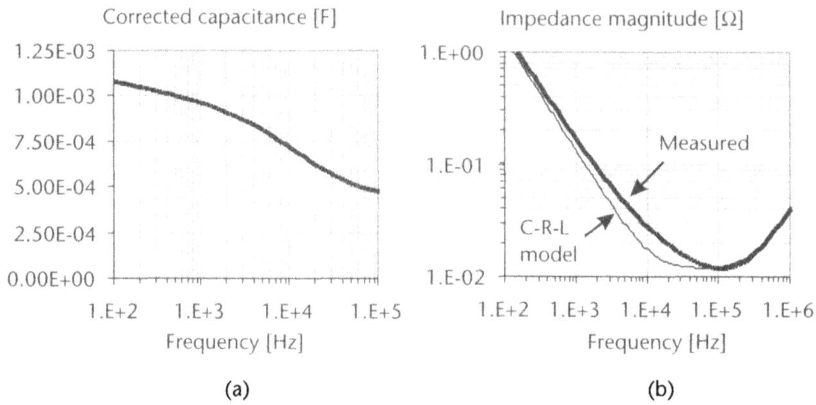

Figure 5.20 (a) Extracted capacitance of the 1,200-μF polymer electrolytic capacitor, using (5.28), with an estimated inductance of L = 6.1 nH. (b) Impedance magnitudes of the measured impedance and that of the approximation with a series *R-L-C* circuit, with frequency independent values of C = 1,200 μF, R = 0.012Ω, and L = 6.1 nH.

usually given at the frequency of impedance minimum. If the real part of impedance varies with frequency, ESR will also depend on the connection inductance, because the loop inductance, together with the capacitance, determines the series resonance frequency.

ESR values are often much smaller than 25Ω; therefore, in simple measurements we may use (5.8), so there is no need to solve for the complex voltage divider of (5.10). Depending on the component and connection geometries, some parts will exhibit significant variation in equivalent series losses, because at higher frequencies the internal current path may change significantly. Also, the real part of the impedance in capacitors will rise at lower frequencies. As an illustration, Figure 5.21 shows the impedance magnitude and real part as a function of frequency for the same 1,200-μF polymer electrolytic capacitor that was used for Figures 5.19 and 5.20.

5.5.3 Extracting Inductance

Capacitors above their series resonance frequencies as well as shorted traces and power/ground planes show primarily inductive impedance. Under these circum-

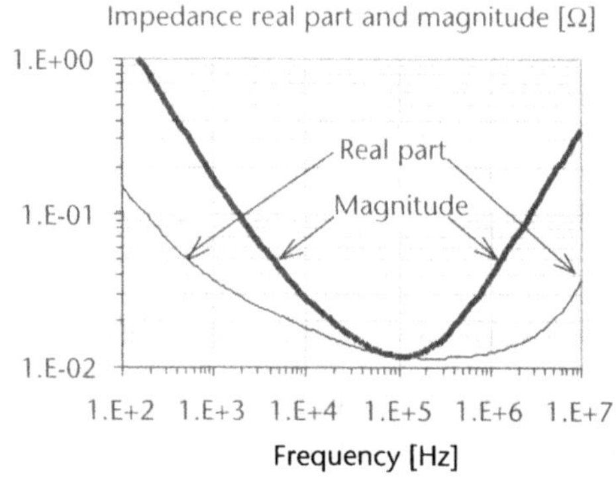

Figure 5.21 Impedance magnitude and real part of impedance of a 1,200-μF polymer electrolytic capacitor.

stances the measured impedance will follow more or less the $X_L = 2\pi fL$ trend. We can then reverse-calculate the inductance from the imaginary part of impedance. As we need the phase information of the measured impedance, a minimum of through calibration is necessary before taking data. For higher accuracy, full two-port calibration is preferred. The inductance estimate is:

$$L = \frac{\text{Im}\{Z_{DUT}\}}{2\pi f} \tag{5.29}$$

where

f is frequency;

Z_{DUT} is the measured impedance of device.

Note that for low-loss inductors, where the phase angle of the impedance is close to 90°, Im$\{Z_{DUT}\}$ can be approximated with magnitude$\{Z_{DUT}\}$.

5.5.3.1 Compensating for Series Capacitance

Figure 5.22 shows the extracted equivalent inductance of the same 1,200-μF polymer electrolytic capacitor that we used for Figures 5.19 through 5.21. Similar to the extracted capacitance, the extracted inductance value will also show a false frequency dependency, as frequency approaches the series resonance frequency: it is an increasing negative error as frequency decreases. As shown in Figure 5.22(a), the inductance value given by (5.29) will exhibit a sharp decrease around 100 kHz. As long as the capacitance at the series resonance frequency in the equivalent circuit can be estimated with reasonable accuracy, a correction can be applied to (5.29) to account for the capacitive reactance. With a series-capacitance estimate of C_S, the corrected inductance can be expressed as:

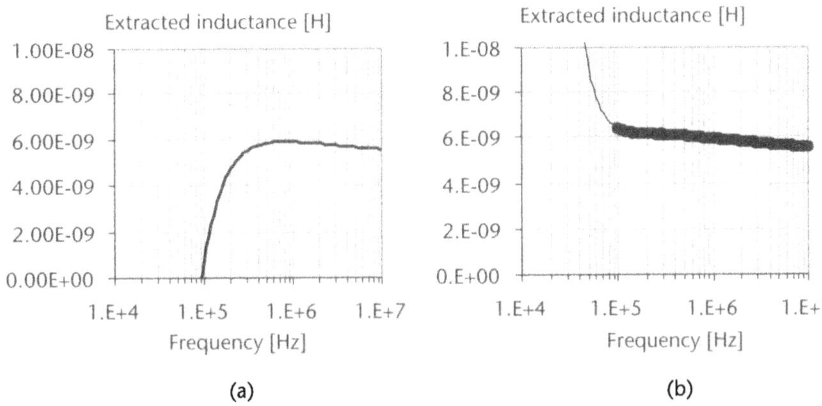

Figure 5.22 (a) Extracted equivalent inductance of a 1,200-μF polymer electrolytic capacitor, using (5.29). (b) Corrected inductance on the same set of data, using (5.30) with C_s = 450 μF.

$$L = \frac{\mathrm{Im}\{Z_{DUT}\} + \frac{1}{2\pi f C_S}}{2\pi f} \tag{5.30}$$

Figure 5.22(b) was calculated with (5.30) on the same measured data, using C_S = 450 μF instead of the nominal 1,200-μF capacitance. This simple correction extends the valid frequency range of the extracted inductance by at least half a decade, all the way to the series resonance frequency. The highlighted heavy trace runs between 100 kHz and 10 MHz, which is the approximate range for valid inductance extraction with this correction. We can use (5.30) also below the series resonance frequency, but we have to expect an increasing error, as shown by the thin line in the graph.

Note that the equivalent inductance after correcting for the series capacitance still shows a frequency dependency above the series resonance frequency, a negative slope. This frequency dependency is now real: it tells us that as frequency increases from 100 kHz to 10 MHz, the current loop shrinks, such that the inductance drops from 6.1 nH to 5.6 nH. The estimation of series capacitance for this correction is covered later in Section 5.5.4.

5.5.3.2 Compensating for Parallel Capacitance

In case of shorted planes, the inductance of the shorting metal is in parallel to the static plane capacitance, which creates a parallel *C-R-L* circuit, similar to that of Figure 5.17(d). The extracted inductance using (5.29) again will show a frequency-dependent error: this time at higher frequencies, as we approach the parallel resonance frequency. Having an estimate on the C_P shunt capacitance, the extracted inductance can be corrected to take into account the susceptance of C_P:

$$L = \frac{\mathrm{Im}\{Z_{DUT}\}}{2\pi f\left(1 + 2\pi f\ C_P\ \mathrm{Im}(Z_{DUT})\right)} \tag{5.31}$$

Figure 5.23 shows the measured impedance of the 3.56 × 25.4-mm (0.14 × 1 inch) plane pair from Figure 5.10, with the capacitor pads short-circuited. Figures 5.24(a) and 5.24(b) show the extracted inductance uncorrected and corrected, respectively. The parallel C_P plane capacitance value is taken from Figure 5.18 as 43.5 pF. Note that we have a similar equivalent circuit when a capacitor is connected to PDN planes, since the capacitor's inductance creates a parallel *L-C* circuit with the static plane capacitance.

5.5.3.3 Approximate Compensation for Series and Parallel Capacitances

When the high-frequency plane short comes from a capacitor mounted on the planes, the inductance of the capacitor is parallel to the static plane capacitance; this results in a series-parallel C_S-*R-L*-C_P circuit similar to that of Figure 5.17(e). The extracted inductance using (5.29) will show an increasing error close to both the series and parallel resonance frequencies. Having an estimate on the C_S and C_P capacitances, we can combine the corrections of (5.30) and (5.31):

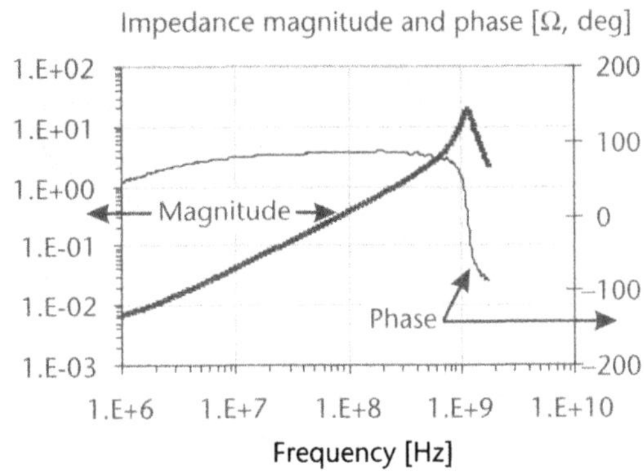

Figure 5.23 The impedance plot shows the impedance magnitude and phase of the small test board shown in Figure 5.10, measured at the test points, with the capacitor pads shorted.

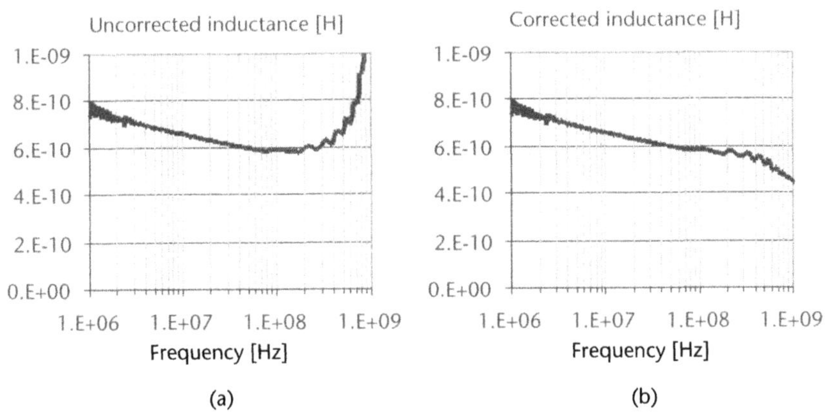

Figure 5.24 Extracted inductance of the 3.56 × 25.4 mm (0.14 × 1 inch) plane pairs (from Figure 5.10) with short on the capacitor site: (a) calculated from (5.29) with no correction for parallel plane capacitance and (b) corrected for plane capacitance, calculated from (5.31) with C_P = 43.5 pF.

$$L \approx \frac{\mathrm{Im}\{Z_{DUT}\} + \dfrac{1}{2\pi f C_S}}{2\pi f\left(1 + 2\pi f\, C_P\, \mathrm{Im}(Z_{DUT})\right)} \tag{5.32}$$

Note that this approximate correction assumes that the series and parallel resonance frequencies are widely separated. We illustrate (5.32) with the measured data on a 1-μF MLCC part, mounted on a small fixture similar to that shown in Figure 5.10. Figure 5.25 shows the impedance magnitude and phase of the capacitor together with the fixture. Note the series resonance frequency of the capacitor is around 4.5 MHz and the parallel resonance between the capacitance of the fixture and the inductance of the mounted capacitor is around 400 MHz. If we take the imaginary part of the measured impedance and extract the inductance with (5.29), we get the inductance curve in Figure 5.26(a). Note the artifact that results: a down slope occurs around the series resonance frequency and a sharp increase of the extracted inductance occurs before the parallel resonance. By applying the compen-

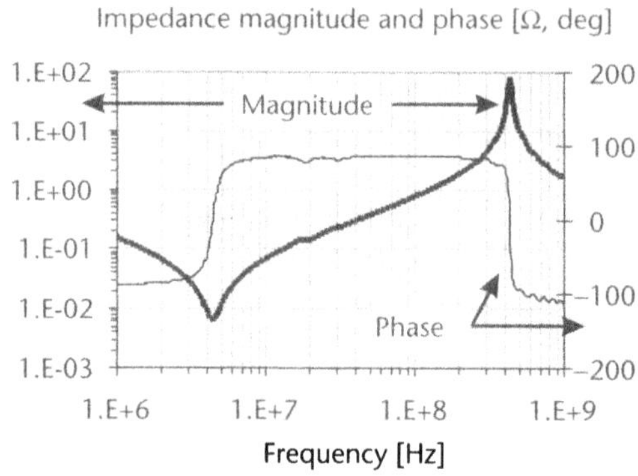

Figure 5.25 The impedance plot shows the impedance magnitude and phase of a small test board, measured at the test points with a mounted 1-μF capacitor.

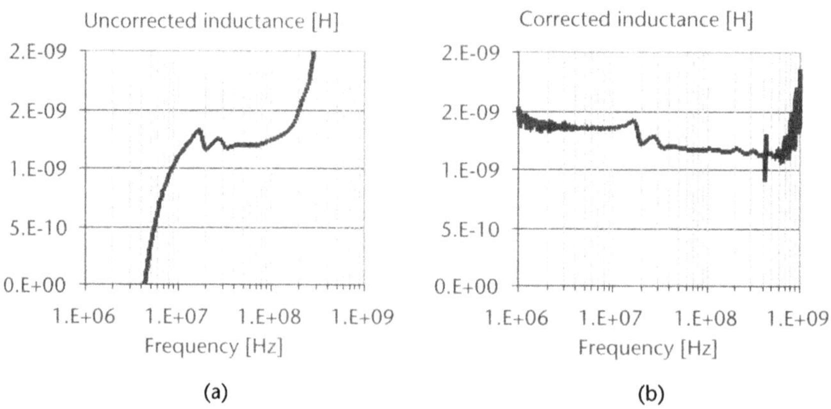

Figure 5.26 Inductance extracted from data of Figure 5.25: (a) calculated from (5.29) with no correction for series and parallel capacitances and (b) corrected for DUT and plane capacitance, calculated from (5.32) with $C_s = 0.97\ \mu$F and $C_p = 115$ pF.

sation of (5.32), these artifacts are removed (for the most part) between the series and parallel resonance frequencies.

5.5.3.4 A More Accurate Compensation for Parallel Capacitance

The compensations for the series capacitance when we extract inductance and the compensations for inductance when we extract capacitance in a purely series *C-R-L* network introduce no errors or approximations as long as the value of capacitance or inductance for which we want to compensate is correctly estimated. These compensations yield the correct answer, because in a series network the real and imaginary parts of the measured impedance do not need further transformations before parameter extraction.

However, it is not the case for the series-parallel networks of Figures 5.17(d) and 5.17(e). Since we have both series and parallel legs in these networks, the real and imaginary components become intertwined in the impedance calculations. For instance, the real and imaginary parts of the equivalent series impedance of the circuit in Figure 5.17(d) become:

$$\mathrm{Re}\{Z\} = R\frac{1}{\left(1-\omega^2 LC\right)^2 - \omega^2 RC} \tag{5.33}$$

$$\mathrm{Im}\{Z\} = j\omega L\frac{1-\omega^2 LC - R^2\left(\frac{C}{L}\right)}{\left(1-\omega^2 LC\right)^2 - \omega^2 RC} \tag{5.34}$$

The interaction between the real and imaginary parts means that the compensation for the parallel capacitance with (5.31) becomes only approximate. With low-ESR parts usually this error is not significant, but, with high-ESR parts, a different solution is required.

The solution is fixture compensation or de-embedding. First, we have to obtain an estimate for C_P (for instance by measuring the bare fixture without the DUT). With this estimate of C_P, we can reverse-calculate the impedance of the series C_S-R-L net and so reduce the problem to the equivalent circuit of Figure 5.17(c). This circuit has no parallel branch so that we can use (5.30) or (5.31).

In dedicated impedance analyzers these steps are usually part of the calibration or fixture compensation process. When we use our own test fixtures with a VNA setup, leaving this compensation process as an optional step increases the flexibility of data mining.

To illustrate this compensation, we use the measured data on a high-ESR capacitor. Figure 5.27 shows the measured impedance magnitude and extracted inductance of a high-ESR ceramic capacitor mounted on a small PCB fixture. The several-ohm ESR of the DUT results in low phase angle up to several hundred megahertz; this makes it challenging to extract inductance. Without the proper fixture compensation, the extracted inductance becomes negative above 50 MHz.

Fixture compensation will be explained in more detail in Section 5.5.5.

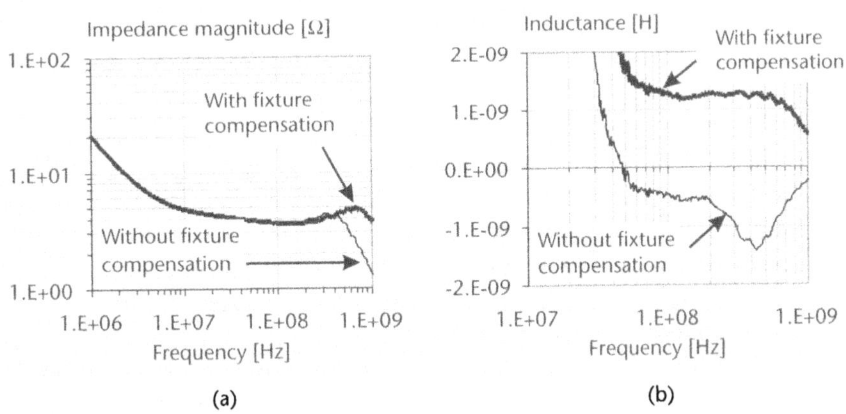

Figure 5.27 (a) Impedance magnitude and (b) inductance of a high-ESR MLCC sample with and without compensating for parallel fixture capacitance.

5.5.4 Estimating Inductance and Capacitance for Compensations

In previous sections we used estimates of inductance for capacitance extraction and estimates of capacitance for inductance extractions. These estimates can be found either by manual iteration or by automated software optimization routines. However, as we show in this section, these estimates cannot be obtained uniquely from the imaginary part of the measured impedance alone, because at any given frequency we have two unknowns (capacitance and inductance) and only one data point (imaginary part of impedance).

Since the inductive and capacitive reactance change in opposite ways as frequency varies, there is a reasonably accurate way to get the capacitance and inductance estimates, as long as they are frequency independent. Assuming this is true, we can estimate the inductance from a frequency point high above the series resonance (so that the capacitive reactance will have negligible influence) and estimate capacitance from a frequency point well below the series resonance (so that inductive reactance has a negligible influence). However, if we assume that the inductance and capacitance are frequency independent, there is little to be gained from extracting the capacitance and inductance versus frequency curves, because they should then be constant (theoretically). Nonetheless, this approach is still useful when the changes in capacitance and inductance with frequency are small (but not zero). There will be some error in the result, but the estimate is still better than without any compensation.

When changes in inductance or capacitance are substantial but gradual, we can create either a model-based estimation or bring in additional constraints to solve for the unknowns. The simplest frequency-dependent models assume that both capacitance and inductance drop monotonically with increasing frequency. Also, as a simple frequency-dependent model, we can assume that inductance below the series resonance frequency will remain relatively constant. Furthermore, we can assume that the rate of capacitance drop with frequency does not change abruptly, so the capacitance slope observed below series resonance will continue through the resonance region. These simple models can be observed when we manually or automatically iterate the inductance and capacitance estimates, such that the compensated capacitance and inductance curves follow our expected models and at the same time yield the capacitance and inductance we assumed for compensation.

Another possibility is to expand the scope of parameter extraction and then to extract capacitance, inductance, and resistance to confirm to a chosen causal frequency-dependent model. This will be shown in Chapter 8.

To illustrate the process of estimating inductance and capacitance only from the imaginary part of measured impedance, we return to the measured data of the 1,200-μF polymer capacitor. The extracted capacitance and inductance of the mounted part was shown in Figures 5.20 and 5.22, respectively. The capacitance curve used a 6.1-nH inductance estimate; the inductance curve used a 450-μF capacitance estimate. To demonstrate how to find correct estimates (and to demonstrate the consequence of an incorrect estimate), a series of extractions were performed on the same set of measured data with capacitance and inductance estimates progressively deviating from the optimum. Starting from the 450-μF and 6.1-nH values, each estimate was changed in 10 steps upward and 10 steps downward, each step representing a 10% change with respect to the previous value. The value of zero represents no change, so we keep the 450-μF and 6.1-nH estimates. The values +1 and –1 correspond to +10% and –10% deviation, that is, the 495-μF and 405-μF capacitance and 6.71-nH and 5.49-nH inductance estimates, for example. Finally, a value at +10 corresponds to a 2.59 times increase in value or 1,165.5-μF capacitance and 15.8-nH inductance; a value at –10 corresponds to a 0.385 times decrease in value or 173-μF capacitance and 2.35-nH inductance.

Figure 5.28 shows the extracted capacitance and inductance values, using (5.28) and (5.30), with three inductance and capacitance estimates. Traces with label B used the optimum 450-μF capacitance and 6.1-nH inductance values. Traces labeled with A used 2.59 times the optimum values. Traces labeled C used 0.385 times the optimum estimates. Note that the figure shows capacitance in the 100-Hz to 100-kHz range and inductance in the 100-kHz to 10-MHz frequency range, respectively.

Figures 5.29 and 5.30 show the absolute value of the relative error between the assumed capacitance and inductance versus the extracted values at the 100-kHz series resonance frequency. The two horizontal axes of the surface plots represent the deviations from the 450-μF and 6.1-nH optimum values. Each integer on the scales represents a 10% change. The plots on the left are 3D surface plots; the graphs on the right plot the same data on a contour plot, which shows the footprint more accurately. If the estimates used for the corrections were ideal, the error would be zero, indicating that the extracted capacitance at SRF is the same value that we use for the corrected inductance extraction and vice versa.

In both figures, we have multiple values of capacitance and inductance steps at which the relative error is close to zero; on the contour plots these values form slanted strips running from upper left to lower right. The strips run over the same points in both figures; this pattern shows multiple solutions with close to zero error (unless additional constraints such as causal models are included).

We return to the extracted capacitance and inductance traces in Figure 5.28 to determine which combinations can be excluded based on further considerations. If we have no additional information about the dielectric material, the only reasonable assumption is that the capacitance will not increase as frequency increases. This eliminates trace C, but traces A and B appear to be equally valid. We can also expect that inductance drops as frequency increases; this eliminates trace C from the

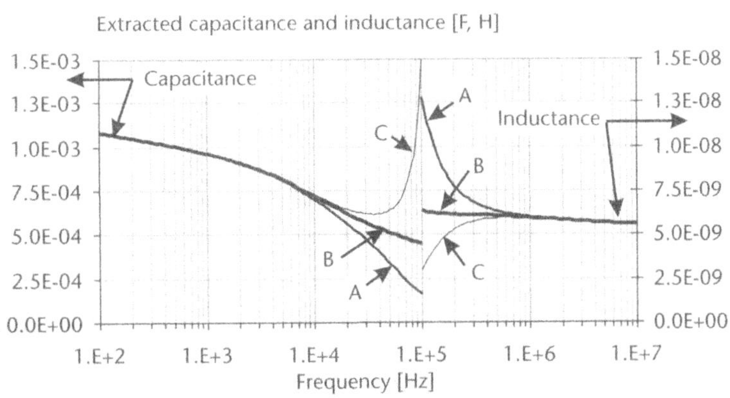

Figure 5.28 Extracted capacitance and inductance curves of a 1,200-μF polymer capacitor, corrected with three different inductance and capacitance estimates.

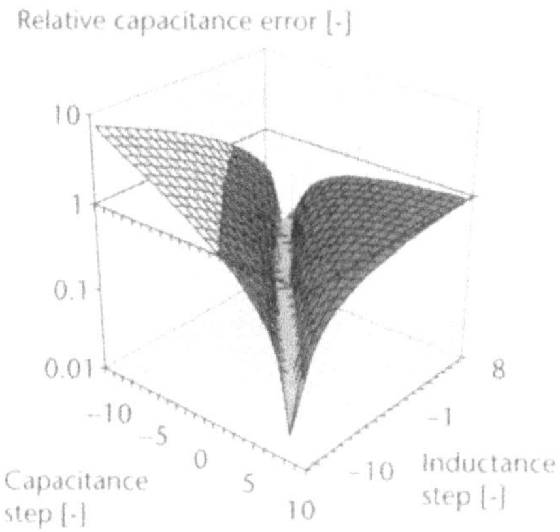

(a)

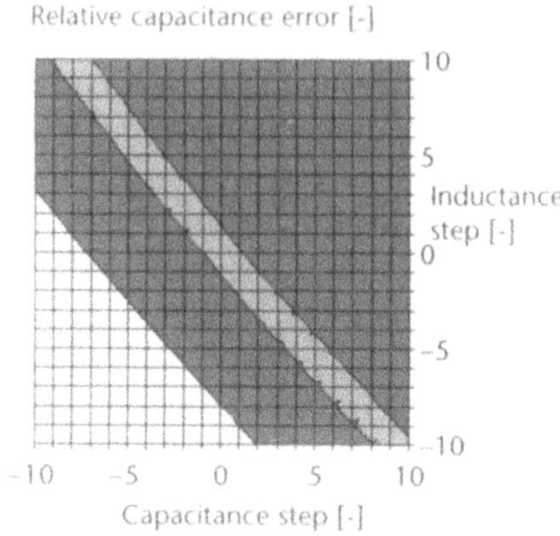

(b)

Figure 5.29 Relative capacitance error as a function of different inductance and capacitance estimates: (a) 3D plot and (b) contour plot.

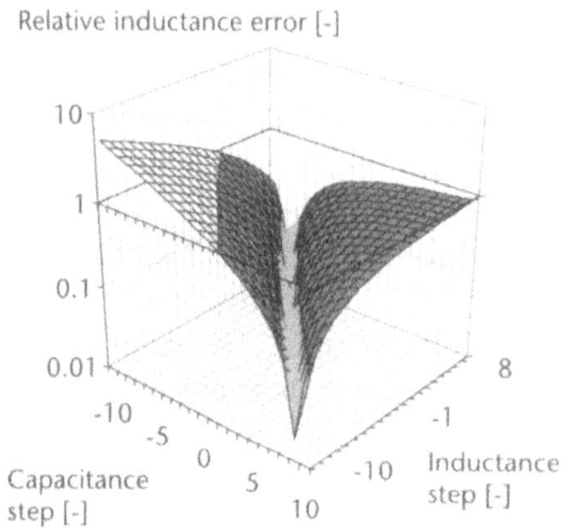

(a)

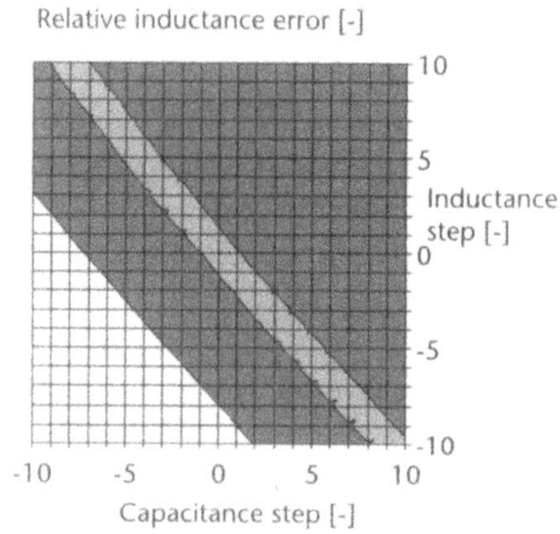

(b)

Figure 5.30 Relative inductance error as a function of different inductance and capacitance estimates: (a) 3D plot and (b) contour plot.

inductance traces. Additionally, it may be reasonable to assume that around the series resonance frequency the inductance is almost constant with frequency. This additional assumption leads us to trace B. We will show later that even in low-ESR and low-ESL MLCCs, inductance stays relatively constant in the close vicinity of SRF. Figure 5.31 plots the slope of extracted inductance at SRF as a function of capacitance estimate. The figure shows a slight negative slope with the chosen 450-μF capacitance estimate, which corresponds to zero on the horizontal axis.

5.5.5 Fixture Compensation, Port Extension, and De-Embedding

Calibration in a VNA establishes a reference plane where its accuracy is specified [7]. Calibration will be covered in detail in Chapter 6. For now we assume that the necessary level of calibration has been performed on the instrument.

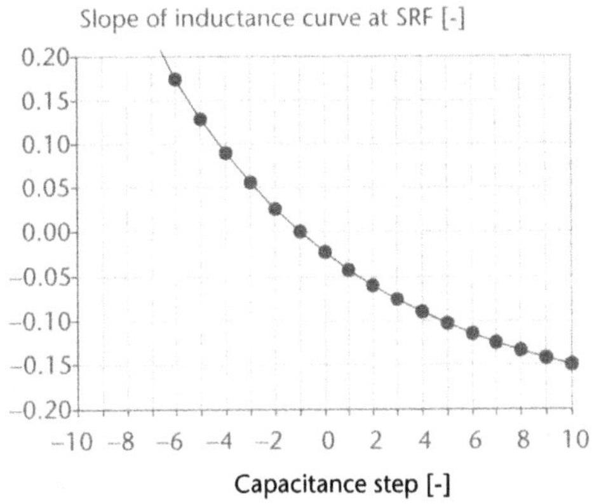

Figure 5.31 Slope of extracted inductance curve at the series resonance frequency.

When cables or probes can be directly connected to a DUT, there is no need for further consideration or data postprocessing. When we have fixtures, the calibrated instrument measures the DUT together with the fixture. With standard fixtures, there is usually an option to do fixture compensation. For low and medium frequencies, where the distributed nature of the fixture can be neglected, the fixture can be modeled with the equivalent circuit shown in Figure 5.32. We assume that the connection points to the instrument are calibrated. During the fixture compensation process, we measure the fixture with a short, which yields Z_s. When we measure the fixture open, the measured value is the sum of Z_s and $1/Y_p$.

The Z_m impedance measured by the calibrated instrument is the convoluted result of the DUT impedance and fixture impedances [7]. With the notations of Figure 5.32:

$$Z_m = Z_s + Z_d \| \left(\frac{1}{Y_p} \right) \tag{5.35}$$

By rearranging (5.35), the Z_d DUT impedance becomes:

$$Z_d = \frac{Z_m - Z_s}{1 + Y_p (Z_s - Z_m)} \tag{5.36}$$

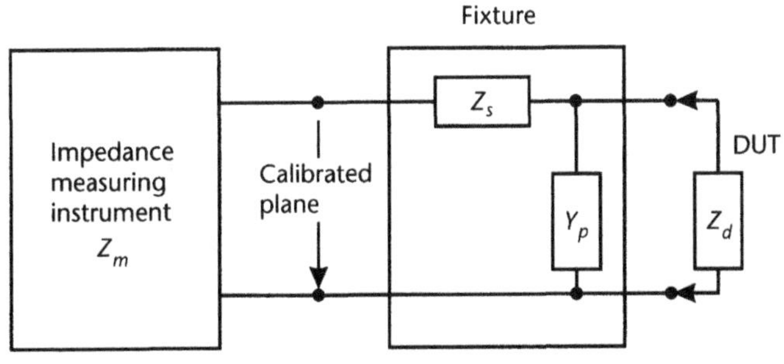

Figure 5.32 Equivalent circuit of DUT measurement through fixture. (*After:* [7].)

Note that in (5.35) and (5.36) all entries are complex numbers. To get additional correlation points, optionally the fixture can also be measured with a known impedance. With very low and very high DUT impedances, there is an additional error: when we measure the fixture with OPEN and SHORT, those reference impedances are not exactly zero and infinite. To account for the finite impedances of the OPEN and SHORT compensation pieces, impedance analyzers usually allow us to enter values for simple equivalent circuits for the compensation pieces: series resistance and inductance for SHORT and parallel capacitance for OPEN.

PCB fixtures are commonly used to measure PDN components: bypass capacitors and inductors. For both through-hole and surface-mount fixtures, we have two options to obtain the impedance of shorted fixtures for fixture-compensation or reference purposes. We can use a regular fixture and measure it with a short attached to it or we can create a shorted fixture where the short is formed on the PCB layers. With aggressive mounting, where the first plane is close [sometimes 0.1 mm (4 mils) or less] below the component, there is a noticeable difference between the two approaches. The geometry difference is shown on the sketches of Figure 5.33. The sketch on the left shows a regular PCB fixture with a shorting plate soldered across the surface-mount pads. The loop height between the bottom of the shorting plate and the top surface of the first plane is the sum of the dielectric thickness, pad thickness and solder thickness. When the short is created from the top metal layer of the fixture, as shown on the right sketch, the loop height is only the dielectric thickness, resulting in lower inductance.

The shorted fixture on the left is a closer representation of the actual usage, since the component body sits above the pad and solder layers. On the other hand, the shorted fixture on the right is more repeatable since the height of the uncertain solder layer is eliminated; this is especially true if the regular and shorted fixtures are created side-by-side on the same PCB.

This is further illustrated in Figure 5.34, where we show a pair of fixtures for an eight-terminal capacitor, with and without a built-in short and the measured loop inductance versus frequency plots for the two sites. The photo shows the two fixtures created on the same multilayer board. Both fixtures have their surface connections to wafer probes in form of horizontal strips. The fixture on the left has four pads formed on the top metal layer and four pads with blind vias connecting to the plane below. The fixture on the right has the built-in short: the top metal layer is solid, connecting all surface pads. The short for the fixture on the left was created by a small-size solid copper sheet covering all 8 terminals. The copper sheet was pushed onto the fixture pads during soldering. Note that the inductance of the fixture with built-in short is lower by about 20 pH, because the shorting metal sheet on

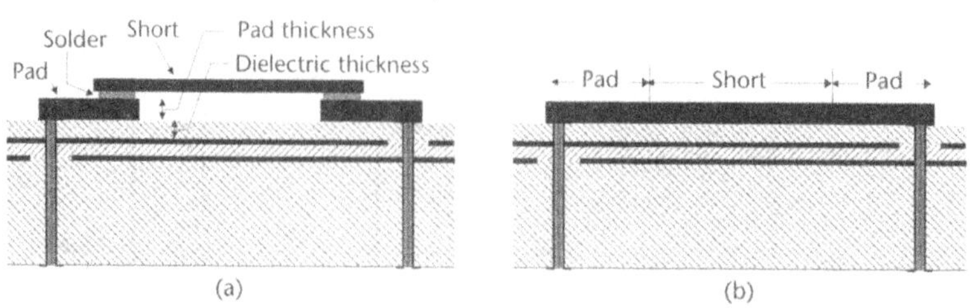

Figure 5.33 Cross-section sketch of the shorted test sites: (a) regular fixture with external shorting plate and (b) short created with PCB metal layer. Not to scale.

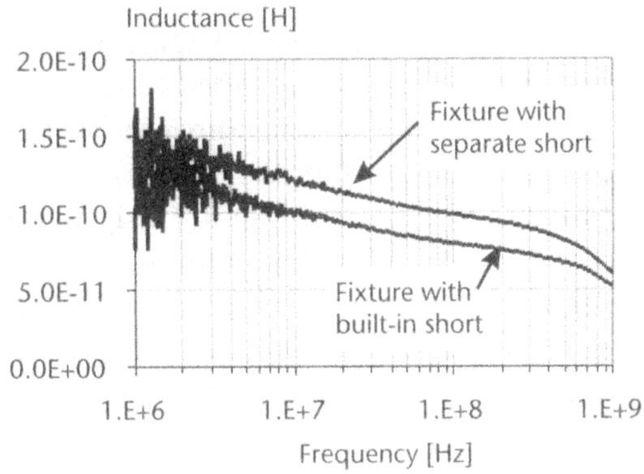

(a) (b)

Figure 5.34 Top-view photo of a pair of fixtures for eight-terminal capacitors (a) with and without built-in short and (b) inductance comparison of the shorted test fixtures. (Fixture courtesy of Samsung.)

the left fixture is sitting on top of the surface pads so it is higher above the return plane.

The fixture compensation process is applicable to both off-the-shelf and home-made fixtures. However, as we will explain in later chapters, for PDN applications we do not necessarily want to apply the compensation.

The fixture compensation process assumes a negligible phase shift across the fixture in the entire frequency range of interest, so that lumped equivalent components can be extracted for the compensation. In some cases, a cable is needed between the calibrated instrument port and the test fixture. In other cases, the frequency range of interest is so high that even short distances on the fixture create

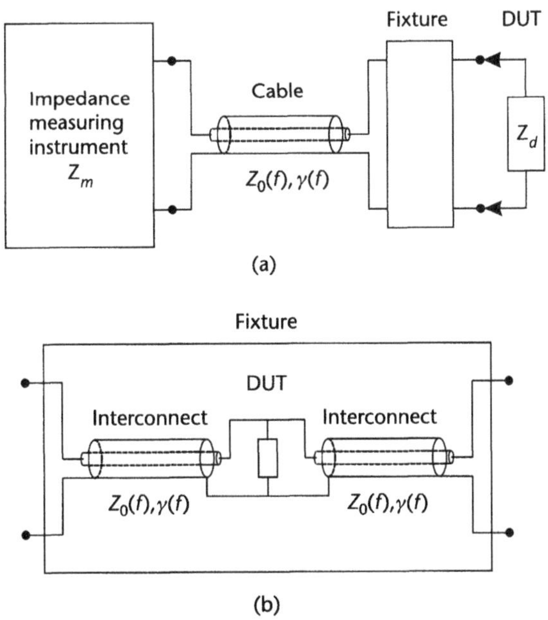

Figure 5.35 Block schematics of (a) port extension and (b) fixture de-embedding.

phase shifts that need to be taken into account. The process to handle this is called *port extension and de-embedding*. The block schematic is shown in Figure 5.35.

Port extension can compensate for the phase rotation created by the phase shift of interconnect, but the attenuation and dispersion of interconnect are either assumed to be zero or they are only crudely approximated. In such cases, it is important to use cables that minimize losses and dispersion. If dispersion cannot be neglected (which may be the case with fixtures working in the gigahertz frequency range) the effect of the interconnect can be de-embedded. The de-embedding procedure moves the calibrated reference plane along the interconnect.

References

[1] Agilent Technologies, *Impedance Measurement Handbook,* December 2003.

[2] Pozar, D. M., *Microwave Engineering*, New York: John Wiley & Sons, 2005, Chapter 4.

[3] Novak, I., "Probes and Setup for Measuring Power-Plane Impedances with Vector-Network Analyzer," *Proceedings of DesignCon 99*, Santa Clara, CA, February 1–4, 1999.

[4] Novak, I., "Measuring Milliohms and PicoHenrys in Power Distribution Networks," *Proceedings of DesignCon 2000*, Santa Clara, CA, February 1–4, 2000.

[5] Agilent Technologies, "Advanced Impedance Measurement Capability of the RF I-V Method Compared to the Network Analysis Method," Application Note 1369-2, July 26, 2001.

[6] Agilent Technologies, "4294A Precision Impedance Analyzer Data Sheet," January 28, 2003.

[7] Agilent Technologies, "8 Hints for Successful Impedance Measurements," Application Note 346-4, 2000.

CHAPTER 6

Connections and Calibrations

There are several possible ways to connect a DUT to the measuring instrument. We always have to weigh and analyze the possible connection methods, cabling, calibration procedures and instrumentation before we decide which one to use. As a generic rule, simpler, less expensive, and easier-to-apply connections, cables, and calibrations tend to be more limited in bandwidth, accuracy, and dynamic range.

As shown in Figure 6.1, we can break down the connection between the DUT and measuring instrument into several blocks: interconnect on the fixture, connector, or probe interfacing the fixture and cable, and cable(s) connecting to the instrument. The location of the calibration or fixture compensation determines which quantity is being measured. Also, various implementations may differ in that some blocks may be combined or left out.

6.1 Port Connections

If the DUT is a single, small-size component, such as a bypass capacitor, we can attach the DUT to a fixture to measure its impedance together with the part. In this case, the calibrated reference plane may be at the end of the cable; so, we may or may not want or need to separate the measured data further by de-embedding or by compensating for the fixture. At the other end of the spectrum is a large system measurement. In a large system, there is probably no fixture; so we connect the instrument to the DUT through cables, with or without probes.

These options will be detailed and illustrated in the coming sections, in which we will proceed from left to right on the blocks of Figure 6.1.

6.1.1 Fixtures

Occasionally one or more of the blocks between the DUT and instrument may be omitted. For instance, an instrument may already contain a fixture that readily takes the DUT without modification or extra interface. In this case, the instrument contains the fixture and its connections to the DUT, as well as all necessary connections to the measuring instrument. As an illustration, Figure 6.2 shows two add-on fixtures to measure surface-mount components.

The fixture is the first (optional) interface around the DUT. Fixtures are used to facilitate the differing geometry between the DUT and test ports. We usually employ fixtures when several or many of the same-size DUTs have to be measured repeatedly. The fixture makes it possible to replace the DUT quickly, while leaving the connections (and possibly also the calibrations) in the measuring instrument

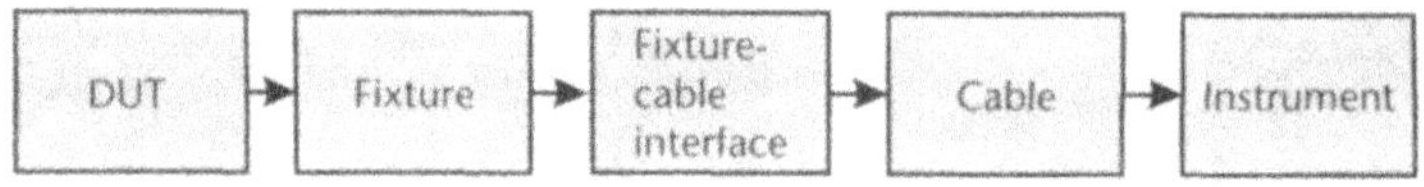

Figure 6.1 Generic scheme of connections between DUT and instrument.

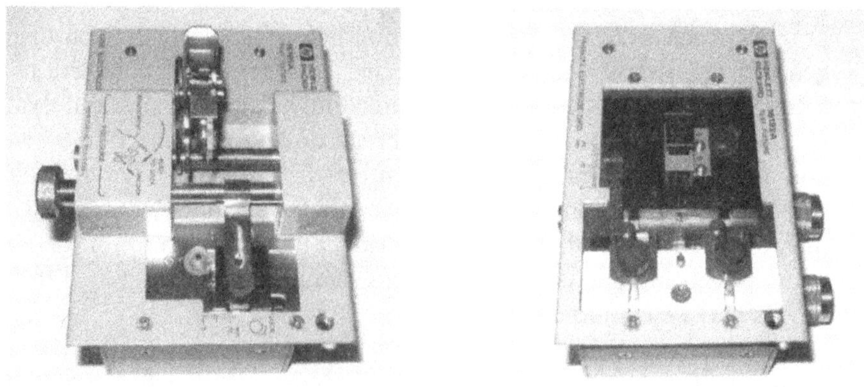

(a) (b)

Figure 6.2 Photos of Agilent 16191A (a) and16192A (b) fixtures for SMD components.

unchanged. We can use fixtures for small-size PDN components, such as bypass capacitors and inductors. We can also use fixtures for larger-size DUTs, such as dc-dc converters or full modules, when an easy connection is important. The connection from the DUT to the fixture can be solder, conductive glue, or conductive (nonglue) paste, dry pressure contact, or a bed of nails on an ICT fixture.

6.1.1.1 When No Fixture Is Needed

Under some circumstances, fixtures are not needed or cannot be used. For instance, there is no need for a fixture when the measurement is taken with calibrated probes on a DUT that has convenient surface connections available for probes.

One such case is shown in Figure 6.3. The DUT is a set of thin-film-capacitor samples on a semiconductor wafer. The metalized surface connections effectively create two capacitors in series [1], as shown in Figure 6.3(b). The two capacitors in series are between top electrode 1 and the bottom electrode, as well as top electrode 2 and the bottom electrode. The probes make the connection across the gap between top electrode 1 and top electrode 2. Full-size production boards can also be probed in a similar fashion, if there are surface features that can be bridged by the small calibrated probes.

Figure 6.4 shows a part of a bare production board of about 50 × 25 cm (20 × 10 inch) size with two wafer probes connected to the land grid array pads of the CPU [2]. The core area of the CPU footprint has a checkerboard pattern of power and ground surface pads. The pads use vias-in-pads to connect to the rest of the PDN through PCB planes. The 1,000-μm wafer probes connect to adjacent power and ground pads along the diagonal of the core area. Populated, and possibly pow-

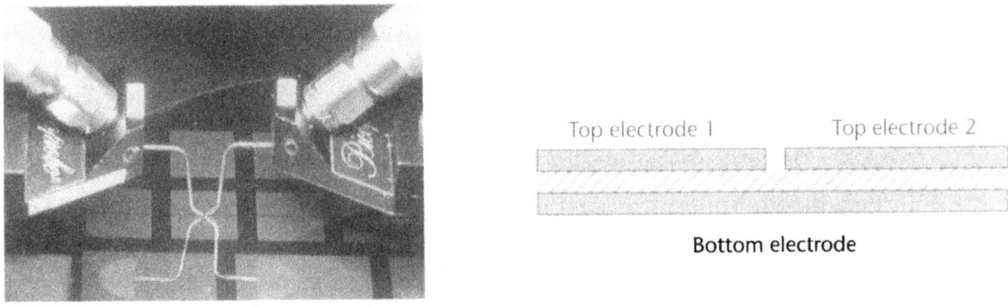

(a) (b)

Figure 6.3 Thin-film DUT measurement. (a) Probe connection from top side with 250-μm wafer probes. Sample courtesy of IWT. (b) Cross-section sketch of sample.

Figure 6.4 Measuring transfer impedance on a large-size bare production board with 1,000-μm wafer probes.

ered-up, boards can also be measured this way, as long as the component height around the connection pads fits into the vertical reach of probes.

Sometimes calibrated probes conveniently match the dimensions of a DUT. For instance, bare laminates can be measured without the need for fabricating a test board where we would need to introduce vertical via connections. We can make the measurements by connecting the probes directly to the edges of the conductive layers. The laminate in Figure 6.5 has a 75-μm dielectric layer with a 36-μm (1 oz) copper on either side. The measured, modeled, and correlated data of this laminate sample was shown in Chapter 4.

For larger modules and systems, we may not need fixtures. In full systems, usually the power-distribution network itself is large enough that it serves the purposes of a fixture. However, we still need connections to probes and instruments. Due to their inherently larger size, connectors very seldom can fit on a high-density digital board, but usually we have room to place test vias. As shown in the next sections,

Figure 6.5 Wafer probe connection to the conductive layers on a bare two-sided 75-μm polyimide laminate.

test vias can provide the necessary connections between the inner plane layers of the fixture or system board and the probes.

6.1.1.2 Fixtures for Large-Size or High-Current DUTs

In high-throughput automated tests, detachable measurement connections to large DUTs require bed-of-nail fixtures. The long pins, however, render many of the implementation options primarily for dc or low-frequency testing.

To connect to high-current devices, such as dc-dc converters, spring-loaded pogo pins can be used. This is shown in the photo of Figure 6.6. The output of the converter has five power and five ground pins, each connecting to an inch-long spring-loaded pin on the fixture board. Note that while the detachable connection is convenient, the long pins add inductance and resistance between the DUT and the base of the fixture.

6.1.1.3 Fixtures and Test Boards Without Probe Connections

As we will see in Chapter 8, the inductance of a bypass capacitor and to a lesser degree its resistance and capacitance, depend not only on the internal geometry of the part but also on the external connections. This makes it important to test parts in a connection geometry that closely resembles the final application. For this purpose, we can use small printed circuit boards with a stackup, surface escape patterns, and via geometry similar to that of the intended application. These test fixtures can be used to test and compare various options for stackups and escapes. We can also use them to correlate simulated data and to validate models.

Figure 5.10 already showed one example of a small test PCB without connectors. The board has several test sites with surface pads for different component packages. The test site shown in Figure 5.10 was for reverse-geometry 0508-size components. It includes a pair of through vias to connect to semirigid probes from the opposite sides to perform two-port shunt-through measurements. The size of the test board is not critical; but the planes should not be unnecessarily large because the increase in area would reduce the parallel resonance frequency between the static capacitance of the plane and the mounted inductance of the capacitor. However, the

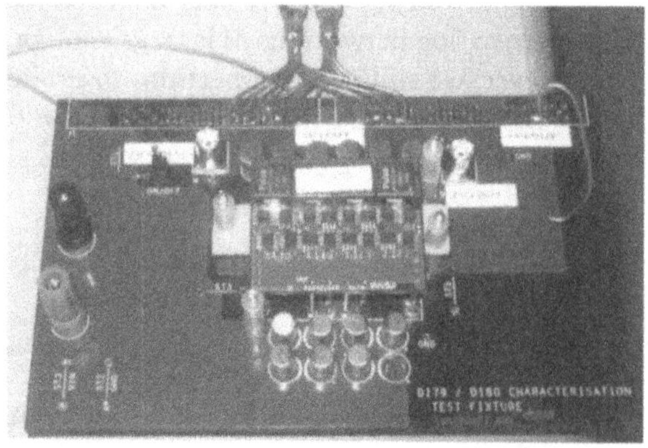

Figure 6.6 Fixture for measuring a 70-A dc-dc converter with multiple output pins.

plane should not be too small either, because that would unnecessarily increase the spreading inductance of connections. A good starting point is to extend the plane around the DUT in all directions by an amount equal to the size of the DUT.

Small test boards can also be made for wafer-probe connections. Figure 6.7 shows the construction of a three-layer small test board for measuring multiterminal capacitors. The board can take 8-terminal or 10-terminal ceramic capacitors; the wafer probes can connect on the bottom to the ends of vias and nearby plane surfaces. The benefit of this construction is that both wafer probes can connect from the same side. This board will be shown again later in Chapter 8 in connection with branched capacitor models.

Test boards can be created also by using surface connections with coplanar transmission lines; this allows the connection of wafer probes to be on the same side where the component is mounted. An example is shown in Figure 6.8. The board has only one conductive layer; the pads are arranged to take three-terminal feed-through capacitors. The G-S-G wafer probes connect to the opposite ends of the strip.

6.1.1.4 Test Boards and Custom Fixtures with Connectors

Connections are more reliably repeatable if we attach the instrument with connectors. Small test boards can be created to implement the two-port shunt-through con-

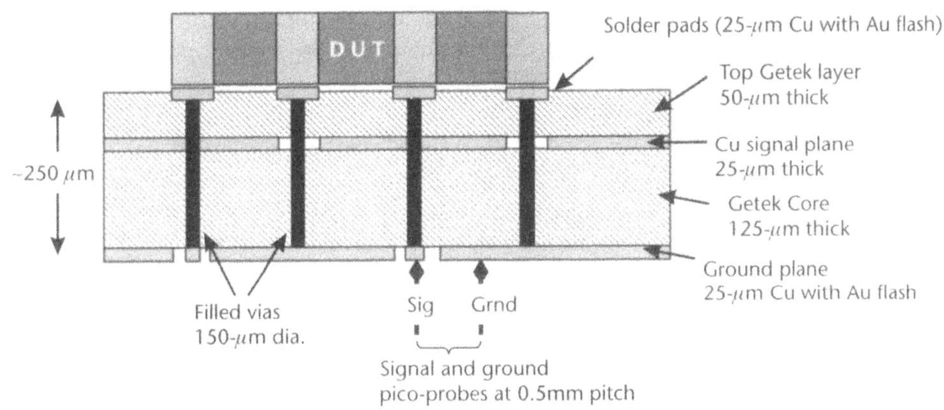

Figure 6.7 Construction of a three-layer test board for measuring 8- and 10-terminal capacitors. (Courtesy of AVX Corporation.)

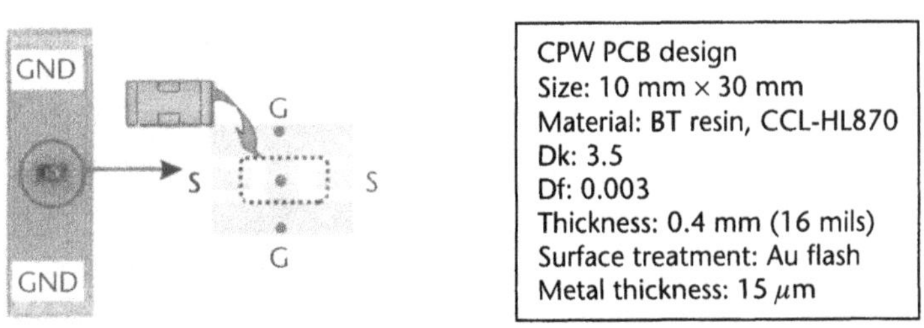

Figure 6.8 Coplanar test board for three-terminal feed-through capacitors. (Courtesy of Taiyo-Yuden.)

nection with a 50Ω through trace terminated in connectors at both ends. We can use microstrip or coplanar trace. With the trace being on the surface of the board, the DUT can be attached conveniently to the trace. For the shunt-through scheme, the DUT connects between the trace and ground. The main benefit of this type of test board is that (with proper design) we can minimize the discontinuity at the trace-connector interface. The drawback is that this connection scheme represents only a part of the PDN applications. All-surface connections are used in low layer count boards or for surface patches; multilayer boards with power planes embedded in the stackup seldom can use this connection scheme.

Figure 6.9 shows a test board with SMA connectors at the two ends. The 50Ω coplanar trace is formed on a two-layer board with ground planes on both sides. The left two photos show the bare test board (front and back view). The photo on the right shows an eight-terminal capacitor mounted on the test board.

The photo in Figure 6.10 shows a custom fixture that takes 7343-size capacitors and connects them to an Agilent 4294A impedance analyzer.

(a) (b) (c)

Figure 6.9 Small-size shunt-through test fixtures with SMA connector at both ends for measuring ceramic multilayer capacitors. Bare test board (a) front and (b) back, as well as (c) front view with mounted capacitor. Test boards courtesy of Samsung.

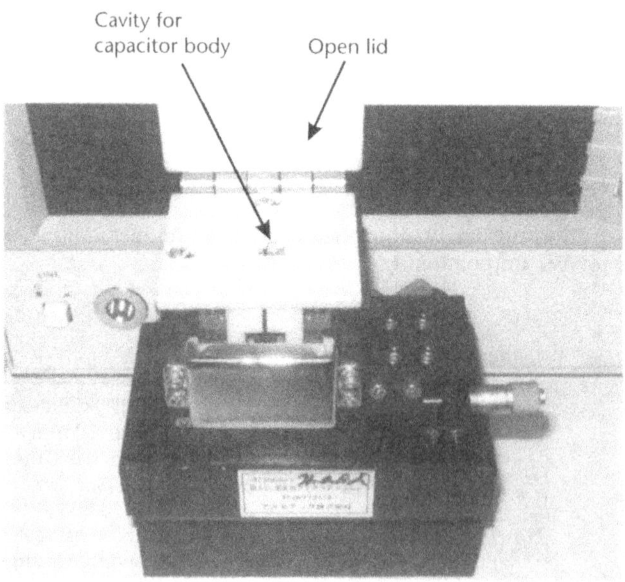

Figure 6.10 Custom fixture for 7343-size three-terminal capacitors, with its lid open. Fixture courtesy of Sanyo.

6.1.2 Test Vias

On production boards, often we do not have space or locations available for connectors solely for the purposes of PDN measurements. Still we can make fairly accurate measurements (up to a couple of gigahertz) by providing test vias on the DUT that can take small-size semirigid probes or can be accessed with wafer probes. On fixtures, too, the ease of establishing or removing the connection can outweigh the drawbacks of somewhat lower accuracy and bandwidth.

If we use a fixture to connect a small-size PDN component to the measuring instrument (as shown above), it is typically a multilayer printed-circuit board where the terminals of the DUT are carried horizontally on different layers. The best characterization is achieved using a test board stackup closely resembling the final usage. Therefore, usually we end up with horizontal connections (planes) in the fixture below the surface; we, then need through-holes or vias to connect to the pins of probes, the tips of wafer probes or to connectors.

As shown in Chapter 5, in two-port shunt-through PDN measurements, the via impedance in series to the 50Ω instrument cables is not a serious limitation of accurate measurements below a couple of gigahertz because we have the via impedance in series to the 50Ω cables and not in series to the DUT.

6.1.2.1 Test Via Geometry

When vias take probe pins, the size and spacing of pins on the probe determine the necessary via geometry. For handheld probing, semirigid coaxial probes with a 2.1-mm (0.082-inch) or 1.2-mm (0.047-inch) outer diameter are convenient: they provide sufficient rigidity for reliable measurement. In addition, the center-wire to sleeve spacing enables us to create probes with 1.25-mm (50-mil) or 1-mm (40-mil) center-to-center separation.

Figure 6.11 is an illustration of possible test-via geometries for semirigid coaxial probes with 2.1-mm (0.082-inch) and 1.2-mm (0.047-inch) sleeve diameters. The figure arbitrarily assumes that the test vias are connecting to the plane pair that is closest to the top. The finished via hole diameter matches the diameter of the center

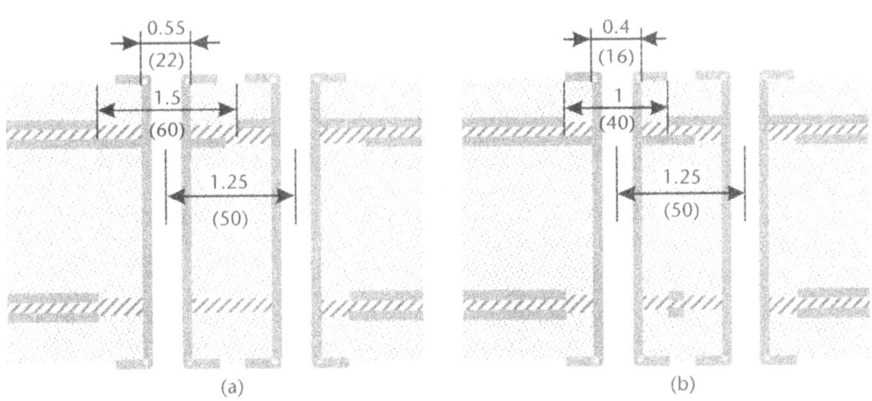

Figure 6.11 Typical test via geometries for homemade semirigid probes: (a) vias for 2.1-mm (0.082-inch) probe; and (b) vias for 1.2-mm (0.047-inch) probe. Dimensions are shown in millimeters and milli-inches in parenthesis.

pin. The antipad sizes for these finished holes depend on the aggressiveness of the PCB technology employed. The figure shows possible values in cross-section views. Note that the 0.55-mm (22-mil) drill-size for the larger probe may require antipads of 1.5 mm (60 mils). With 1.25-mm (50-mil) center-to-center separation of the power-ground vias, this antipad size will leave a small area between the two via barrels where both planes are removed. When we use smaller probes, requiring 0.4-mm (16-mil) drill size and 1-mm (40-mil) antipads, the same 1.25-mm (50-mil) center-to-center via separation will guarantee that we have no void in the plane overlap.

Figure 6.12 shows the correct and incorrect probe connections to the test vias from the opposite sides. The correct connection is shown in Figure 6.12(a): the probe pins fit snugly in the holes and all four probe pins make connections to the via walls without touching each other. If hot-air level solder (HASL) finish is used on the PCB, the soft solder material in the via hole will help the compliance of the pins. Incorrect probe positioning results in a loose connection in the via barrels or the pins being pushed so far into the holes that the tips of a probe pins from opposite sides touch each other. If the pins touch each other instead of making the connection to the via barrel first, the direct pin-to-pin path will increase the S_{21} reading; hence, the measured impedance will be increased erroneously.

Unless the rim of the sleeve is insulated, we need to be careful not to push the probe all the way to the surface of the board, as it may short to nearby surface traces or pads. At the same time, we need a minimum length of the probe pins to guide them into the holes manually. This also sets the range of convenient and appropriate pin length of the probes: pin lengths shorter than 1 mm (40 mils) are hard to guide and position, even when using the smaller probe. The maximum length is about half of the board thickness to be probed, so that we can avoid touching probe tips from opposite sides.

Through vias with probes attached from the opposite sides allow us to measure the transfer impedance across the distance of the via spacing; using handheld semirigid probes gives us the closest measure of true self-impedance (which ideally would require the probes to be connected to the same exact location). Unfortu-

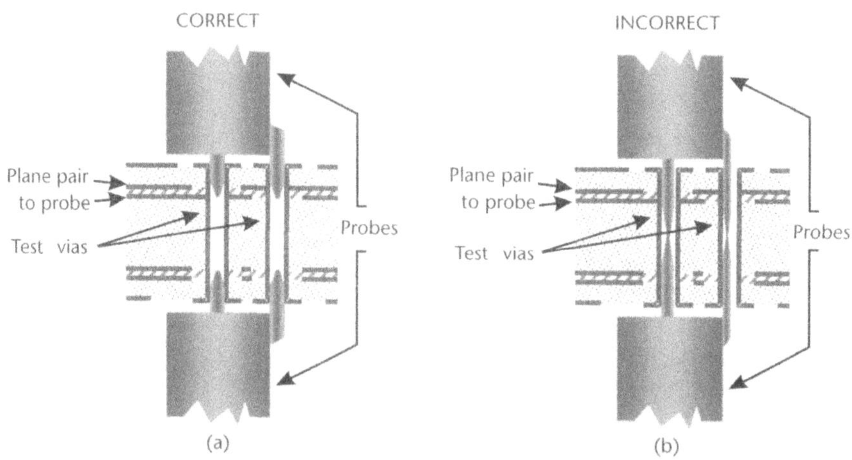

Figure 6.12 Connection of two-pin probes to test vias from opposite sides. (a) Correct connection, where the probe pins make a tight fit in the via barrels, and the pins are short enough not to touch each other. (b) Incorrect geometry, where the probe pins are either loose in the via barrel and/or the pins from the opposite sides touch each other.

nately, attaching the probes from the opposite sides is not always easy; the best method is to position the DUT vertically, so that the two probes are horizontal. This can be done easily with small fixtures. With larger production boards, especially if the measurement has to be done in a chassis, accessing both sides may be impossible.

As it was shown in Figures 5.5 and 5.6, self- and transfer impedances do not differ much if we have the connection points up to a small fraction of the wavelength of the highest frequency of interest. This allows us to use two pairs of test vias within a short distance, so that we can connect both probes from the same side. While this technique simplifies the connections of the probes, it requires us to create space for a second pair of vias and consider the possible coupling between the two loops of probes.

Figure 6.13 shows some of the possible connection permutations for self-impedance measurements. The sketches show a PDN with two plane pairs (we assume the test vias connect to the upper plane pair). Option (a) is the best connection, electrically, for self-impedance measurement, because the loops of the probes are lined up so that coupling between them is minimal. Also, the planes will further isolate the probes at high frequencies. This option requires probes to be connected from the opposite sides: this is easy if the connection is soldered for low-frequency measurements; but it can be difficult with probes on large DUTs for high-frequency measurements. Option (b) is the preferred solution when test-point pairs are close to each other. If the two probes are electrically close, we would still measure self-

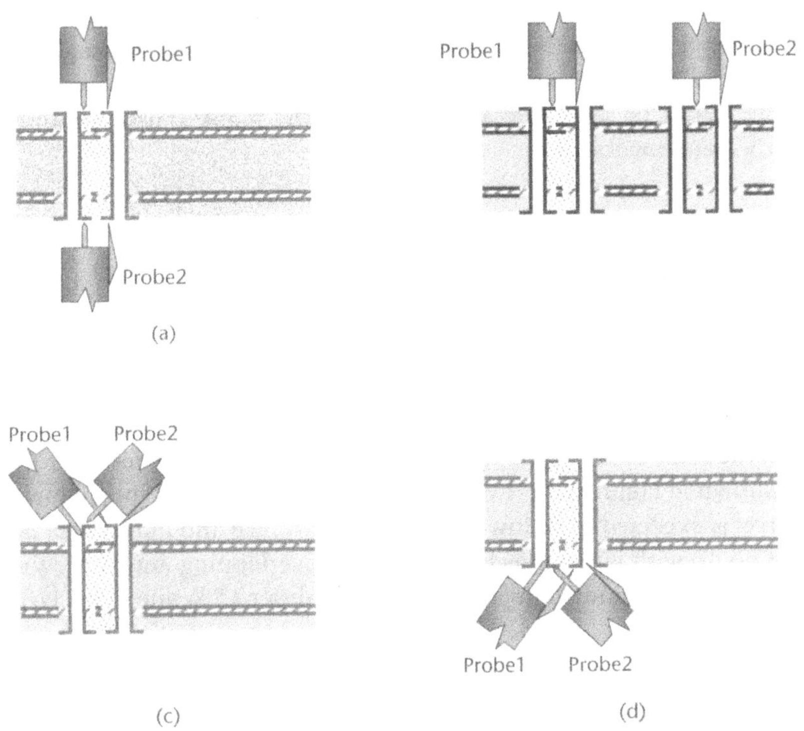

Figure 6.13 Sketches of some of the permutations of probe connections for self-impedance measurements from same versus opposite sides: (a) probes from opposite sides, (b) probes on same side attached to separate via pairs, (c) probes on same side attached to via pair close to planes, and (d) probes on same side attached to via pair far from planes.

impedance without the need to attach the probes from the opposite sides. Since via discontinuities are in series to the 50Ω interconnect impedance, we can connect through the shorter or longer parts of the vias. The probe loops, however, are parallel unless the test points are arranged orthogonally; but this arrangement increases the chances for high-frequency errors due to probe coupling. Option (c) or (d) is the preferred solution when we have only one test-point pair and we cannot connect the probes from opposite sides. We can connect from either side, but the impedance of the portion of the via connecting from the probes to the planes appears as an error in the measurement result; so, connecting through the shorter via portion results in reduced error.

To capture the self-impedance properly through test points separated by a distance of d, we have to satisfy two conditions: both the phase shift at the highest frequency of interest and the lumped attenuation due to the series impedance between the test points should be kept small. To illustrate the effect of the phase shift, we can use the data from Figure 5.5. The frequency of the first impedance minimum around 100 MHz showed no noticeable deviation until the separation reached about 3 mm (0.12 inch). With the assumed dielectric constant of 4, the wavelength at 100 MHz is 1.5m (59 inches). With these numbers, the separation-to-wavelength ratio is 0.2%, or a 0.72 degree of phase shift.

Figure 5.7 also illustrated the options of Figure 6.13 on a low-frequency low-impedance set of measurements, where self-opposite, self-top, self-bottom, and transfer 20 mm referred to the connection geometry of Figures 6.13(a), (c), (d), and (b), respectively. The deviation of the transfer 20 mm curve of Figure 5.7 is an illustration of the possible error stemming from lumped attenuation due to the series impedance between the two test-point pairs. At 100 kHz the phase shift through a 20-mm (0.8-inch) distance is only 0.005°; the consistently lower reading of the curve is the result of the attenuation between the plane resistance and low-impedance PDN components.

6.1.2.2 Spreading Inductance of Antipads in Test Vias

The larger through-holes are more convenient for manual probing, but typical PCB manufacturing processes may not allow us to make small enough antipads to avoid overlapping of voids on the facing planes. This increases the residual inductance as the current spreads out from the test vias. This is illustrated in Figure 6.14 with the measured resistances and inductances of two shorted test fixtures. The two fixtures had the same exact geometry, except the diameter of the test vias and antipads were as shown in Figure 6.11. Even though the resistance and inductance of the larger via barrel is expected to be lower, both the resistance and inductance of the shorted test site go up with larger vias because of the overlapping antipads. With the particular geometry, spreading inductance increases about 15% due to the larger through-hole antipad.

6.1.2.3 Thermal Relief in Test Vias

The through-hole test via connections should have no thermal relief connections at the planes, since that would increase the residual resistance and inductance of the

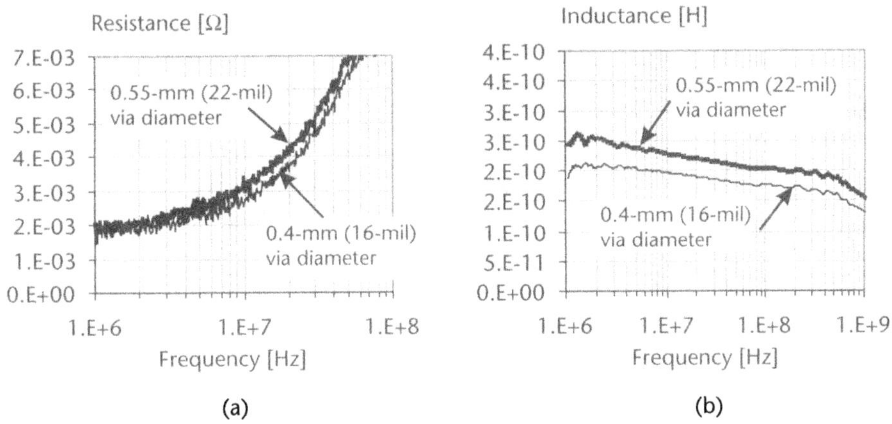

Figure 6.14 Extracted resistance (a) and inductance (b) of the same shorted-capacitor fixture with test vias of 0.55-mm (22-mil) and 0.4-mm (16-mil) diameters.

connections, too. A thermal-relief pattern is used primarily to improve the solderability when the hole takes soldered component leads. Since test via through-holes take probe pins but no soldered component leads, a thermal-relief pattern is not necessary. Figure 6.15 shows a plane layer with two vias; the via on the left connects to the plane through the thermal relief pattern. The via on the right does not connect to the plane, so that we can see the size of pad and antipad around the same size via barrel.

Figure 6.16 gives the measured ac resistance and inductance of a pair of vias shorted on a plane, with and without thermal relief. The via barrels had a 0.3-mm (12-mil) diameter, the shorting plane was at 0.5 mm (20 mils) from the surface, and the thermal-relief shape from Figure 6.15 was used. Note that larger cutouts and narrower spokes in the thermal-relief pattern will result in larger extra resistance and inductance.

6.1.3 Using Component Pads or Component Bodies as Test Points

To connect probes or solder cables, we may be tempted to use some of the usually numerous component pads or component bodies on an assembled board or package. While this type of connection eliminates the need for dedicated test points, at the same time it may severely alter the measurement result. Therefore, it should only be employed with a full understanding and acceptance of the implications.

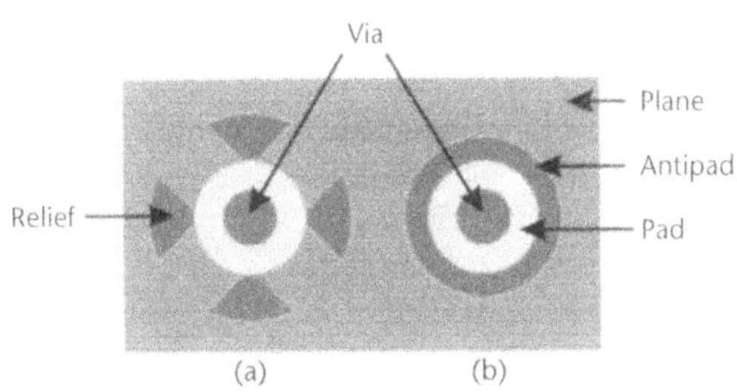

Figure 6.15 (a) A thermal-relief pattern and (b) its pad and antipad. The geometry of the thermal pattern may vary with CAD package and thermal requirements.

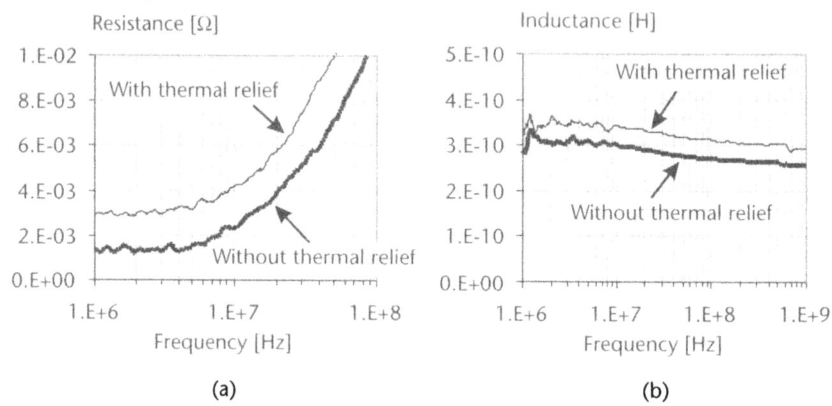

Figure 6.16 Comparison of ac resistance (a) and inductance (b) of pairs of vias with 0.3-mm (12-mil) diameter, shorted on a plane 0.5 mm (20 mil) below the surface. One via pair had solid hits on the plane, the second via pair had a thermal relief pattern as shown in Figure 6.15.

Figure 6.17(a) shows the equivalent circuit when we connect the instrument to a populated capacitor site. The bypass capacitor is represented by the series C_{cap}, R_{cap}, and L_{cap} elements, where L_{cap} represents inductance of the portion of the loop associated with the capacitor body. The vias and traces/pads are represented by series R-L circuits; they connect the capacitor to the rest of the PDN. The instrument is assumed to connect across the capacitor body. The circuit can be simplified by combining the series inductors and resistors of vias and pads. This results in the simplified equivalent circuit of Figure 6.17(b).

As was shown earlier, the series via and trace/pad impedances alone do not create significant errors up to a couple of gigahertz frequencies. However, significant error may result from the voltage divider formed by the via and trace/pad impedance

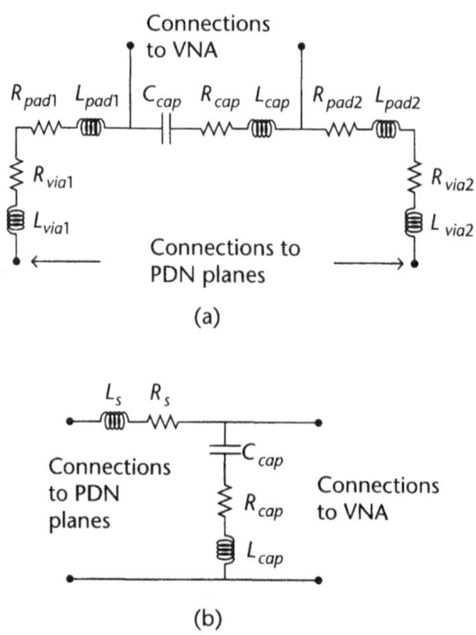

Figure 6.17 Equivalent circuit, showing the coloring of results when we connect to a bypass capacitor body: (a) full schematic and (b) simplified schematic.

and the impedance of the capacitor. With low-ESR parts, we get a frequency-dependent error around and above the series resonance frequency of the part. This approach is viable only if the ESR of the part is much higher than the series impedance of vias/pads (which may be the case for high-ESR aluminum and/or tantalum capacitors) or if we restrict the measurement to low frequencies, where the shunt impedance of the capacitor body is much higher than the series impedance of vias and pads.

The coloring of results can be expressed with the voltage transfer function from the planes to the VNA connections. The voltage transfer function is used, because in typical situations the PDN impedance is much lower than the impedance shown by the capacitor loop. Also, the impedance of the capacitor is usually much less than 50Ω. The voltage transfer function can be expressed as:

$$\frac{V_{out}}{V_{in}}(\omega) = \frac{R_{cap} + j\omega L_{cap} + \dfrac{1}{j\omega C_{cap}}}{R_s + j\omega L_s + R_{cap} + j\omega L_{cap} + \dfrac{1}{j\omega C_{cap}}} \tag{6.1}$$

When this expression is simplified, it reduces to the ratio of two second-order complex terms:

$$\frac{V_{out}}{V_{in}}(\omega) = \frac{1-\left(\dfrac{\omega}{\omega_{cap}}\right)^2 + j\omega\tau_{cap}}{1-\left(\dfrac{\omega}{\omega_1}\right)^2 + j\omega\tau_1} \tag{6.2}$$

where

$$\omega_{cap} = \frac{1}{\sqrt{C_{cap}L_{cap}}}, \quad \tau_{cap} = R_{cap}C_{cap}, \quad \omega_1 = \frac{1}{\sqrt{C_{cap}\left(L_{cap} + L_s\right)}}, \quad \tau_1 = \left(R_{cap} + R_s\right)C_{cap}$$

It is easy to find the values of the transfer functions at the extremes: at dc, the ratio is one; at infinite frequencies, the ratio corresponds to that of the inductances:

$$\frac{V_{out}}{V_{in}}(0) \equiv 1, \quad \frac{V_{out}}{V_{in}}(\infty) \equiv \frac{L_{cap}}{L_{cap} + L_s}$$

At interim frequencies the transfer function can take on any value, depending on the resonance frequencies and damping factors resulting from the interaction of the impedances of capacitor and vias and pads.

As an illustration, we show the measured frequency- and time-domain data on a power rail, where the measurements were done at a test-via pair connecting to the PCB planes and across the body of a nearby bypass capacitor. The supply rail had a multiphase dc-dc converter, bulk capacitors, and a number of 22-μF MLCC parts. Only components belonging to this supply rail were populated on the board. The bypass capacitor with two probes connected to its body is shown in Figure 6.18. The connection shown was used to measure the self-impedance across the capacitor body. There were also test vias connecting to the planes; the two square pads of the test vias are immediately on the right and below the capacitor, approximately 1.5 mm (60 mils) from the capacitor. The impedance at the test vias was measured with the same two probes, connected to the test vias from the opposite sides. To measure the time-domain noise across the capacitor or at the test vias only one of the probes was connected.

Three frequency-domain and four time-domain measurements were done on the rail. The frequency-domain characteristics were measured in the 100-kHz–1,800-MHz frequency range, with the probes connected:

- To the test vias from opposite sides;
- Across the capacitor body;
- One probe connected to the test vias, the other probe connected across the capacitor body.

The measured frequency-domain data is shown in Figure 6.19. Figure 6.19(a) shows the self-impedances at the test vias and across the capacitor. Note the 30-times difference at 1 MHz. Above the series resonance of the capacitor and below the lowest plane resonance, the two curves run in parallel; the impedance

Figure 6.18 Measuring probes across a 1210-size ceramic bypass capacitor.

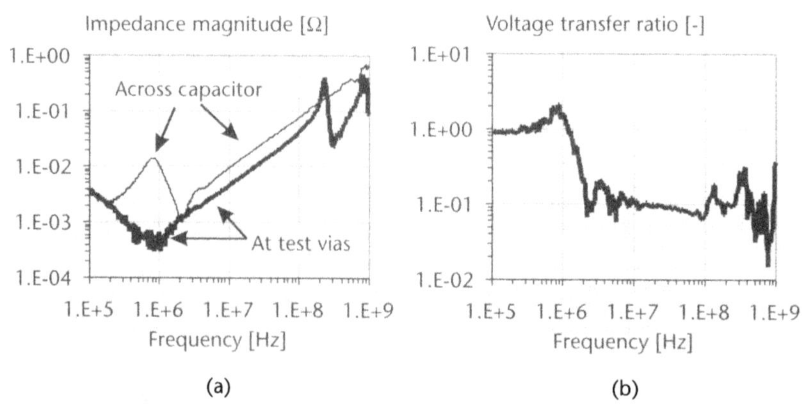

Figure 6.19 (a) Measured impedances and (b) extracted voltage transfer ratio on the board shown in Figure 6.18.

measured at the vias is lower by a factor of 2. Figure 6.19(b) shows the voltage transfer ratio calculated with (5.24). As expected, the transfer ratio settles at 1.0 at low frequencies. At 1 MHz the voltage transfer ratio reaches a peak of 2; the voltage across the capacitor body is twice as high as the voltage on the planes. Above 5 MHz, except around the structural resonances, the transfer ratio is approximately 0.1; the small slope is the result of inductances changing differently with frequency. This means that when the measurement is done across the capacitor body, noise is overestimated at 1 MHz and underestimated anywhere above 2 MHz.

To show the implications of the above frequency-dependent transfer function in the time domain, a low-frequency and a high-frequency test signal were measured and compared across the test vias versus across the capacitor body. Instead of using test signals from laboratory sources or using the random noise generated by the logic circuit, two signatures of the dc-dc converter's output signal were utilized. The switching ripple of a dc-dc converter was measured with a 20-MHz bandwidth setting of the oscilloscope, and the 220-MHz ringing escaping the converter was measured with a 500-MHz bandwidth setting. Figure 6.20 shows the four time-domain waveforms.

With the 20-MHz oscilloscope bandwidth, we remove the high-frequency ringing and measure the switching ripple. At the test vias we measure 4 mVpp, whereas across the capacitor the ripple is only 1 mVpp. Though the waveform is not sinusoidal, the ratio of peak-to-peak values shows good agreement with the transfer-function value at the 3-MHz fundamental frequency of the ripple waveform. With the 500-MHz bandwidth, we compare the ringing waveforms. At the test vias we measure 28 mVpp, whereas across the capacitor body the ringing is 6.2 mVpp; a ratio of 0.22. This correlates well to the measured voltage transfer ratio, which is 0.1 at 200 MHz and 0.27 at 300 MHz.

We can always use unpopulated pads to land wafer probes or solder cables. On unpopulated boards, we can use any component pad for probe connection. If extra pads are available on the board already (so that we do not need to take components off the populated board), we can simply redesignate those pads as test points. If there are no vacant pads on a populated board, we can remove parts to free up pads. If we want to connect to pads only (and not to vias, in order to avoid the time-consuming and always questionable de-embedding of pad and via impedance), we need

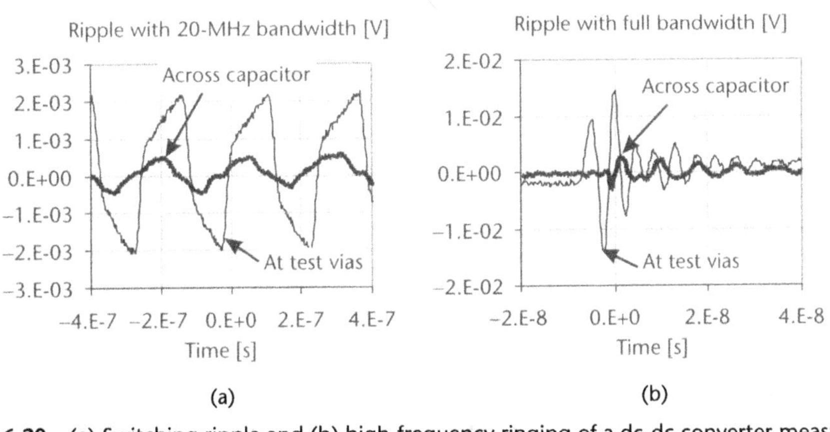

Figure 6.20 (a) Switching ripple and (b) high-frequency ringing of a dc-dc converter measured at the test vias and across the capacitor body.

to free up two pairs of pads to get the connection scheme of Figure 6.13(b). When we have a PDN rail with numerous bypass capacitors, often we can find two components that we can depopulate without significantly altering the PDN impedance. However, on supply rails with only a few components, this approach is not viable because the missing components may grossly impact the measured results. When we can connect to vias from the opposite sides of the board and find a bypass component that can be safely removed, we can depopulate this single component and create a connection scheme according to Figure 6.13(a).

6.1.4 Location of Test Points

To determine the required number and proper assignment of test points is really a design question, and as such we do not cover it in this book. Some of the consequences of test-point selections were discussed in the previous sections. For instance, the gradual transformation of transfer impedance into self-impedance as a function of test-point spacing was shown in Section 5.3. The impact of using existing components or component pads was covered in the previous section. Some further illustrations are also shown in Chapter 9, where we discuss full-system measurements.

In Section 6.1.2.3, we showed that the connection of the test point to the horizontal planes or patches is important at high frequencies; thermal relief or unnecessarily large antipads will negatively impact the results.

Figure 6.21(a) shows the impedance magnitudes on a 25 × 35-cm (10 × 14-inch) plane pair in a large multilayer board, measured at two different test points. One test-point pair was in the middle of the plane. The second pair of test vias was placed at a distance of 1 mm (40 mils) from to the plane edge, as shown in Figure 6.21(b). With the size of necessary antipads around the vias, there is only 0.15 mm (6 mils) of plane left between the antipad and edge of plane. This increases the measured impedance significantly: above 1 MHz the increase is approximately fourfold. Since incorrect location also increases resistance, the impedance increase starts to manifest itself as low as 10 kHz in frequency.

The equivalent inductance changes over a pair of planes gradually: it is lowest in the middle, rises as we approach the plane edges, and even rises more sharply as we

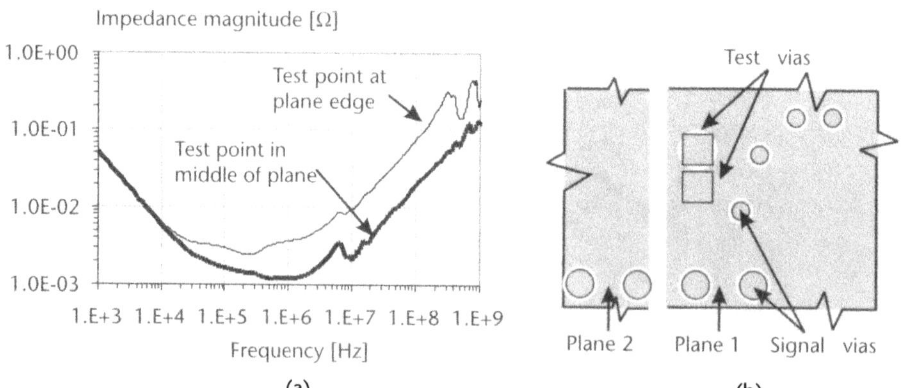

Figure 6.21 Illustration of impedance increase with test points located very close to plane edge: (a) measured impedance magnitude at plane edge versus middle of planes and (b) layout detail with test-point pair close to plane edge.

approach corners. When measuring bare plane pairs, at low frequencies this inductive behavior is masked by the large capacitive reactance; at high frequencies it is usually masked by modal resonances. We can capture this behavior either with plane pairs shorted at the edges or with bare plane pairs with very thin dielectrics so that modal resonances are suppressed. This technique was explained in detail in Chapter 4. In Figure 6.22, we use an illustrative impedance plot measured at 100 MHz on a 25.4 × 12.7-cm (10 × 5-inch) plane pair with very thin dielectric layer. The impedance was measured at test points located on a 2.54-cm (1-inch) grid over the board. The floor of the graph represents the top view of the board. The outermost test points were 6.35 mm (0.25 inches) from the plane edge. The board construction was similar to that shown in Figure 4.44.

Similar to plane edges, any cutout or larger antipad in close proximity to the test vias will change the observed impedance. Therefore, test vias have to be kept away from such disturbances.

6.2 Probes, Connectors, and Cables

At low frequencies, simple hand-soldered cable connections to the DUT or its fixture may suffice. With increasing frequencies, probes and connectors are necessary to achieve the required repeatability and accuracy.

6.2.1 Soldered Connections

At low frequencies, soldering the open pigtails of cables to the DUT is often acceptable; no connector or probe may be necessary. An example is shown in Figure 6.23. The error related to the open pigtail connections is due primarily to the small uncertainty of overall length, which translates to uncertainty (error) in the measured

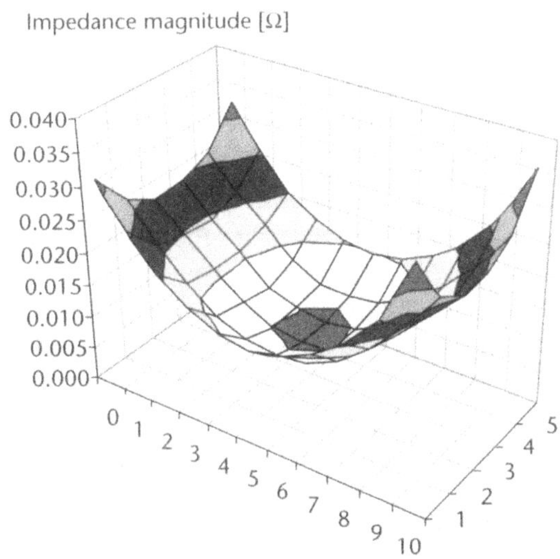

Figure 6.22 Illustration of self-impedance magnitude changing over a rectangular bare plane pair at 100 MHz.

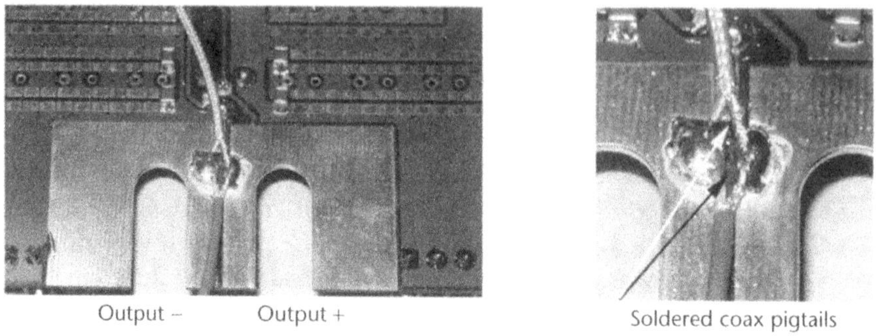

Figure 6.23 (a) Open pigtail test-cable connection to a dc-dc converter evaluation module and (b) an enlarged view of the connections.

phase. Assuming an uncertainty of length Δl corresponding to a Δt_{pd} uncertainty in propagation delay, the $\Delta\Phi$ phase error is:

$$\Delta\Phi = 2\pi f \Delta t_{pd} = \omega \Delta t_{pd} \quad (6.3)$$

At 1 MHz, we get 0.1° phase error from Δt_{pd} = 1/3.6 ns uncertainty of delay, which corresponds to about 5 cm (or about 2 inches) of cable length uncertainty. This shows that as long as we can maintain less than a centimeter (few hundred mils) of pigtail length, the phase error below 1 MHz is negligible for all practical purposes.

Typical low-frequency PDN measurements are done on active power sources, such as switching-mode dc-dc converters, voltage regulator modules, linear regulators, or low dropout regulators, and on low-frequency bulk capacitors or inductors.

If both cables are soldered on one side of the DUT so that plated through-holes are connecting the DUT to the cables [as shown in Figure 6.13(c, d)], the resistance and inductance of vias are in series to the unknown DUT impedance. The resistance and inductance of a via may be several hundred micro-ohms and probably a few hundred picohenries. Via resistance will be a limitation for dc-dc converters with high dc gain. The inductance of the series via creates an added inductive reactance in series to the measured impedance. An assumed 1 nH via inductance represents 6.3-mΩ inductive reactance at 1 MHz. For quick measurements of slow dc-dc converters, this error may be acceptable.

If the two cables are soldered from the DUT from the opposite sides or there is no via connection involved, the error comes from the uncertainty of the length (and delay) of the open pigtails. If we assume 100 ps/inch propagation constant, a 1-mm (40-mil) long open pigtail of a coaxial cable represents an approximately 4-ps delay. At 1 MHz with 4-ps delay (6.3) evaluates to 25 microradian phase error, which is safely negligible.

6.2.2 Homemade Probes

The highest bandwidth (up to tens of gigahertz) and greatest accuracy can be achieved with wafer probes, which come with matching calibration substrates.

Wafer probes will be covered in Section 6.2.3. Wafer probes achieve high bandwidth and good accuracy because they are made of small and flexible parts. As their name suggests, wafer probes are primarily targeted for semiconductor and package measurements; they are not well suited for large, rigid boards. Large and heavy DUTs require more robust probes; fortunately, for board applications, lower probe bandwidths are usually acceptable.

There are several companies offering different styles of rigid probes made of semirigid or machine-made coaxial cables with fixed or spring-loaded contacts. These probes may be viable up to several gigahertz, but usually they do not have a matching calibration substrate.

Homemade probes can be created easily of standard semirigid cables. Semirigid cables are available in various standard sizes (outer diameters) and cable types (braid material and style, dielectric material). Since these homemade semirigid probes are usually short and we do not use them for frequencies greater than a few gigahertz, the loss of the cable portion of the probe is largely irrelevant. Since we are not concerned about the cable braid material and dielectric losses, we can choose the semirigid coaxial cable based on the mechanical properties that are of interest.

For instance, we can use 2.1-mm (0.082-inch) or 1.2-mm (0.047-inch) outer diameter semirigid coaxial cables. With the 2.1-mm (0.082-inch) outer diameter cable, the center wire's diameter is approximately 0.55 mm (22 mils). If we solder a piece of the center wire to the sleeve to make the ground pin, the center-to-center pin spacing will be approximately 1.25 mm (50 mils). Figures 6.24 and 6.25 show two homemade probes with these dimensions. Figure 6.11 showed matching test vias for these probes.

When connecting two probes to closely spaced locations on a DUT, the coupling through the finite surface-transfer impedance of cables and probes may limit the dynamic range. Isolation calibration may help, but it assumes that the probe and cable geometry is exactly the same during calibration and measurement. We can also improve isolation by putting absorbing ferrites around the probes (shown in Figure 6.24) and around cables (shown later in Figure 6.32). The ferrite material can be chosen to maximize absorption in the frequency range of interest.

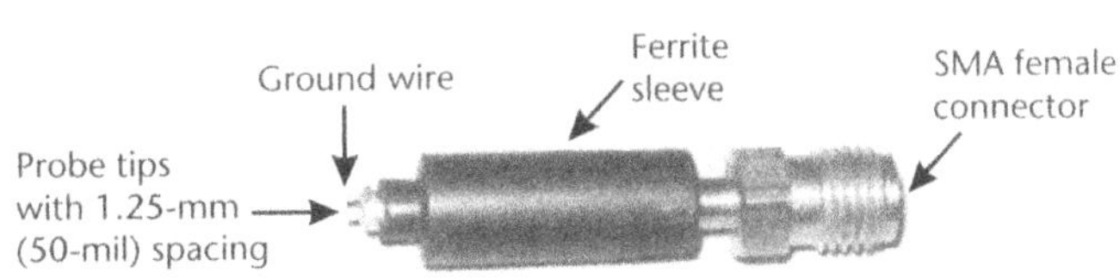

Figure 6.24 PTH semirigid probe made of 2.1-mm (0.082-inch) semirigid coaxial cable, with female SMA connector.

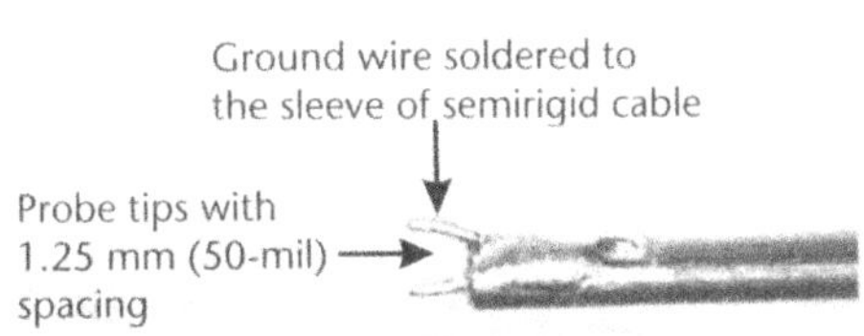

Figure 6.25 End view of a 1.2-mm (0.047-inch)-diameter semirigid probe with 1.25-mm (50-mil) pin spacing.

The optimum length of the probe is determined by the geometry of DUT. For probing bare boards or PCB test fixtures with low-height components, we want to use the shortest probe that we can make and use conveniently. This will reduce the errors from the probes that we cannot remove by calibration. On the other hand, for probing populated boards with potentially tall components, at least one of the probes has to be long enough that it can reach down among them, unless we have enough clearance around the test vias to accommodate not only the diameter of the probe, but also the connector at the end of the probe. Note that when we use the two-port shunt-through connection, the two cables and two probes do not have to be the same length. Most boards usually have tall components on top and low-profile components on the bottom side. When this is true, we can use a short probe on the bottom side, and a longer probe for the top.

We can buy semirigid cables with SMA connectors already assembled at one or both ends. Most connecting flexible cables have male connectors, but high-performance cables are available with either male or female connectors. If the connecting cables have a male connector, semirigid probes with a female connector (as shown in Figure 6.24) are preferable, because we can connect them directly to the cable without an adapter. However, when we calibrate to the end of the cables, we will need to connect and disconnect the probes several times to perform the calibration. That is, because we do not have calibration standards with a connection geometry matching that of the probe tips, we have to do reflection calibrations at the end of the cable with SMA standards. Unfortunately, screwing and unscrewing the probe several times may cause the center pin on the cable's SMA connector (which is attached to the center wire of the flexible cable) to break. To avoid this, we can use an adaptor piece that we leave screwed on the cable. In accurate high-frequency measurements, usually we want to minimize the number of connectors and adaptors; but, since the semirigid probes are rapidly losing their accuracy above a couple of gigahertz, the adaptor does not create a problem (i.e., at frequencies where we use the probes). This solution is illustrated for cables with male SMA connectors at both ends in Figure 6.26. At both cable ends a female SMA slug serves as a thread and cable saver. In this case the probe has a male SMA connector.

The rigid tips of the probe pins can be best used if inserted into through holes with matching center-to-center spacing. If we have to probe surface pads instead of through-holes, the rigid probe pins make it difficult to ensure proper connections of the four probe tips at the same time. To ease this difficulty we can replace the rigid ground pin with a miniature spring-loaded pin. These spring-loaded pins are available, for instance, as an oscilloscope-probe accessory. While we can make the connections more easily using this method, the spring-loaded pins will add to the ground-path inductance. Figure 6.27 shows a 2.1-mm (0.082-inch) homemade semirigid probe with spring-loaded ground. The zoomed picture on the right shows that the uncompressed

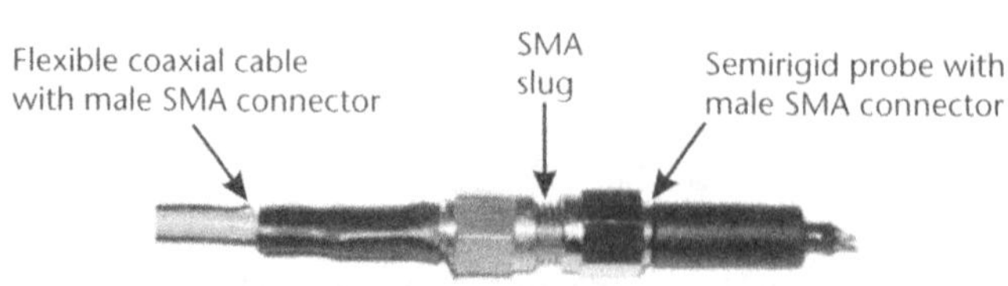

Figure 6.26 Homemade semirigid probe with cable and thread-saver slug.

(a) (b)

Figure 6.27 Photo of a 2.1-mm (0.082-inch) semirigid probe with spring-loaded ground pin: (a) full picture and (b) an enlarged view of the probe tip.

ground pin extends beyond the center pin. When the spring-loaded pin is compressed, the ground-pin tip will come in line with the center pin.

The pins create discontinuities at the end of the semirigid probes. When inserted into through-hole barrels, the depth of insertion is not well defined. Also, when we use spring-loaded ground pins, the spring attachment creates extra length and extra discontinuity. It is hard to measure this extra discontinuity in the frequency domain because we usually do not have calibration substrates for these probes. However, we can get an overall comparison by measuring the TDR response with the various shorting options. Figure 6.28 shows the time-domain responses. The trace labeled "shorting cap without probe" is the response with a shorting cap directly screwed on the sampling head, with no cable. The trace labeled "probe shorted at base" was measured with a 2.5-mm (1-inch) long semirigid probe directly screwed on the sampling head; and its pins were shorted with a solid metal sheet in line with the end of the sleeve. The trace labeled "probe shorted at 1 mm" was measured with the same setup as above except the solid metal sheet was 1 mm (40 mils) away from the end of the sleeve, leaving the 1-mm center pin and ground pin exposed. The trace labeled "probe with spring-loaded ground shorted at base" had the spring-loaded ground as shown in Figure 6.27, and the solid shorting metal sheet was in line with the end of sleeve.

The numbers in parenthesis show the 10–90% rise times of the responses. Note that the semirigid coax cable of the probe and its associated discontinuities at the connector and shorted pins creates an about 8-ps rise-time degradation. This

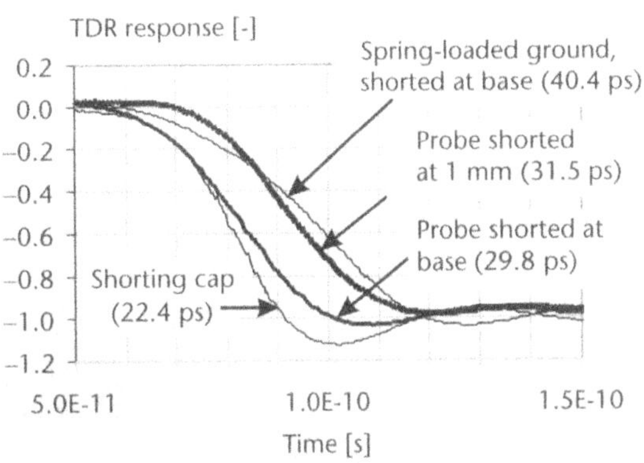

Figure 6.28 TDR responses of various shorted probes.

rise-time degradation increases only by 1.7 ps as we move the shorting plane 1 mm (40 mils) away from the base of the semirigid sleeve. However, adding a spring-loaded ground pin increases the rise-time degradation from 8 ps to 18 ps.

6.2.3 Wafer Probes

PCBs with through-hole test vias can be conveniently probed with semirigid probes. There are structures, however, where we do not have the luxury of through-hole test vias. Complex packages, for instance, may be too dense or expensive to add through-hole test vias for measurement purposes. In these situations we may need to connect the probes to surface pads, which require compliance, otherwise the probes would either break or not make connection. Interestingly, the reverse is also true: with a few exceptions, we do not want to probe a large and heavy board with tiny wafer probes. Though the wafer probes are compliant, it is likely that the large board will distort or sag beyond the compliance threshold of the wafer probe, in which case we damage the probe, often beyond repair.

Wafer probes are short, tiny, semirigid coaxial cables with an SMA, 3.5-mm, or 2.9-mm connector attached to a mounting base at one end and a short center pin and a compliant grounding blade at the other end. The photo of Figure 6.29 shows the end of a wafer probe with 450-μm (18-mil) pin spacing. These wafer probes are supposed to land on the target at a low angle; then the small deflection and wipe of the flexible ground blade and the (somewhat) more rigid center pin ensure the proper connections.

Today wafer probes come in different pin pitches anywhere in the range of 50 μm (2 mils) to 1,500 μm (60 mils) and in various shapes. Straight, low-angle versions are good for unpopulated DUTs with gold-plated pads, so that a low contact force is enough to make good connection. Longer probes are also available to reach between components on a populated board. Figure 6.30 shows deep-reaching probes connected to the back side of a ceramic LGA package and to a large rigid board.

6.2.4 Probe and DUT Holders and Probe Stations

To hold a large DUT, we might use a simple vise. If the DUT is held vertically, the probes can easily reach the DUT from the opposite sides. DUTs for wafer probing have to be mounted very firmly, so that we do not accidentally crush the probes.

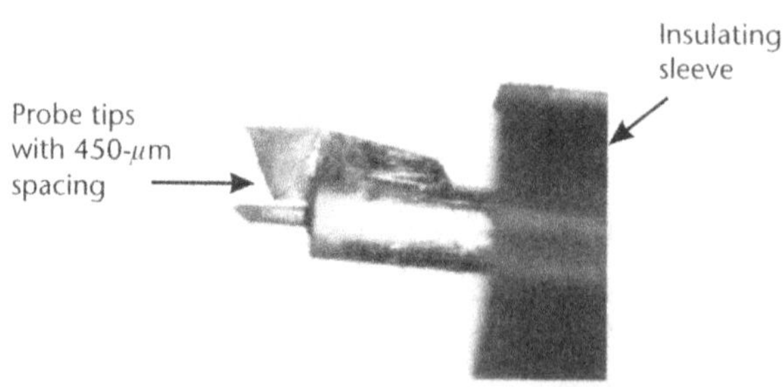

Figure 6.29 Tip of a wafer probe with 450-μm (18-mil) pin spacing. The large black area on the right is the end of the protective plastic sleeve around the tiny semirigid coaxial cable.

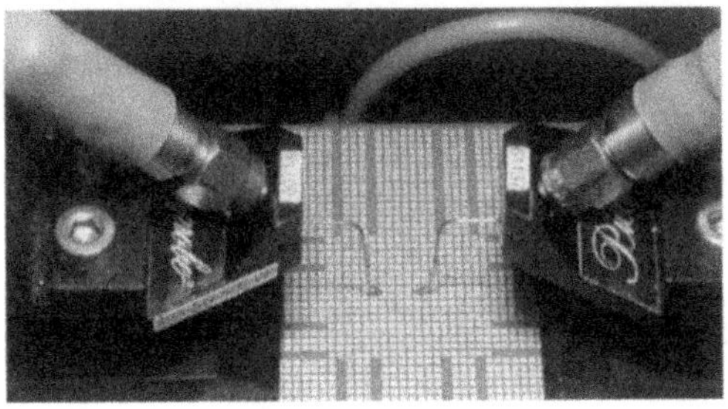

(a)

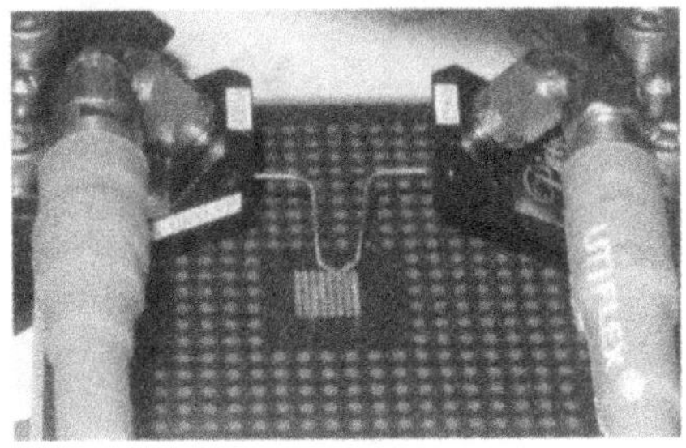

(b)

Figure 6.30 (a) 500-μm wafer probes on an LGA ceramic package to measure transfer impedance and (b) 1,000-μm wafer probes connected to a via array on a large rigid board.

Semirigid probes can be handheld for quick and brief measurements. To attach the probes from the opposite sides, adjustable probe arms may be used. Wafer probes need probe positioners with adjustability along all three axes. Figure 6.31 shows an adjustable probe arm with a semirigid probe and a 3D probe positioner.

6.2.5 Cables

Unless the component to be measured can be brought to the measuring instrument's calibrated terminals, we need cables to connect to the instrument and to the DUT, or to the probe reaching to the DUT. For a matter of convenience, the characteristic impedance of the cable should match the terminal impedance of the measuring instrument; for most VNAs it is 50Ω. The length and type of the cable depend on the application geometry and the highest frequency to be tested. If the cable had no losses and dispersion, its length would not matter. In real-world application, the limit comes from the need to calibrate the frequency-dependent losses and impedance mismatches. The greater the cable losses, the less dynamic range remains for the measurements after calibration. The per-unit-length loss of a cable depends on the cable size and material. Larger diameter and better dielectric material mean lower losses (and usually higher cost).

Figure 6.32 shows three representative cables; their main parameters are compared in Table 6.1. On the left a low-cost and high-loss cable is shown with a

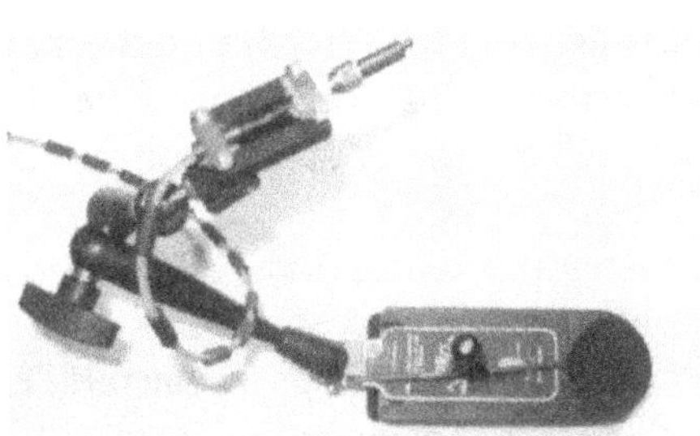

(a)

(b)

Figure 6.31 Photo of (a) an adjustable probe arm and (b) a probe positioner with 3D adjustment capability for wafer probes.

(a) (b) (c)

Figure 6.32 Cables for PDN measurements: (a) low-cost flexible cable; (b) ferrite-loaded cable with semirigid probe, the purpose of the ferrite beads on the cable is explained in Section 7.1; and (c) high-performance cable for microwave frequencies.

Table 6.1 Parameter Comparison of the Cables from Figure 6.32

Cable type	RG178B/U	Tensolite 461 Q-Flex	MegaPhase
Loss	0.17 dB/m at 10 MHz	1.08 dB/m at 2 GHz	1.64 dB/m at 10 GHz
Outer diameter	1.9 mm (75 mils)	1.9 mm (75 mils)	15.88 mm (625 mils)

crimped-on SMA connector. This is a thin and flexible cable, good for low-frequency measurements to check dc-dc converters and bulk capacitors. The cable in the middle has a somewhat higher diameter and soldered connectors instead of a crimp-on attachment. This cable is convenient for measurements up to a few gigahertz. The cable on the right is a high-end cable; in spite of its large diameter, it is still relatively flexible, and its outer conductor construction provides good surface transfer impedance. This cable is good for measurements up to many gigahertz. The large diameter and the weight of the cable require a connectorized DUT or a probe station; it is not convenient for handheld probing.

The connector at the cable ends affect results mostly at high frequencies. At low frequencies, below a couple of gigahertz, the connector type can be changed with an appropriate adaptor without sacrifice of accuracy and stability. We need to be careful, however, with the cable/connector termination. Crimped or unsoldered screwed cable terminations are acceptable only at low frequencies; above 100 MHz we may see an increasing repeatability error arising from the unstable connections.

Semirigid cables provide high performance, but their limited flexibility makes them useful only for those applications where the endpoints of the cable do not change frequently.

At high frequencies, the surface transfer impedance of the cables also needs to be considered. When input and output cables run close to each other or close to other aggressor lines, good isolation is essential. The ferrite beads on the middle cable provide this isolation. This will be further illustrated in Section 7.1.

6.3 Calibrations

Calibration of high-frequency instruments create a reference plane, where the specified accuracy of the instrument is guaranteed. Measuring instruments require periodic calibration to reduce the errors due to short-term and long-term environmental changes, such as supply voltage, temperature, humidity, and aging of components. Some instruments perform the calibration in the background without interrupting the measurement process. Typically, these are instruments without the need for cables connecting to the DUT or where the cables are not included in the calibration loop. Note also that fixture compensation, port extension, and de-embedding are different from calibration. Those topics were covered in Section 5.5.5.

With vector network analyzers, typically DUTs are connected with cables of arbitrary length and type. Even if the internal circuitry of the VNA could be calibrated in the background, the removal of cable effects still requires a set of calibration steps each time the cabling is changed.

VNA calibration is well documented in the literature (see [3]); as long as we follow the prescribed procedure, we can leave the details to the VNA vendor. In this section, we include practical information that the user may find useful for VNA measurements in various frequency ranges.

We will illustrate the various calibration options with data on different VNA models, covering the frequency range from 10 Hz to 20 GHz.

In terms of calibration need, we can divide the frequency range into three broad categories:

- Low frequencies: dc–1 MHz;
- Medium frequencies: 1 MHz–1 GHz;
- High frequencies: 1 GHz and up.

The selected boundaries are somewhat arbitrary, but may serve as easy-to-remember guidance.

6.3.1 VNA Calibrations in the Low-Frequency Range

In the low-frequency range, the wavelength of the test signal is so long that a cable in most laboratory setups would introduce only a small phase shift. At 1 MHz the wavelength in a typical cable is around 200m. So, for a 1-m-long cable, from (6.3) we get 1.8° of phase shift. This maximum phase shift is so small that the reflection ripple due to cable and probe mismatch is negligible. In the low-frequency range, the errors arise mostly from nonideal flatness of source and receive responses of the VNA.

Figure 6.33 shows the uncorrected (uncalibrated) through response of an Agilent 4395 VNA model with a short cable connecting the RF out connector to the B input connector. Model 4395 covers the 10-Hz–500-MHz frequency range; it does not have built-in directional couplers. The RF source and tracking receiver inputs are accessible at the front panel. Without calibration, the magnitude reading stays within ±10% of the ideal 1.00 reading in the 100-Hz–100-MHz frequency range.

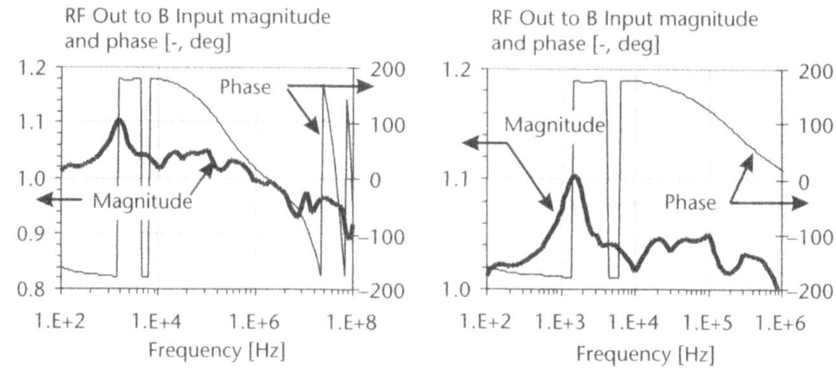

(a) (b)

Figure 6.33 Uncalibrated through response of an Agilent 4395 VNA: (a) full frequency range and (b) 100-Hz–1-MHz frequency range.

In the low-frequency range, assuming that the source and receive flatness of the VNA is good, quick magnitude measurements can be done with no calibration at all. Approximate DUT impedance magnitude can be obtained from (5.8) with reasonable accuracy as long as its value is much less than 25Ω.

Figure 6.34 shows the measured impedance profile of a 1,200-μF polymer capacitor. The data was taken with a 4395 VNA in the two-port shunt-through configuration with no calibration and no ground-loop isolation, connecting the DUT with short pieces of coaxial cable between the RF out and B in connectors.

Figure 5.19(a) showed the impedance magnitude and phase of the same part, measured with proper calibration and ground-loop isolation. The impedance magnitude traces in Figures 6.34 and 5.19 look similar at first sight, but the phase curves are very different. The correct data in Figure 5.19(a) exhibits a phase plot smoothly changing from −90° to +90°. The phase from the uncalibrated measurement has artificial slopes and a large offset. Note that because without calibration the phase information is totally irrelevant, the capacitance and inductance could be extracted only in an approximate way from the impedance magnitude. The graph of Figure

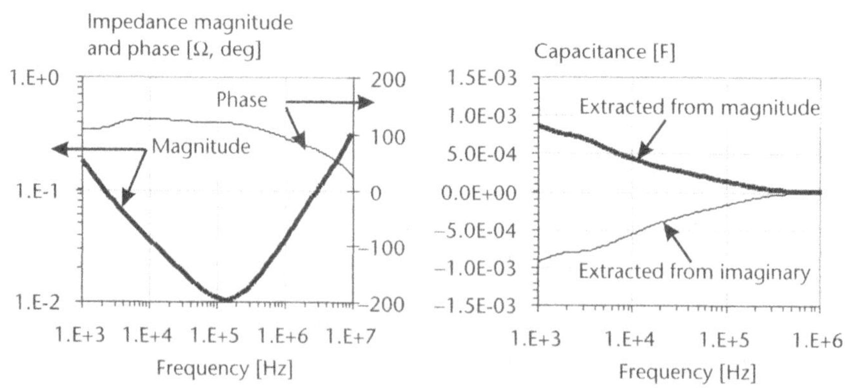

(a) (b)

Figure 6.34 Uncalibrated measurement data on a 1,200-μF polymer capacitor: (a) magnitude and phase and (b) extracted capacitance.

6.34(b) shows the capacitance versus frequency extracted in two different ways from the uncalibrated measurement data: extracted from the impedance magnitude, and extracted from the imaginary part of impedance. When we attempt to extract the capacitance from the imaginary part of the uncalibrated data, we get negative capacitance. When extracted from the uncalibrated impedance magnitude, the capacitance curve is closer to the correct one shown in Figure 5.20(a). Note that the accurate process requires the extraction of capacitance or inductance from the imaginary part of the impedance measured in a calibrated setup.

At low frequencies, a simple through calibration is usually sufficient to get accurate result. The data on Figure 5.20 was taken with 30-cm (12-inch) long cables with through-calibration only. This is the frequency range where soldered connections are acceptable, so often we do not need probes or connectors.

Full two-port calibration is not offered on low-frequency instruments, because mismatches do not affect results. We have to remember, though, that the cable should be electrically short. If we were to do a through-calibration-only measurement at 10 MHz with 5-m-long cables, the phase-shift through the cables would be 90°.

6.3.2 VNA Calibrations in the Mid-Frequency Range

Model HP4396 covers the 100-kHz–1,800-MHz frequency range. Similar to model 4395, it also has direct access to the RF source and tracking receiver inputs. As an option, transmission/reflection kits or *S*-parameter test kits can be attached to this model, to measure *S*-parameters directly.

Figure 6.35 illustrates the source and receive flatness of a Agilent 4396A VNA with direct RF out to B input connection and with an 85046 *S*-parameter test kit, measured with a short, low-loss cable.

In the mid-frequency range, both the flatness errors and the potential magnitude offsets are usually larger than those in the low-frequency range. Therefore, uncorrected (uncalibrated) measurements are not meaningful, even as approximations. If we are confident that our cables and probes are close enough to the nominal impedance, we may forego the calibration related to reflections and perform only a simple

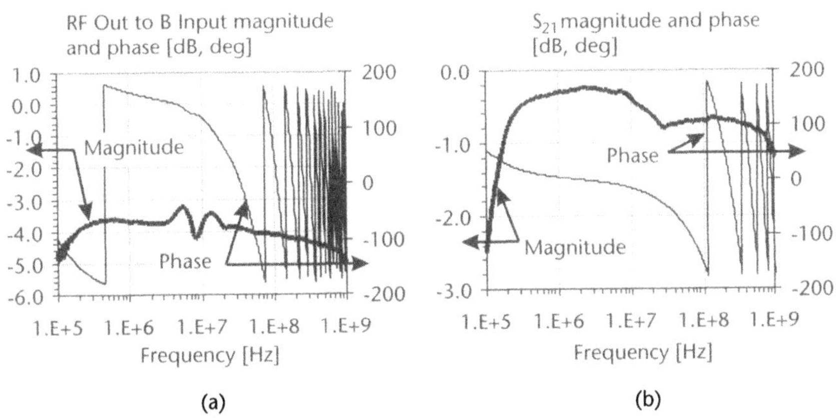

Figure 6.35 Uncorrected through response of an Agilent 4396 VNA: (a) RF out to B input and (b) S_{21} with 85046 *S*-parameter test kit.

through calibration. In this case, the errors related to the impedance mismatches will show up as a small periodical ripple superimposed on the measured data. The difference between through-calibration only versus full two-port calibration is illustrated in Figure 6.36.

With semirigid coaxial probes, full two-port calibration can be performed up to the end of the two coaxial cables leading to the probes. Standard SMA open, short, and load elements can be used for the reflection calibrations. For the through-calibration part, a semirigid SMA-SMA cable is used with a total length equaling the two semirigid probes. The only uncalibrated error is due to the impedance mismatch in the short semirigid probes; the mismatch and attenuation of the connecting cables are removed.

Figure 6.36 compares data taken on a reverse-geometry 1-μF 0612-size capacitor with through-calibration only versus full two-port calibration. Note that with through-calibration only, the magnitude and phase above 100 MHz show a periodical ripple error, originated from the cable mismatches.

In the mid-frequency range, we can use semirigid probes to connect to test vias on packages, boards or test fixtures. Not having calibration substrates for semirigid probes with rigid pins, we cannot perform full calibrations to the ends of probe pins, only to the end of cables. In this case, short probes with good impedance continuity at the connector are essential. Alternately, we can use wafer probes with matching calibration substrates and calibrate to the tip of the probes (as shown in the next section).

6.3.3 VNA Calibrations in the High-Frequency Range

In the high-frequency range, the approximate calibrations that we used in the low-frequency and mid-frequency ranges are no longer viable. Source and receive flatness errors are usually so pronounced that through-calibration is mandatory. Even with high-performance cables the periodical mismatch error would create unacceptably large fluctuations on the measured data; furthermore, losses and mismatches of semirigid probes would alter the result significantly; therefore, reflection

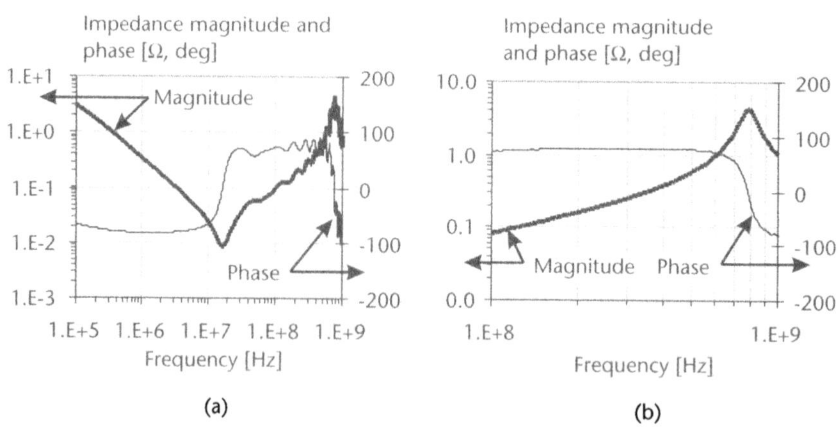

Figure 6.36 Impedance magnitude and phase, measured on a 1-μF MLCC part mounted on a small fixture: (a) measured with through-calibration only and (b) measured with full two-port calibration, with an enlarged view focused on the horizontal scale.

calibrations are required as well. Isolation calibration can be foregone if the reading does not approach the noise floor of instrument and cable crosstalk. When we perform the isolation calibration to reduce error due to probe or cable coupling, the calibration should be performed with the same or similar probe and cable geometry that we use later for measuring the DUT. In the high-frequency range, a full two-port calibration up to the connection points is required. This means we either need connectorized DUTs or wafer probes.

With miniature two or three-pin probes, such as wafer probes, full two-port calibration can be done with the calibration kit available on ceramic substrates. Figure 6.37 shows probes with 450-μm-pin pitch. The probes are positioned over the shorting pads of the calibration substrate. With the calibration substrate, calibration can be performed up to the tips of the probes.

The graphs in Figure 6.38 illustrate the necessity of full two-port calibration in this frequency range. In the 1-MHz–10-MHz frequency range, the impedance magnitude traces on both graphs appear to be straight slopes, suggesting the capacitive nature of the DUT. The absolute values, however, are visibly different even on this crude scale: at 1 MHz the magnitude from the corrected data is above 100Ω, while

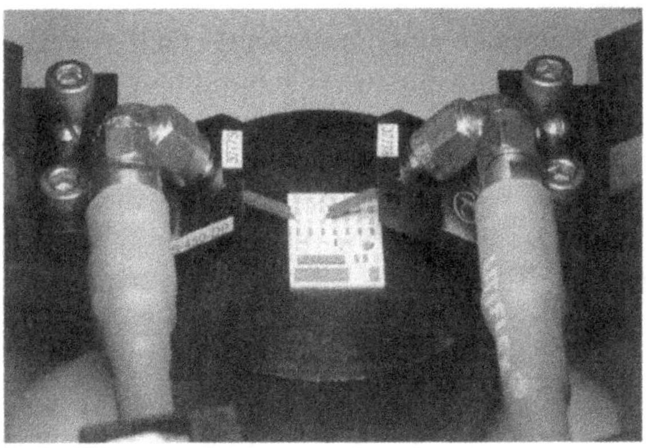

Figure 6.37 SG-style 2-pin 0.45-mm (18-mil) wafer probes on ceramic calibration substrate.

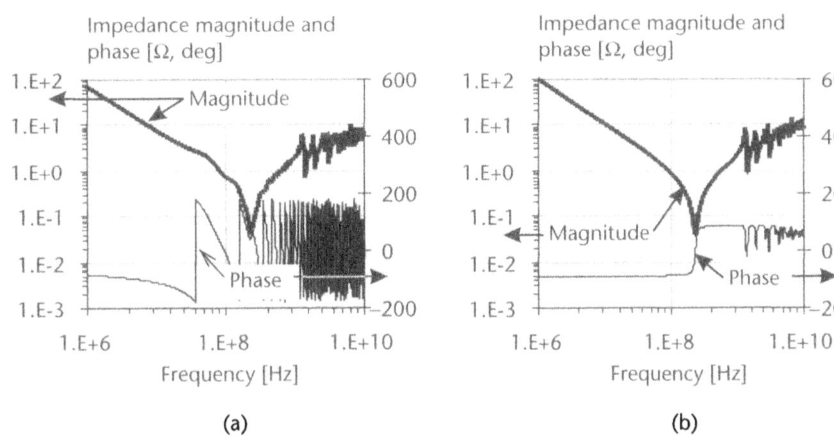

Figure 6.38 A 50 × 50-mm (2 × 2-inch) PCB laminate measured with an Agilent N5230A VNA and wafer probes, with correction turned (a) off and (b) on.

from the uncorrected data it is below 100Ω. Without calibration, the phase shift of the cable appears in the measured phase; the phase rotates significantly faster with frequency in the uncorrected dataset. This phase error in the uncorrected data is present even when the cable impedance is not mismatched. With the same periodicity, we also see a ripple superimposed in the magnitude trace of the uncorrected data, suggesting a slight mismatch in the cables.

6.4 Stability and Accuracy of Measurements

Calibrations are only as reliable as our calibration standards. Moreover, calibrations are useful only as long as the resolution, stability, and repeatability of the instrument, cables, and connections are of sufficiently high quality. We also need to remember that calibration and measurement setups are never the same. When we switch from calibration to measuring a DUT, not only does time elapse, but probes and cables are moved; also, depending on the frequency range, we sometimes need to desolder and resolder connections (in the low-frequency range) or move probes and cables or change adaptors or connectors (in the mid- and high-frequency ranges). These changes occur outside the calibration loop, so we need to be careful to introduce a minimum amount of change (and associated error).

The potential errors due to cable-braid loops in the low-frequency range, and due to finite surface-transfer impedance of cables in the mid-frequency and high-frequency ranges will be covered in Chapter 7. Here we show a few additional illustrations of effects we need to be careful about: drift with time (and temperature), instrumentation settings, probe placement, and quality of probe and DUT connections.

6.4.1 Response Drift with Time

The VNA performance drifts with time and temperature, therefore periodic recalibration is necessary. The maximum time interval between recalibrations depends on the instrument and on our applications needs. It is best to consult the instrument's manual and to make a series of reference measurements which track the actual performance of the unit over time. Immediately after calibration, the accuracy is determined primarily by the accuracy of calibration standard, repeatability of instrument, cables, and connections. Under these conditions, the accuracy can be significantly better than the guaranteed accuracy of the instrument.

As an illustration, Figure 6.39 shows a series of measurements taken at 1 minute, 10 minutes, 100 minutes, and 1,000 minutes after calibration. The room temperature was held within ±1°C for the time of the experiment. The DUT was a bare pair of planes with approximately 20-nF static capacitance. In the 100-Hz–1-MHz frequency range, the expected impedance plot is a straight slope.

Below 10 kHz, where the impedance magnitude is higher than 1 kΩ, there is a significant difference in the measured impedance with time. One minute after calibration, the impedance ceiling is around 40 kΩ. This ceiling drops with time; after 100 minutes, the ceiling has dropped to only 4 kΩ. The next day, 1,000 minutes after calibration, the impedance ceiling is only 1.5 kΩ.

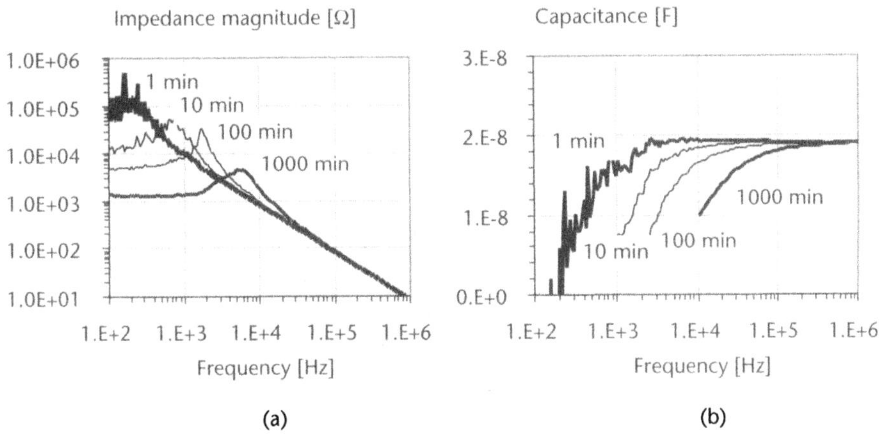

Figure 6.39 Illustration of VNA response drift with time. (a) Measured impedance magnitude and (b) extracted capacitance on a pair of bare power-ground plates.

Above 50 kHz, where the measured impedance is below 100Ω, the impedance-magnitude traces appear to very nearly coincide. The capacitance extracted from the same data set still shows noticeable difference up to about 300 kHz. In Chapter 4 we showed that the capacitance of PCB materials typically drops with a constant percentage over a logarithmic frequency scale. In this case, we get this expected trend only up to a few minutes because we are stressing the accuracy limit of the setup by measuring high impedances in two-port shunt-through connection. Note that, in this example, we intentionally stressed the setup because we wanted to show that immediately after calibration the resolution and accuracy of the instrument can be significantly better than the specifications. The performance gradually degrades with time.

6.4.2 Instrumentation Settings

When we change any of the sweep-related parameters of the VNA or impedance analyzer, we should recalibrate. Under some circumstances instruments can and will interpolate (or extrapolate) when we change the number of frequency points or change the start and stop frequencies. Unfortunately, this interpolation (or extrapolation) will introduce errors almost inevitably.

Similarly, if we decide to change the filter bandwidth of the instrument after calibration, some additional errors may result. In well-maintained instruments, the error due to changing the filter bandwidth is usually very small. However, this may not be guaranteed on the data sheet, so we need to track the performance of each instrument. Figure 6.40 shows the measured impedance magnitude of a DUT with different analog bandwidth settings, without recalibration. The IF bandwidth of the VNA was stepped through 2 Hz, 10 Hz, 30 Hz, 100 Hz, 300 Hz, 1 kHz, 3 kHz, 10 kHz, and 30 kHz values. The DUT was a high-power 12V rail: a set of low-ESR bulk capacitors distributed over a large PCB. The graph shows the impedance magnitude over the 100-Hz–10-MHz range. The impedance real part at 1 MHz was approximately 1 mΩ. In spite of the small impedance reading, note that the data traces show little difference, as bandwidth is switched in the 2-Hz–300-Hz range.

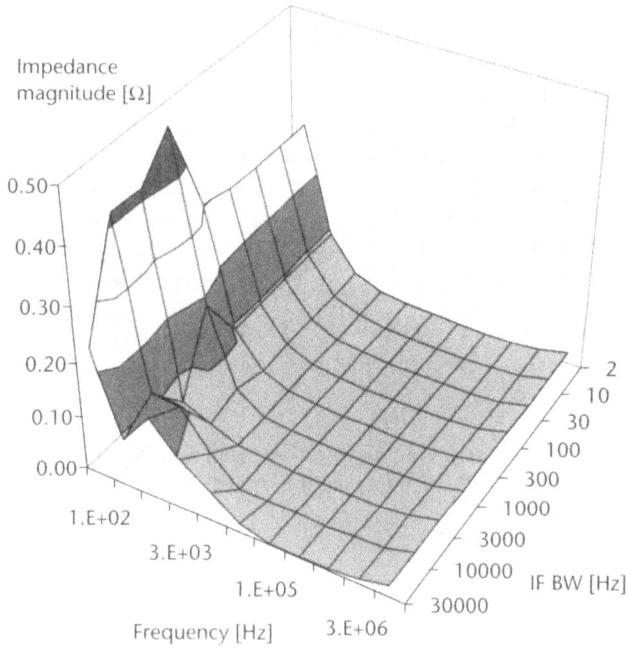

Figure 6.40 Effect of the IF bandwidth of VNA on impedance magnitude.

With wider IF bandwidth, the trace noise increases and also at low frequencies the selectivity curve of the instrument causes a rise in the impedance-magnitude plot.

The noise floor of the instruments is the lowest with the narrowest IF bandwidth. However, the settling time, and with it the total sweep time increases as bandwidth is reduced. For a 201-point sweep on the VNA used to collect data for Figure 6.40, one sweep took 270 ms with 30-kHz BW, versus 189 seconds of sweep time with a 2-Hz bandwidth. It is best to perform the calibration with the same bandwidth that we use later for measurements. If the bandwidth we will use is unknown, but we are confident that the instruments response is stable as we change the bandwidth, we should perform the calibration with the narrowest bandwidth for which we can afford to wait.

Output level on the test port is another parameter to watch carefully. Usually, we have the option to set the signal level at the test port. For instance, in the Agilent 4395A VNA, the test-port power can be set in the –50-dBm to +15-dBm range. In the Agilent 4294A impedance analyzer the oscillator voltage can be set in the 5- to 500-mVrms range. To maximize the available dynamic range, we may want to set the output level to the highest value that will not drive the instrument's receiver into saturation. When we measure DUTs over several decades of frequencies, the DUT impedance (together with the ac voltage and current through the DUT) may change considerably. Some PDN components, notably ceramic capacitors with high dielectric constant and thin layers are sensitive to the test level; so, the reading will change as the ac level changes. This will be explained and illustrated in more detail in Chapter 8. Here we show two illustrations in Figure 6.41. The same DUT, a 100-μF MLCC part was measured with two different instruments, with different test-power settings. We compare the impedance magnitudes and the capacitances extracted from the imaginary part of impedance. In the frequency range shown, the DUT is

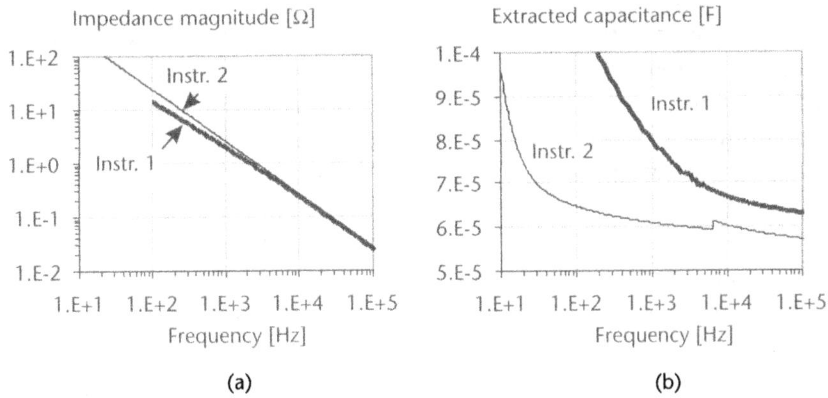

Figure 6.41 Impact of test-power level on measured DUT data. (a) Impedance magnitude and (b) extracted capacitance of a 100-μF MLCC with two different VNAs.

capacitive and the impedance magnitude and the test voltage across the part changes four decades.

Instrument 1 used higher test power and the source power was kept constant during the sweep. This resulted in lower measured impedance, hence higher extracted capacitance. Instrument 2 used lower power and to achieve higher dynamic range, the power was changed at 6 kHz. The lower power resulted in higher impedance and lower extracted capacitance. While the impedance magnitude curves from the two instruments are somewhat different, the more significant difference is in the extracted capacitance. As we will show in Chapter 8, many thin-layer MLCC parts exhibit increased capacitance as the ac bias across them increases. The same DUT with the same two instruments showed identical impedance and extracted capacitance curves as soon as the test power in both instruments was reduced to a level where the capacitance did not change any further with ac bias.

6.4.3 Probe Placement

When we use semirigid probes inserted into test through-holes from the opposite sides, the interaction of the small loops formed by the probe pins will be minimal and negligible. When we measure low-inductance DUTs at high frequencies, such as packages or PCBs with thin laminates, the interaction of probes may still create noticeable errors. A convenient way to eliminate those errors is to put lossy ferrite tubes around the probe sleeves, as shown in Figure 6.24.

Wafer probes create different placement challenges. Though wafer probes can be positioned on the opposite sides of the DUT by using either vertical probe arms or 3D flippable probe stations, a more convenient way of using wafer probes is to land both probes on the same side of the DUT for the two-port shunt-through connection. For self-impedance measurements at high frequencies, we want to position the probes close to each other. In such cases, the coupling between the loops formed by the wafer-probe pins cannot be neglected. If we know beforehand the precise probe separation on the DUT, we can try to perform the calibrations with the same probe separations. The calibration substrates offer a series of calibration features arranged with different separations; depending on the increments of spacing, we may get a good approximation of the probe separation that we will need in the actual measurement. This will remove a large part of this error, but requires us to recalibrate each time the probe spacing changes on the DUT.

To illustrate this error, Figure 6.42 shows the measured results on a shorting plane with different probe spacing, without recalibration.

The data was taken with 1,000-μm wafer probes on a matching calibration substrate. The reflection calibration was done on calibration features 1,000 μm apart. After the calibration, the probes were placed back on the shorting features. The result is shown in Figure 6.42(a). The phase is negative, indicating capacitive impedance. At 1 GHz, the impedance reading corresponds to –7-pH inductance. The graph in Figure 6.42(b) was measured with the same calibration, but the probes were shorted on the same features 500 μm apart. The impedance changed to inductive: at 1 GHz it corresponds to 20-pH inductance.

6.4.4 Quality of Probe and DUT Connections

As we saw earlier, the two-port shunt-through VNA connection scheme is very robust and forgiving about minor discontinuities in the connections when we measure low-impedance DUTs. In fact, it is so forgiving that sometimes bad connections may not be apparent simply by looking at the S_{21} magnitude plots. This problem is more likely to arise with nonconnectorized attachments (where the quality of connection may vary). When we do measurements with semirigid or wafer probes, or when we use nonsoldered fixtures, the contact pressure, wipe, and surface conditions of the contacts all have an influence on the result.

Figure 6.43 shows the measured impedance and extracted capacitance of a bare capacitor test fixture with good and poor probe connections during calibrations and measurements. The probes were homemade semirigid probes as shown in Figure 6.24. With clean contacts, the impedance magnitude and capacitance follow straight lines. With dirty contacts, dependent on which or how many of the four connections are not good, there will be error at different frequency ranges. At low

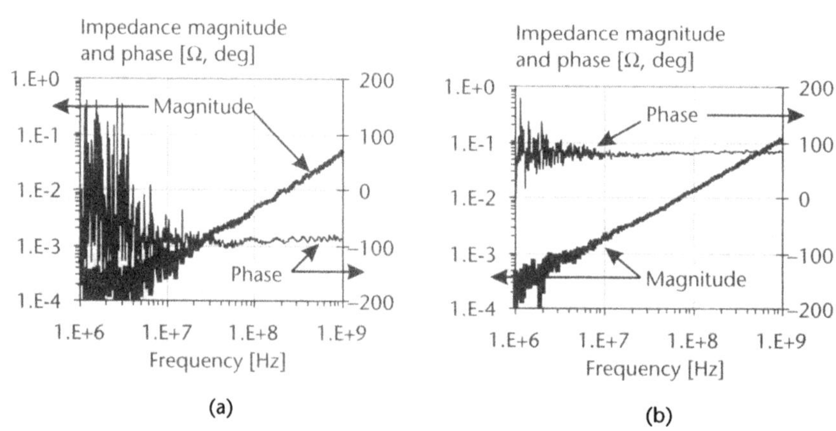

Figure 6.42 Effect of probe spacing on short measurement. Probe spacing during measurement: (a) 1,000 μm and (b) 500 μm.

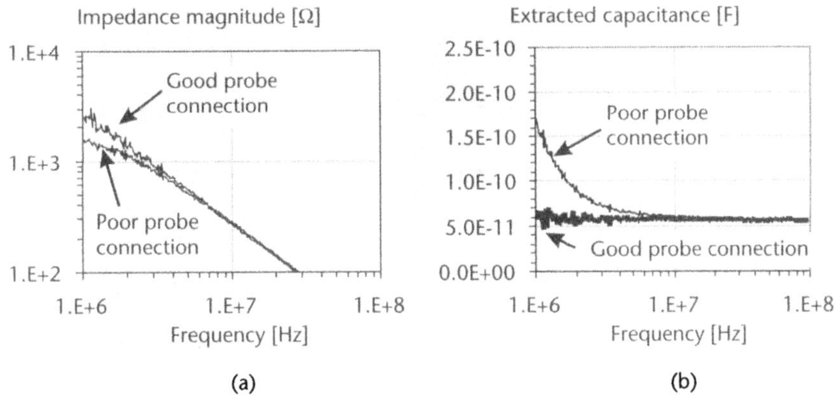

Figure 6.43 Comparison of (a) impedance magnitude and (b) extracted capacitance of a DUT with proper and improper probe connections.

frequencies the illustration in the figure shows seemingly lower impedance and higher capacitance.

Figure 6.44 shows another potential problem with probes: unintentionally, we may reverse the polarity of one probe with respect to the other, so that the center conductor of one probe connects to the ground side of the other probe and vice versa. The figure shows the measured impedance magnitude and phase of an odd-shaped power-ground plane pair on a large-size production board. Figure 6.44(a) shows the result with proper probe polarity. Figure 6.44(b) is the result on the same DUT with cross-connected probes. Note that there are two major problems created when the probes are reversed: the impedance magnitude saturates below 1 MHz, and the phase is inverted. However, if we look only at the impedance magnitude above 1 MHz, we may not realize that there is a problem.

Unreliable connections can also occur in fixtures. We show the impedance real part measured on a monolithic multiturn ferrite-bead inductor, measured with an Agilent 4294A impedance analyzer, 42942A port extender and 16193A fixture. Part of the fixture, where the DUT is placed, is shown in Figure 6.45. The fixture contacts are gold plated, and there is a thin plastic rod with adjustable spring force

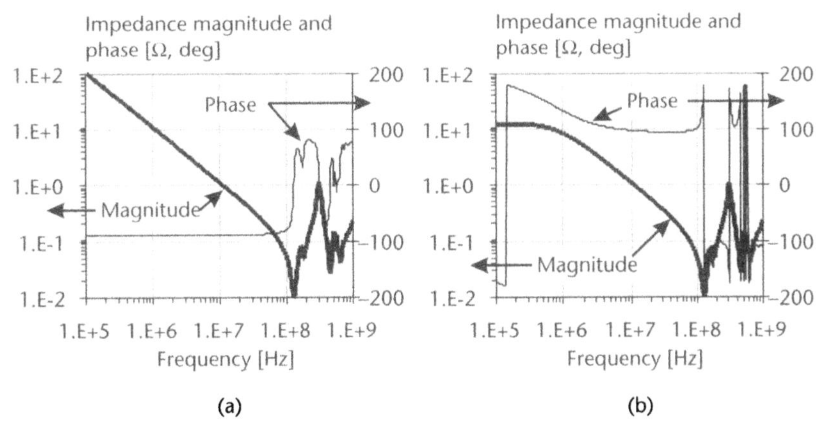

Figure 6.44 Impact of cross-connected probes: (a) proper connections and (b) cross-connected probes.

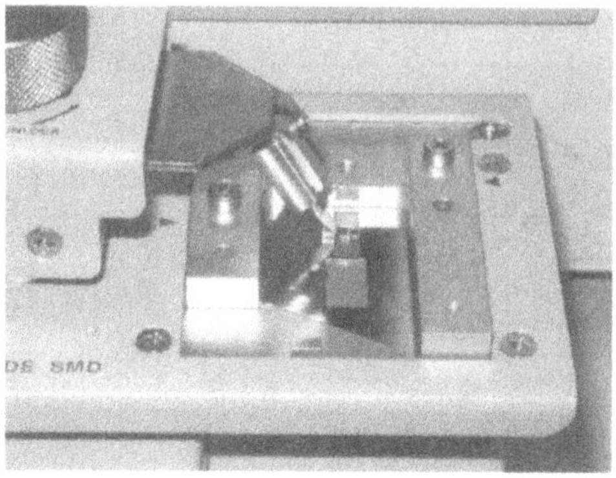

Figure 6.45 Ferrite bead placed on SMD fixture.

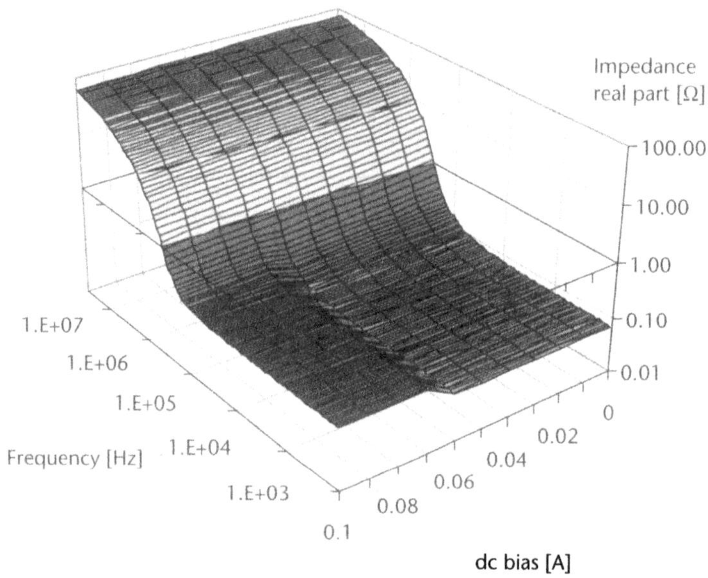

Figure 6.46 Impedance real part measured with the fixture and DUT shown in Figure 6.45.

to hold the DUT in place. With this setup, the impedance of the ferrite bead was measured with different dc–bias currents, starting with 0A and stepping up to 0.1A in 10-mA increments. The result is shown in Figure 6.46. At low frequencies, we see a step increase around 0.06A dc bias current; the impedance real part jumps from 13 mΩ to 67 mΩ. The jump in impedance was caused by an unintentional tap of the fixture; this indicates that the adjustable contact force was set to an insufficiently low value that did not enable a reliable connection.

References

[1] Ulrich, R. K., and L. W. Schaper, *Integrated Passive Component Technology*, New York: IEEE Press, Wiley-Interscience, 2003.

[2] Novak, I., "SUN's Experience Ultra-Thin Laminates for Power Distribution Applications," *Proceedings of DesignCon 2006*, Santa Clara, CA, February 6–9, 2006.

[3] Agilent Technologies, "Specifying Calibration Standards and Kits for Agilent Vector Network Analyzers," Application Note 1287-11, 5989-4840EN, 2006.

CHAPTER 7

Measurements: Practical Details

In Chapter 5 we covered the measurement basics, and in Chapter 6 we looked at calibrations and connections. In this chapter, we look at the practical details that are necessary to measure low impedances of PDN over wide ranges of frequencies.

7.1 Making the Proper Connections

The proper choice of calibration and probes is a necessary first step, but further considerations are needed to make sure that the connections between the VNA and probes do not impose limitations on the measurements.

7.1.1 Eliminating Cable-Braid-Loop Error at Low Frequencies

The ground returns of port 1 and port 2 in most VNAs are connected together inside the instrument. When we measure a low-impedance DUT, for instance, an active dc-dc converter output, or a metal short, the current of port 1 will create a voltage drop across the parallel equivalent of the braid resistances of the two cables. This is shown in the simplified schematics of Figure 7.1.

When we have a very-low-impedance device connected in the two-port shunt-through connection, the source impedance, cable center-wire resistances (R_{c1} and R_{c2}), and braid resistances (R_{b1} and R_{b2}) create a voltage divider, which produces a V_e error voltage and an erroneous impedance reading floor. This divider is created as the source current flows through the parallel equivalent of the two braid resistances, raising the potential of the lower node of Z_{DUT}. The measured low-frequency impedance value, instead of Z_{DUT}, thus becomes

$$Z_{measured} = Z_{DUT} + R_{b1} \| R_{b2} \tag{7.1}$$

where || indicates parallel equivalent. Though VNAs do not operate at dc, the very-low-frequency response can be approximated with the dc resistance values. Note that in (7.1) we have the two braid-resistance values in parallel, so if one of the connections is very short or has very low dc resistance, this limitation becomes negligible. We typically encounter the cable-braid loop problem when the VNA connections need to be brought over longer cables to the DUT. When we measure small-size components, which we can easily move to one of the VNA ports, one cable can be shortened significantly or be eliminated altogether. On the other hand, to reach into large-size DUTs, we require longer, flexible cables, so the cable-braid loop has to be eliminated by other means.

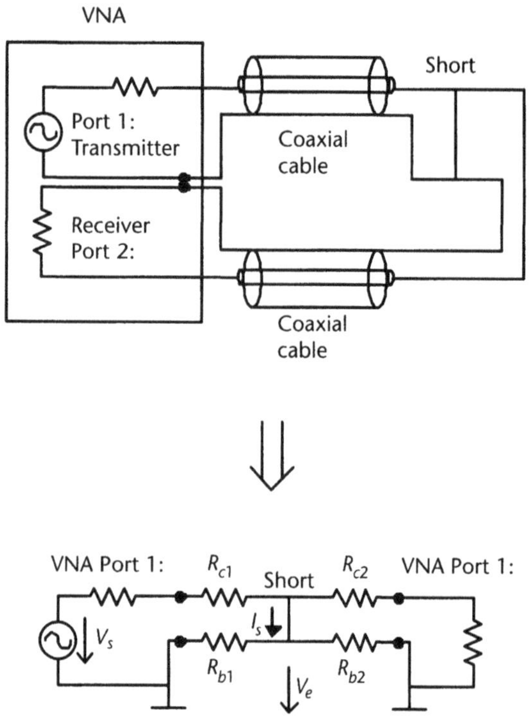

Figure 7.1 Simplified schematics, showing the cable-braid loop at low frequencies.

7.1.1.1 Ferrite Clamps/Sleeves

At high frequencies, the cable-braid ground loop will not create this error, because the cable inductance introduces a –20-dB/decade roll off of this residual reading above the corner frequency determined by the braid resistance and braid inductance. By placing ferrite clamps or ferrite beads around the cables, the inductance of the cable can be increased. The increased inductance reduces the corner frequency, beyond which the error drops.

Figure 7.2 shows a sheet-metal short that is used to illustrate the effect of cable-braid loop in various configurations. Figure 7.3 shows the impedance magnitude measured with different cable configurations with the sheet-metal short across the test points. All cables for this illustration were 61-cm (24-inch) long, type RG174 flexible coax and 2.1-mm (0.082-inch) semirigid coax. The plots show the residual error with and without ferrite clamps around the cables. The measurements were done with an HP4395 VNA, between RF out and input B.

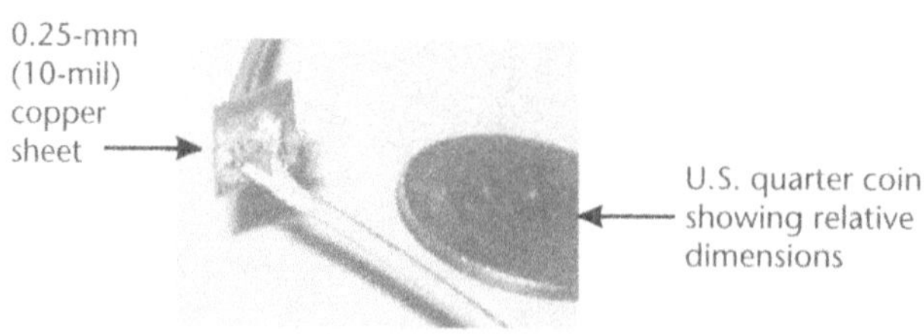

Figure 7.2 Sheet-metal short used for the measurement data shown in Figure 7.3.

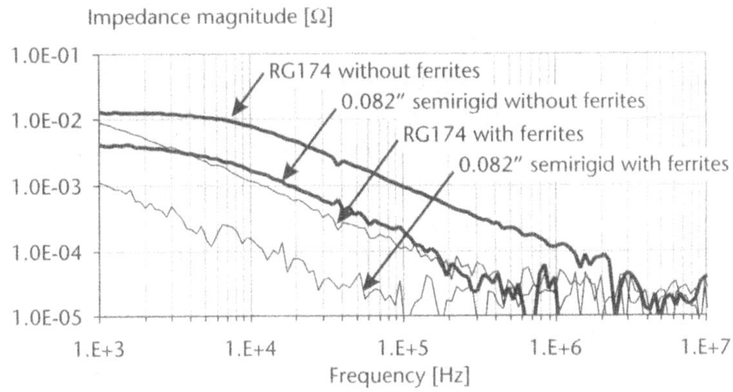

Figure 7.3 Measured impedance magnitudes of residual error on shorts with different cables, with and without ferrite sleeves. For this illustration, no isolation transformer or isolation amplifier was used (see in the next sections).

A cable with ferrite beads and one low-frequency clamp is shown in Figure 7.4. With the ferrite clamps the residual error still has a –20 dB / decade roll off, but the corner frequency gets lower. For measuring capacitors, reducing the corner frequency may be sufficient, as the capacitor's impedance curve follows a similar trend at low frequencies. In those cases, it is sufficient to reduce the residual error to at least 10 dB below the expected impedance of the DUT. However, eliminating the cable- braid-loop error by ferrite clamps does not work for very low dc resistances combined with a low series inductance, such as shorted planes, shorted via loops, and active voltage regulator modules and dc-dc converters.

The ferrite material can be selected based on the frequency range of interest and type of cables. For low-frequency measurements, high permeability is useful to increase the common-mode braid inductance. For high-frequency resonance suppression purposes (explained in Section 7.1.3), we need a specific loss of the ferrite material in the frequency range of the resonances.

To break the cable-braid loop, we can use capacitive, inductive, or electronic isolation. Capacitive isolation requires sufficiently large capacitance in series to both the hot and return connections of at least one of the ports. The capacitance should be large enough that the occasional dc bias voltage from a DUT and the ac bias from the VNA do not change the transfer gain significantly in response to potentially changing capacitance. Inductive and electronic isolations will be described in the next sections.

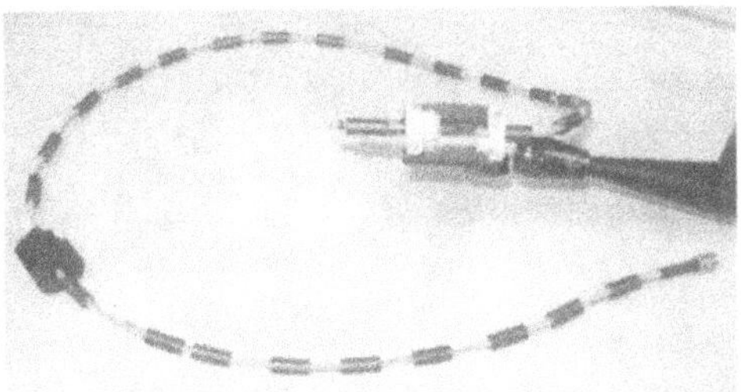

Figure 7.4 Flexible coaxial cable, with ferrite sleeves along its jacket. The short semirigid probe at the end of the cable is held by a probe positioner.

7.1.1.2 Isolation Transformer

Another way of eliminating the ground loop of cables is to use an isolation transformer. Figure 7.5 shows the photo of a homemade transformer created on a Phillips ferrite toroid core, type TX51/32/19-3F3. The 52-mm ferrite core has 2 × 50 turns with AWG20 wires. The large-size core provides low distortion for a 0-dBm VNA measurement level. The large size is also useful in measuring active devices, like VRMs, where the dc output current of the VRM may drive a smaller core into saturation. A series capacitor would block dc current, but it would introduce secondary resonances. The number of turns is a compromise between the main inductance (which sets the lower corner frequency) and the winding capacitance (which sets the upper corner frequency of the transformer response). The transformer described here had its –3-dB points between 50Ω terminations at 500 Hz and 3 MHz (see Figure 7.6). The main and stray inductances and the winding capacitance of the transformer create a frequency-dependent transfer response, which cannot be completely removed by VNA calibrations for a wide range of DUT impedances. Therefore, the calibration should be performed with a low-value standard (1Ω or 0.1Ω), which approximately matches the upper end of expected range of measured DUT impedances. This frequency-dependent error can also be reduced by limiting the measurement bandwidth: for the given transformer, in the 1-kHz to 1-MHz frequency range, the residual error is negligibly small for any DUT impedance.

7.1.1.3 Isolation Amplifier

Isolation amplifiers can also be used to break the ground loop. The ground loop is broken at the floating differential input of the amplifier. The amplifier can be placed either in front of port 2 of the VNA or at the output of port 1 before the DUT. As opposed to isolation transformers, low-frequency response is usually not a concern, as operational amplifiers are dc coupled. However, the high end of the frequency range is still limited by the frequency response and maximum slew rate of the amplifier. The circuit shown in Figure 7.7 had a flat frequency response with nonsaturated slew rates up to at least 10 MHz. For measuring passive components, the placement of the amplifier within the measurement loop is not critical.

The amplifier can be placed between port 1 of the VNA and the cable leading to the DUT, as shown in Figure 7.7, or it can be placed after the DUT. Placing the

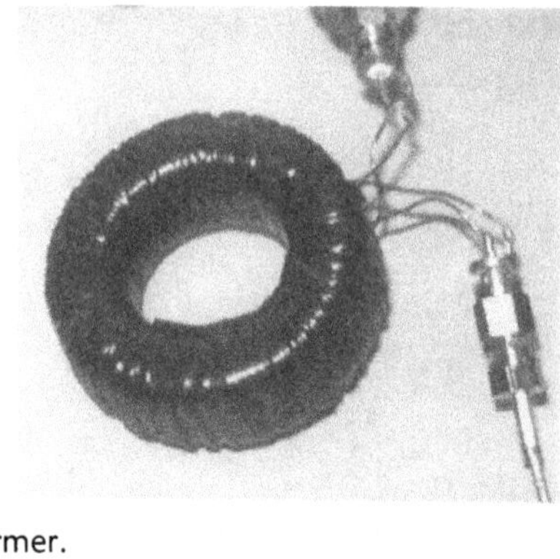

Figure 7.5 Isolation transformer.

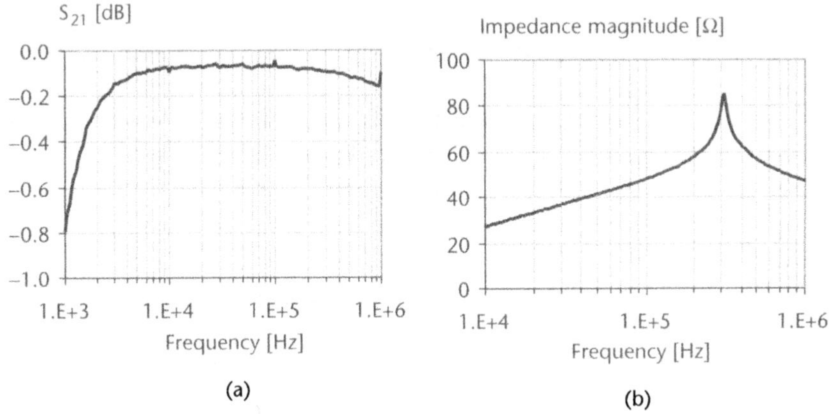

Figure 7.6 Transfer response (a) and impedance (b) of the isolation transformer shown in Figure 7.5.

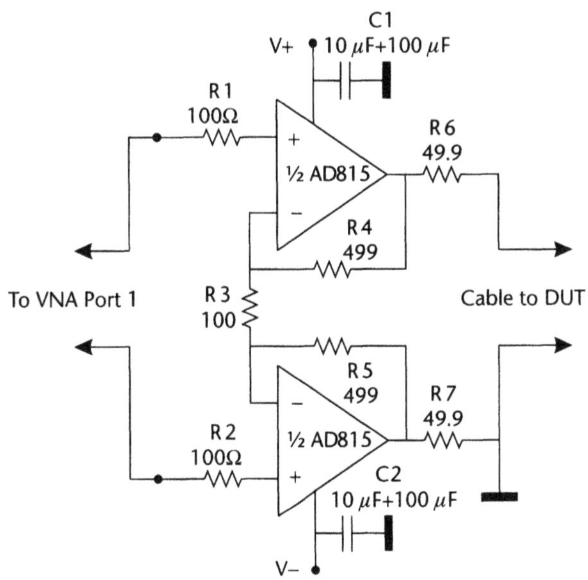

Figure 7.7 Differential-input amplifier with a dual operational amplifier.

amplifier directly at port 1 of the VNA is advantageous with active DUTs or when using long cables. With the differential input facing port 1, the amplifier's common-mode voltage range is not stressed; only the low voltage drop across the cable braids has to be absorbed. When active devices (powered PDNs, dc-dc converters) are measured, the source voltage of the DUT drives a current through the amplifier's output resistance; but, as long as the amplifier can tolerate that dc current in its linear range, the dc output voltage of the amplifier will stay close to zero.

If we place the amplifier between the DUT and port 2, there are two additional considerations. If the amplifier is close to the VNA and we use long cables to connect to the DUT, we need a differential termination resistor across the amplifier input to match the cable. Moreover, when measuring active DUTs, the maximum dc input voltage is limited by the maximum unsaturated output voltage of the amplifier divided by its differential gain. For the circuit shown, it is slightly above 1V dc.

Figure 7.8 shows the frequency response and supply-voltage sensitivity curves. Note that the gain is very stable below 1 MHz; also it varies less than ±0.1 dB at 10 MHz as the supply voltage varies in the ±6- ... ±8-V range. In this simple homemade setup, both the supply-voltage sensitivity and the absolute frequency response of the amplifier limit the application to frequencies below 10 MHz.

The two photos in Figure 7.9 show the construction of the homemade amplifier. The amplifier was hand-built on a small piece of two-sided PCB material. The floating differential input connects to a board-mount female SMA connector, which can be screwed directly on the N-to-SMA adaptor at port 1 of the VNA.

A drawback of the isolation amplifier is the need for an external power supply. The supply voltages must be selected such that the connected VNA input should survive the maximum saturated output voltage when transients or circuit shorts occur. The lowest value of the supply voltage is limited by the specified minimum rail voltage of the operational amplifier. With the circuit shown in Figure 7.7, stable operation was achieved at and above ±4V supply voltage. For most low-frequency VNA models, the maximum input dc voltage is at least 7V; this suggests a ±7V maximum supply voltage for the differential amplifier.

Instruments may have built-in low-current power supplies to power probes. In the Agilent 4395A VNA, the probe power connections offer +15V and −12V raw voltages. We can use small analog voltage regulators to drop the voltage to the ±7V level, but we have to realize that the common point of the built-in power supply is tied to the instrument's ground internally, which recreates the ground loop we had hoped to eliminate. To avoid reestablishing the ground loop, we have to use an isolated dc-dc converter in the supply path. One possible circuit is shown in Figure 7.10. The benefit of this circuit is that it eliminates the separate bench supply and thereby reduces the size of the power-supply loop.

When using an external power supply to feed the differential amplifier, resonances and noise from the power-supply loop may enter into the measured data. Figure 7.11 shows the error created when the supply feed of the isolation amplifier has resonances. The DUT was a shorted capacitor test site measured with two con-

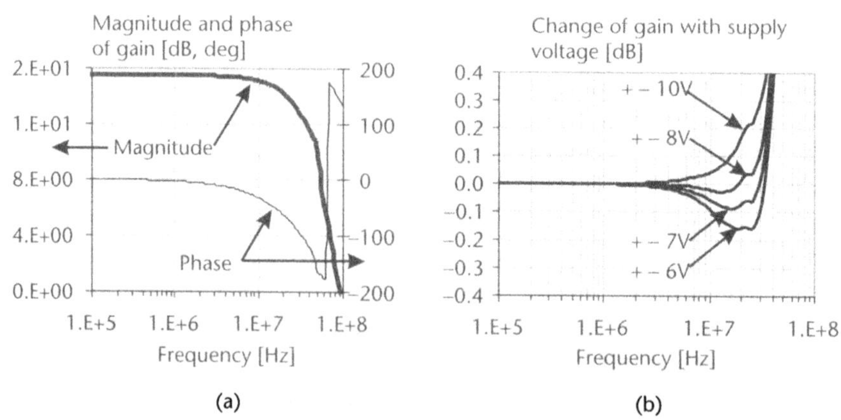

Figure 7.8 Performance graphs of the differential amplifier: (a) gain versus frequency and (b) change of gain response with supply voltage.

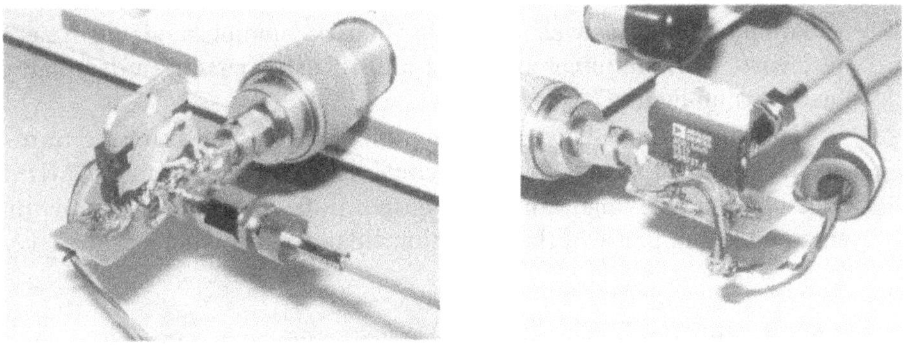

(a) (b)

Figure 7.9 (a) Front and (b) back photos of the amplifier, connected to the RF-out output of the 4395A VNA. Note that the three supply wires are fed through a ferrite ring.

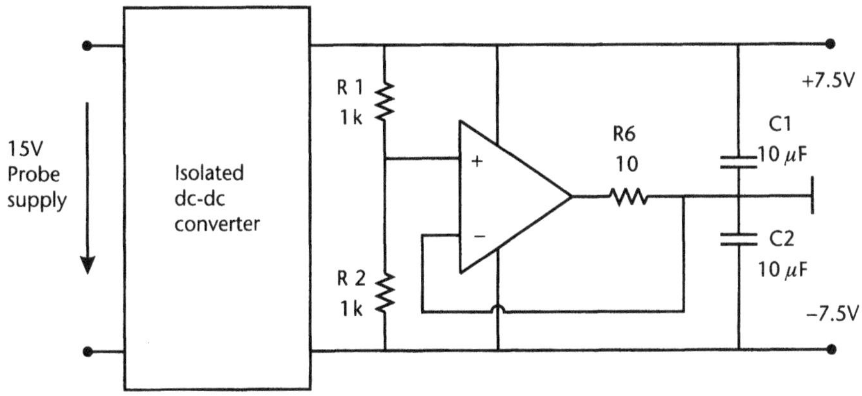

Figure 7.10 Isolated dc source for the differential amplifier, generated from the VNA probe source.

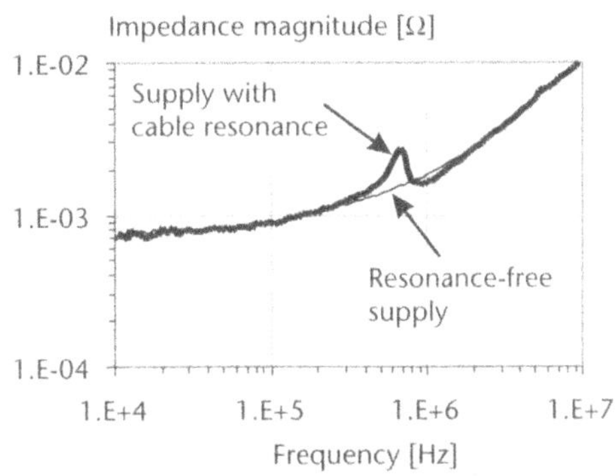

Figure 7.11 Illustration of resonance in the isolation amplifier feed.

figurations of isolation-amplifier feed. The first setup had large loops formed by the VNA and bench-supply power cords and power wires with only 10 μF bypassing at the isolation amplifier. The second setup had an external bench supply feeding the isolation amplifier as well, but it also had a common-mode ferrite on the twisted feed wires of the amplifier and 10 μF and 100 μF bypassing at the isolation amplifier (as shown in Figure 7.9).

The output impedance of the amplifier is close to 50Ω in the entire 0–10-MHz frequency range. Active DUTs (such as dc-dc converters and VRMs) up to a few volts of source voltage can be connected to its output without any problem. The magnitude and phase of the output impedance is shown in Figure 7.12.

7.1.1.4 Other Possibilities to Eliminate Cable-Braid Loop Errors

Instead of homemade accessories connected to a VNA, instrument-grade probes or different instruments altogether can also be used to open the cable-braid loop.

Active differential probe is one of the possible solutions. The key is differential input: regardless of its input impedance, a single-ended probe will not break the cable-braid loop on the ground side. Contemporary differential probes are designed primarily for oscilloscopes, which also provide the power and other controls for the probes. For some of the probes, we can get an adaptor box, which feeds the probe, takes care of the controls, and makes the output voltage of the probe available at a connector. The probe amplifiers usually attenuate the signal; the output stage is not designed to drive large output voltage swings. For this reason, we cannot use these probes between port 1 and the DUT; instead, we have to place the probe between the DUT and port 2. The probe head usually connects directly to the DUT with the shortest pins.

These probes can achieve multigigahertz bandwidth and, therefore, can be very convenient to measure passive DUTs over a wide frequency band. However, the common-mode voltage range of wideband probes is typically limited to a few volts; therefore, active devices can be measured only within those limits.

Dedicated systems for low-frequency floating measurements are also available. The Venable Model 3120 covers the 10-MHz–2.4-MHz range with two isolated floating sense channels (see later in Figure 7.19). The AP Instruments Model 200 Parallel Frequency Response Analyzer covers the 11.6-MHz–15-MHz frequency range with two receive channels, and comes with optional injection isolator and differential probe.

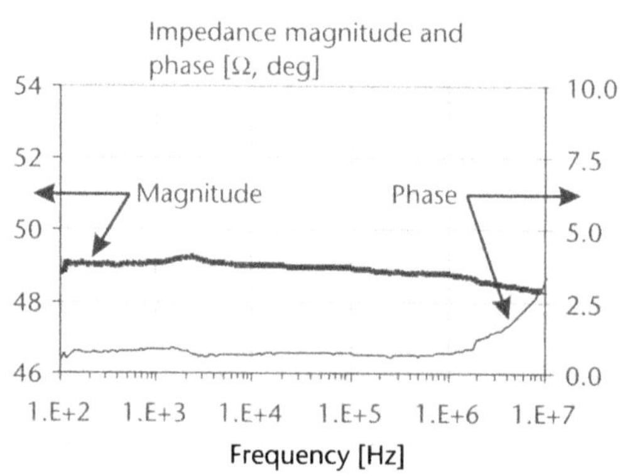

Figure 7.12 Output impedance of the isolation amplifier.

7.1.2 Examples of Correct Connections

In this section, we examine scenarios where the proper connection is important to get accurate low-frequency results.

7.1.2.1 Example: Measuring Low-ESR Bulk Capacitors

As was shown earlier, the impedance magnitude of some low-ESR bulk capacitors may be so low that the cable-braid loop resistance could produce noticeable errors in the measurements unless we break the loop. For lower capacitance values, placing many ferrite clamps or ferrite beads around the cable sleeve may reduce the residual error sufficiently below the measured impedance values. For large capacitance values combined with low ESR, isolation transformer or isolation amplifier is needed.

The two-port shunt-through VNA connection with isolation transformer or isolation amplifier helps to eliminate the measurement error in DUT connections, but the connections of the probes to the device still need to be carefully selected. For through-hole parts, the probes can be soldered or pressed against the capacitor's leads. However, the actual connection points on the leads and their distance from the capacitor body will eventually influence the measured value of resistance and inductance.

Usually we are interested in the performance of the part in the actual environment. The capacitor is connected through vias and optional traces to planes or plane shapes that connect the capacitor to the rest of the PDN; the value we need to determine is the impedance presented by the capacitor to the rest of the PDN. The best way to determine this impedance is to create a small test board with the same stack up and PDN-plane allocation that we have in the final application. The capacitor can be soldered onto or pressed into the test board; then, at a nearby via pair, the impedance can be measured. The additional impedance of the plane connection from the probe points to the capacitor site can be established by measuring the same test board with the capacitor site shorted. Note that this also means that the inductance, and, to a smaller degree, also the ESR of the device will depend on the geometry of the application.

The stack up of the small printed-circuit board in Figure 7.13 has six copper layers, nominally with 2.4-mm (0.093-inch) total board thickness. Every copper layer is nominally 35 μm (1 oz). There are two thin cores with 50-μm (2-mil) dielectrics, 0.2 mm (8 mils) below the surfaces on either side. Only one of the plane pairs is connected to the capacitor pads and test vias. Therefore, the side from which we insert a through-hole part determines whether the plane will be near or far.

The setup used an isolation transformer. The transformer limited the upper bandwidth to 1 MHz; only through-calibration was performed, with a 1Ω resistor soldered across the capacitor site. After calibration, the test boards resistance and inductance were measured by inserting a shorting wire into the through-hole test points.

Separate impedance plots are shown in Figure 7.14 for the short inserted from the near and far sides. The measured impedance profiles for the bulk capacitor inserted from the near and far sides are shown in Figure 7.15 with the extracted ESR and inductance values labeled. The resistance and inductance values measured on

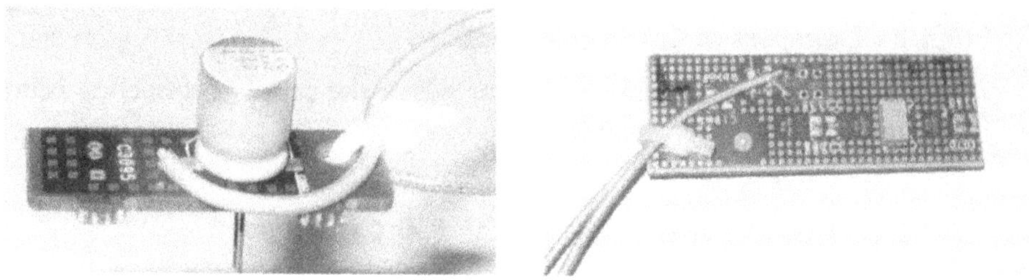

(a) (b)

Figure 7.13 Small multilayer test boards for measuring (a) through-hole and (b) surface-mount low-ESR bulk capacitors.

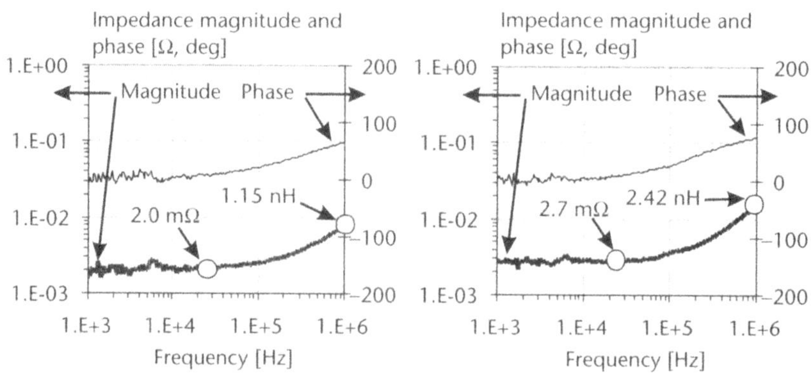

(a) (b)

Figure 7.14 Impedance reading of reference short at the through-hole test points on the small test board shown on Figure 7.13(a). (a) Short inserted from the near side, where the plane pair is 0.2 mm (8 mils) below the surface and (b) short inserted from the far side, where the plane pair is 2.2 mm (85 mils) below the surface.

the shorted capacitor site can be subtracted from the measured capacitor data. ESR on the near side is 7.6 – 2.0 = 5.6 mΩ. ESR on the far side is 8.5 – 2.7 = 5.8 mΩ. Note that the difference can be attributed to the resistance of the lead-length difference between the near and far sides.

The extracted added inductance (added meaning beyond the inductance corresponding to a shorting wire in the given fixture) can be calculated in a similar way. Added inductance on the near side is 3.56 – 1.15 = 2.41 nH; inductance on the far side is 4.72 – 2.42 = 2.3 nH. This inductance is about the same (within measurement errors) from the near and far sides, because the inductance is dominated by the capacitor body. The various inductance definitions will be detailed in Section 8.3.1.

7.1.2.2 Example: Measuring the Output Impedance of DC-DC Converters

The dc-dc converters are switching-mode regulators that convert the incoming (e.g., 48V, 12V, or 5V) dc either to interim bus voltages or to the final regulated voltages. These converters should provide the specified dc current; also, at low frequencies,

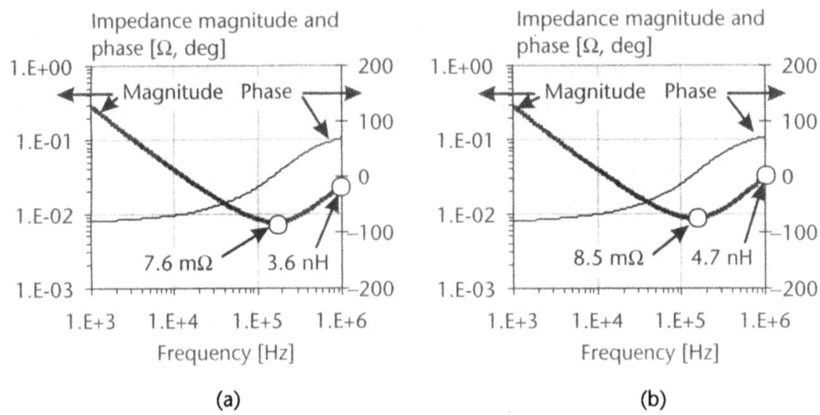

Figure 7.15 (a) Through-hole low-ESR bulk capacitor in the small test board, inserted from the near side. (b) Through-hole low-ESR bulk capacitor in the small test board, inserted from the far side.

we expect them to provide low impedance. The bandwidth in which the low impedance should be maintained by the converter depends primarily on the switching frequency, circuit topology, and loop compensation. Currently, the mainstream switching frequency is several hundred kilohertz per phase; this achieves much greater than 1-MHz effective switching frequency with multiphase converters. The output impedance in the fast point-of-load converters can be as low as a fraction of a milliohm up to at least 10 kHz. On the other hand, VRM response can be significantly nonlinear for large load-current steps. Also guaranteeing the stability of the control loop could be a challenge for widely varying load currents and bypass-capacitor impedances.

For the purposes of PDN design, the most convenient solution is to have the converter remain in the linear region over the entire specified line and load range. When the simulated or measured output impedance is maintained under all operating conditions, there may be no need to consider the large-signal conditions separately.

Measuring the output impedance of a VRM or dc-dc converter requires considerations similar to that of measuring low-ESR bulk capacitors. Since dc-dc converter control loops tend to have high gain at dc, the typical output impedance response is *R*-*L*-like, achieving very low impedances at low frequencies. When using standard VNAs, the cable-braid ground loop must be opened by either transformer isolation or by differential-input amplifiers; ferrite loaded cables are insufficient usually if we want to measure the VRM impedance below a few kilohertz.

If we use the differential-input single-ended output isolation amplifier with its input connected to the VNA port 1 output, the series 50Ω resistor protects the amplifier's output from the dc voltage of the regulator. The proper dc load current of the VRM can be set either with a separate resistive load or by using an active electronic load device.

Isolation transformers can be used to measure an active converter as well. In this case, however, we need to consider further factors. First, unless we use a dc-blocking capacitor, the transformer winding's resistance creates a dc load to the VRM. When we decide to use a dc-blocking capacitor, it has to have large capacitance; therefore, it probably will be bulky. The large capacitor will create secondary resonances with the transformer that will further limit the usable bandwidth.

The isolation transformer shown on the previous figures has a few milliohms dc resistance; this would overload most VRM outputs if connected directly to them. To limit the dc load current, we can insert some length of thin (RG174 or RG178) coaxial cable; the dc resistance of the inner wire will limit the dc load current, while the resistance of the wire is removed from the measured data by calibration.

Unless we use two transformers, to isolate both VNA ports from the VRM, the nonisolated VNA port will get the full output voltage of the VRM. We always have to check the VNA specifications for maximum allowable dc input voltage: usually a few volts is not a problem.

Of course, if the VRM is not powered up, its output impedance can be measured similar to measuring any low-frequency bulk capacitor.

Figure 7.16 shows the setup to measure the output impedance of a small-size socketed converter. The dc input power was connected through wires on the left. The output voltage appeared at multiple pins on the socket. To maintain the low output impedance, a small double-sided thin printed-circuit board was soldered to the appropriate pins: all positive output pins were soldered to the top copper plane, all negative output pins were soldered to the bottom copper plane. The solid copper planes across a thin dielectric separation ensure that the converter output pins are connected to the measurement points through a low-resistance and low-inductance path. The heavy isolated wire on the right shows the connection to the electronic load.

The VNA probes were soldered in the middle of the two-sided plane connection. The remote sensing wires were soldered to the same location.

The dc-dc converter can be measured without input power (inactive loop) or powered (active loop). Figure 7.17(a) shows the output impedance with no input power. It follows the impedance of its output capacitors. Note that if the loop stability requires or assumes external capacitors, those capacitors should also be connected to the output while we measure its powered output impedance.

Figure 7.17(b) shows two versions of the same converter with different loop compensations. All data was taken with no external bypass capacitors. The two versions of active output impedances illustrate the potential problem of instability. One of the curves maintains less than 1-mΩ up to 20-kHz frequencies, but at 85 kHz this

Figure 7.16 Connections to measure the output impedance of a dc-dc converter.

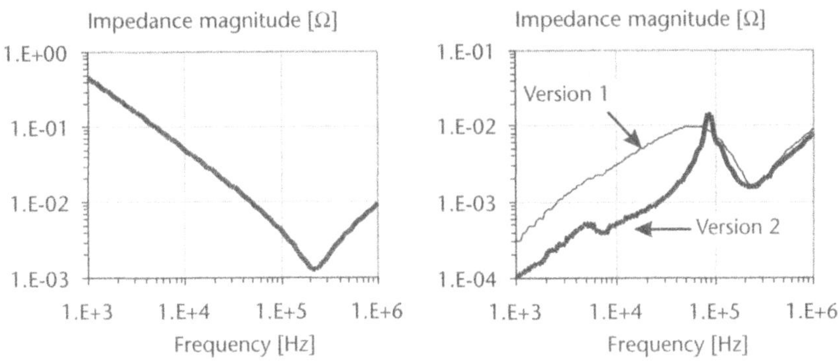

(a) (b)

Figure 7.17 Measured low-frequency output impedance of a VRM with (a) no input power and (b) with input power, for two different loop-compensations.

version has a sharp impedance peak. While the magnitude of the impedance peak itself (near 15 mΩ) may not be a problem, the sharp peak suggests the possibility of a low phase margin, as well as the risk of self-oscillation under a slightly different operating point. In fact, separate analysis of the phase margin of the control loop indicated a low phase margin. This impedance peak is similar to the classic intercapacitor antiresonance peaking: that peak occurs when a bypass capacitor's inductive impedance creates a peak with another bypass capacitor's capacitance. In this case, the inductive behavior comes from the decreasing loop gain in the VRM with increasing frequencies.

The trace labeled "version 1" exhibits no sharp peaking. However, as the relative positions of the two traces show, unless the switching frequency is increased or the control-loop compensation is carefully optimized, the increase of phase margin usually comes with a lower loop bandwidth. In this particular case, the output impedance at 10 kHz increased from 0.4 mΩ to 3 mΩ. Finally, above the peak, starting at about 200 kHz, the output impedances of both versions follow the impedance curve of the output filter capacitors.

Note that while the output impedance profile alone does not provide any direct measure about the loop stability, this data does provide an easy and quick way to qualitatively check stability. Note also that measurement of loop stability usually requires opening up the control loop, but the measurement of the output impedance can be performed without any modifications to the VRM.

The next example shows the output-impedance measurement of a small nonisolated surface-mount dc-dc converter with two different instrumentations. The converter had a nominal current rating of 6A; it had one pin each for input, ground, and output connection. The data sheet of the converter recommends a minimum of 100-μF input capacitance, but the external output capacitor can be omitted. The converter was measured with a 4395A VNA with isolation amplifier in a setup similar to that of the previous figures. The same converter sample was measured with a Venable Model 3120 analyzer also. As shown in Figure 7.18 schematics, the Model 3120 analyzer requires a current-monitoring resistor in series to the unknown impedance. The setup photo is shown in Figure 7.19.

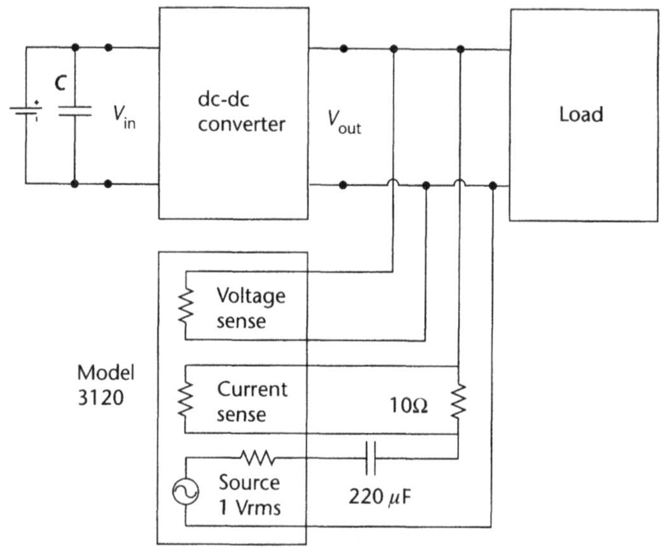

Figure 7.18 Schematics of output-impedance measuring setup with Venable Model 3120.

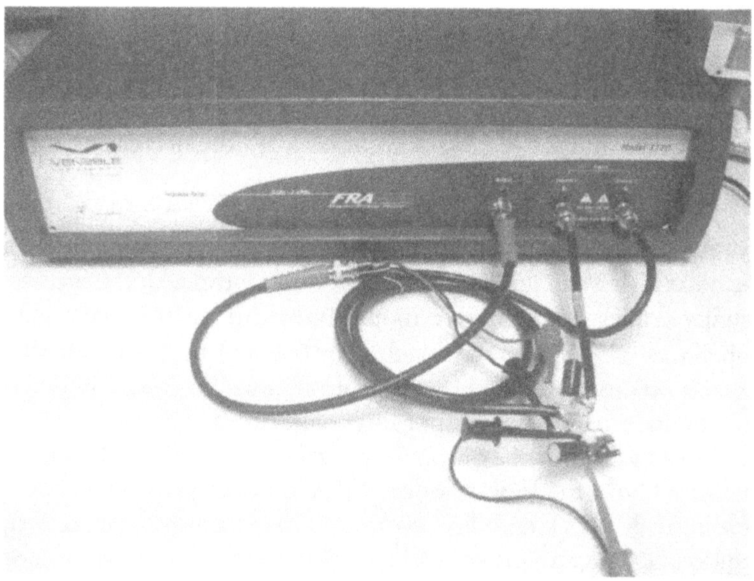

Figure 7.19 Test setup for output impedance measurement with Venable Model 3120. (Courtesy of Texas Instruments.)

The close-up photos of Figure 7.20 show two ways of connecting the cables. The arrows labeled with *R* indicate the current-sensing resistors. Setup 1 has a few millimeters of wire between the output pin of the converter and where the voltage-sense cable is soldered to it. This location is marked by an unlabeled arrow. The resistance and inductance of the wire will appear in series to the measured output impedance. Setup 2 had the voltage-sense cable connected very close to the converter pin, so the extra voltage drop is minimized. The measured impedance magnitudes with 3.3-V

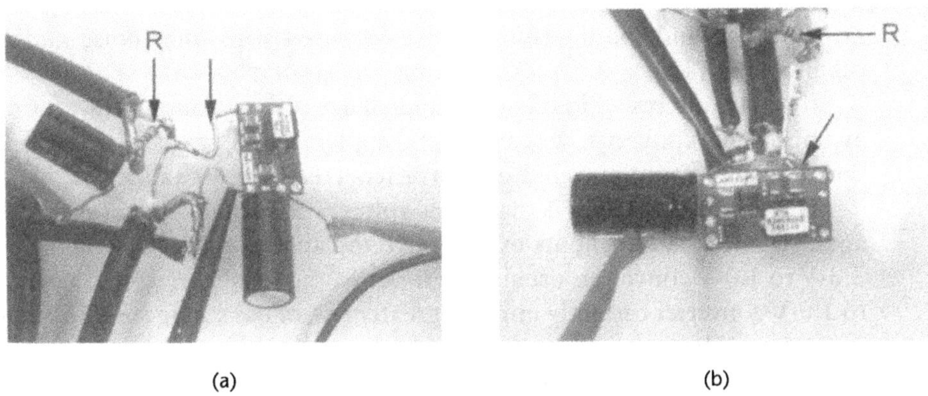

Figure 7.20 Connection geometry: (a) setup 1 and (b) setup 2. (Courtesy of Texas Instruments.)

input, 0.9-V output voltages, 0-A and 5-A dc load current are compared for the three setups in Figure 7.21. The trace marked setup 3 was taken with a 4395A VNA.

Traces from setup 2 and setup 3 have good agreement, though setup 2 still produces a fraction of a milliohm higher impedance below 1-kHz frequency. The trace from setup 1 has errors both at low frequencies from the resistance of the extra wire and at high frequencies from the inductance of the extra wire.

We will show the 3D output-impedance surface and model correlations for this converter in Chapter 9.

7.1.2.3 Measuring the Input Impedance of DC-DC Converters

In frequency-domain analysis, the primary concern for dc-dc converters is their output impedance. In converter designs where the input and output has sufficient built-in filtering, we can neglect the mid-frequency and high-frequency interaction between the input and output. With higher densities and reduced on-module filtering, the converters become more transparent, so it may become necessary to measure their input impedance and transfer function also from input to output in both directions. This need is amplified further in applications in which dc-dc converters

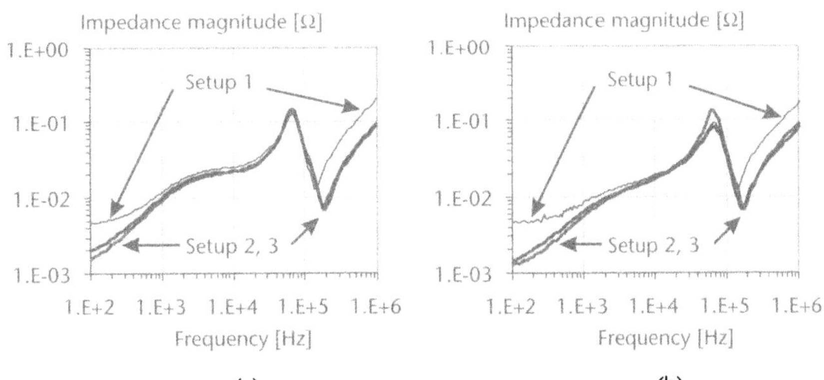

Figure 7.21 Output impedance of a nonisolated dc-dc converter, measured in three different setups: (a) 0-A load current and (b) 5-A load current. (Data for setup 1 and setup 2 courtesy of Texas Instruments.)

are cascaded and the interim voltages are used with tight noise and regulation requirements [1].

If we are interested in the input impedance or input-output transfer function of an unpowered converter, we can directly apply the findings of Section 5.4. However, the need to measure powered-up converters creates new challenges.

One of the challenges is that input voltages are usually higher than output voltages, so not all VNA inputs may survive the applied dc voltage. When we have a 3.3-V to 1.0-V converter, usually the input voltage is not an issue. However, a 12-V to 1.0-V converter certainly imposes greater risk of damaging the VNA input; inputs of 24V or 48V are beyond safe limits for most VNAs.

A second challenge is the connection. For output-impedance measurements, the load can be a resistor or current source. Due to the regulation loop, at low frequencies the output behaves like a voltage source with much lower resistance than the load. The input of the converter, however, cannot be fed with a high-impedance current source. Instead, it must be fed by sufficiently low impedance that is in parallel to the input impedance of the converter we want to measure. A low-impedance dc source with good bypassing would shunt out most of our test current. The solution is to have a floating separation between the source and receiver; this separation not only eliminates the cable-braid loop but also allows us to inject a known test signal in series to the input. If the power source feeding the input of the DUT is floating, we can inject the test signal on the low side, without stressing the isolation. When the power source has a common return with the VNA return, we need to inject the test signal on the high side.

Figure 7.22 shows one possible solution. We use a low-frequency VNA with a source and two receivers, for instance, Agilent 4395A or Ridley AP200 parallel frequency response analyzer or Venable Model 3120. There is an *N*:1 step-down transformer to inject the test signal with low source impedance after the dc source is bypassed with capacitor C. There is a small-valued series resistor, *R*s, in the primary path of the step-down transformer, which allows us to measure the current entering the transformer. The two inputs of the VNA measure the voltages across the input of the dc-dc converter and across the current-sensing resistor. Assuming an ideal transformer and neglecting the receiver input impedance in parallel to the DUT, the input impedance is:

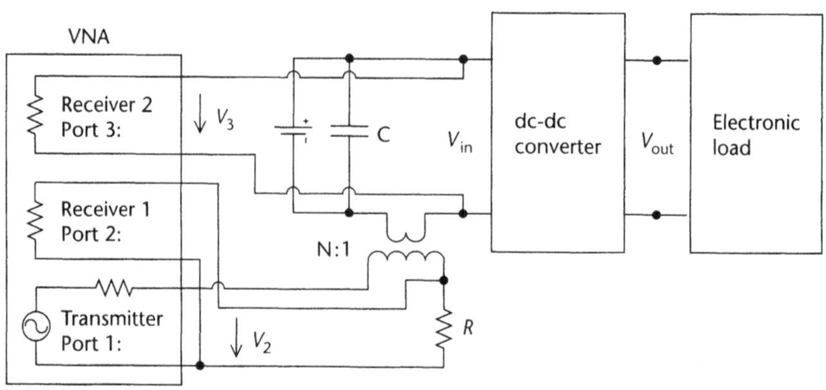

Figure 7.22 Injecting test current to measure input impedance.

$$Z_{in} = \frac{V_3}{V_2}\frac{R}{N} \tag{7.2}$$

A homemade 50:1 transformer with a small-size dc-dc converter is shown in Figure 7.23.

The nonideal response of the step-down transformer can be characterized separately and its effect can be removed from the measured data. Note that the low-frequency, input impedance of regulated dc-dc converters with constant load is a negative resistance [2]. This negative resistance simply follows from the fact that the regulation keeps the output voltage constant; so, if the input voltage increases, the input current decreases. The input impedance of the converter from Figure 7.23 is shown in Figure 7.24. The input dc voltage and load current were 3.3V and 3A, respectively.

7.1.2.4 Measuring Transfer Functions of DC-DC Converters

In addition to the small-signal output and input impedances of dc-dc converters, the transfer functions may be equally useful and important to measure. By knowing the input and output small-signal impedances and by measuring the S_{21} small-signal transfer parameter, we can calculate the transfer impedance from input to output and from output to input, as in Section 5.4. Since the converter is an active and nonreciprocal device, the transfer impedances in the two directions are not necessarily the same.

The typical converter parameters and application circumstances, however, make other forms of description more attractive. As we saw above, the input of the converter has to be fed with sufficiently low impedance, whereas its output can be approximated as a voltage source for typical loads. When considering the noise propagation from input to output, we can assume that the noise source creates a certain noise voltage across the input terminals; we are interested in the noise generated at the output. This requires an input-voltage–to–output-voltage transfer ratio,

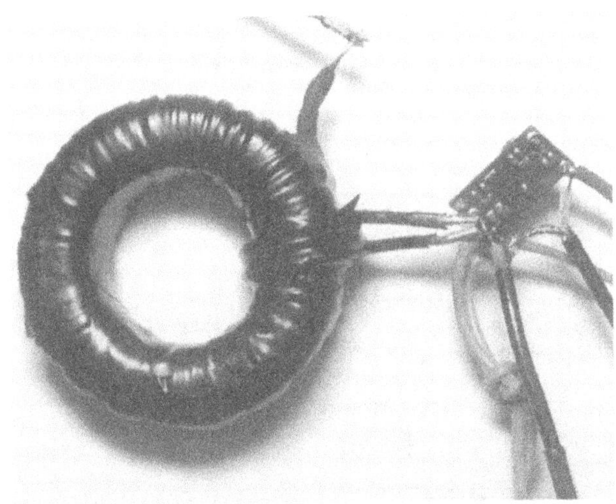

Figure 7.23 Homemade toroid 50:1 transformer with small-size dc-dc converter for input-impedance measurement.

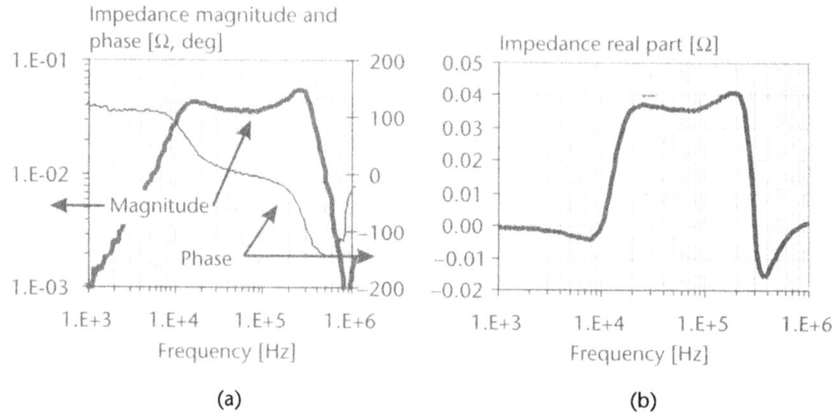

Figure 7.24 Input impedance of the dc-dc converter from Figure 7.23: (a) impedance magnitude and phase and (b) real part of impedance.

rather then transfer impedance. When we consider the propagation of noise from the output toward the input, it is more illustrative to obtain the output-voltage to input-current transfer function, which is transfer admittance.

Figure 7.25 shows the setup scheme for measuring the input-to-output voltage transfer ratio.

At low frequencies, both the output impedance and the input and source impedances are typically much lower than 50Ω; therefore, we do not need to de-embed the VNA impedance. For the same reason, either floating differential inputs or isolated outputs are required to open up the potential cable-braid loop. This stresses the dynamic range of receiver 1; and if there is attenuation from the input to the output, it stresses even more the dynamic range of receiver 2. For this reason, the measured voltage-transfer ratio may be noisy. If the two inputs track together, and the cable attenuations to the two inputs are similar, we do not need calibration (since we are measuring a ratio). The input-to-output voltage transfer ratio then simply becomes:

$$V_{transfer} = \frac{V_3}{V_2} \tag{7.3}$$

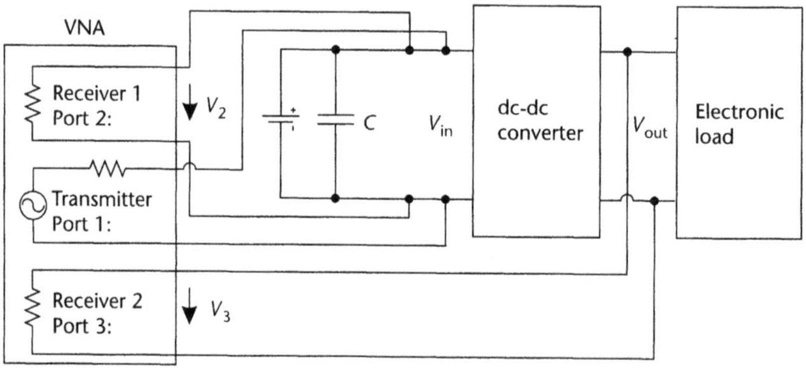

Figure 7.25 Measurement connection scheme for input-to-output voltage transfer ratio.

An illustrative result is shown in Figure 7.26. A small-size nonisolated dc-dc converter was measured in the setup of Figure 7.25. The voltage transfer ratio was measured under four conditions: converter without input power (off) and input power applied with three different dc load currents, namely, 0A, 3A, and 6A (on). The three on traces run on top of each other; this indicates that the voltage transfer ratio has only a very weak dependence on the dc load current.

The port 1–port 2 connection in Figure 7.25 appears to be similar to a two-port shunt-through connection scheme; so it may appear to be suitable to measure the input impedance of the converter. In fact, if we are interested in the input impedance of the converter in parallel to the impedance of its dc source, this is the setup we should use. However, as it was mentioned in Section 7.1.2.3, converters require low source impedance: so, we would actually be measuring the input impedance of the converter in parallel to the low source impedance. We could measure the impedance at the converter's input with the source and converter present but disabled; then we could de-embed this value from the impedance measured at the same point with the converter enabled. However, the difference measured is usually small, and the result is noisy.

To measure the output-to-input transfer admittance, we have to apply an ac test voltage across the converter's output and then measure the corresponding ac current at the input. To measure the current at the input, we can use a step-up transformer similar to one we used to measure the input impedance in Figure 7.22. The setup is shown in Figure 7.27. With a 50:1 turns ratio the 50Ω VNA impedance is transformed into $50 / 50^2 = 0.02\Omega$ in series to the converter input.

The resistance R represents the load resistance to the current-sensing transformer. If receiver 2 is a high-impedance input, we need to connect a separate load resistor across the transformer's output. When we use a VNA with a 50Ω input impedance, it can serve as a load resistance to the transformer. Unless the instrument has floating inputs, we need an isolation transformer or amplifier to break the cable-braid loop between port 1 and port 2. With the instrument, we need to measure the ratio of V_2 and V_3. If the two inputs track together and the cable

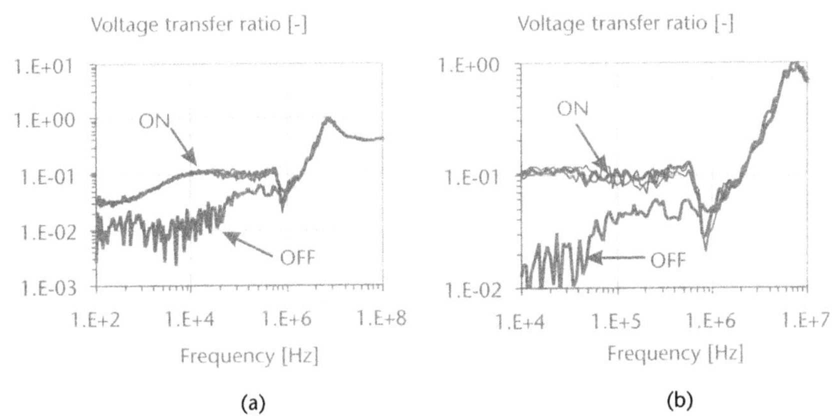

Figure 7.26 Measured input-to-output voltage transfer ratio of a nonisolated dc-dc converter: (a) full 100-Hz–100-MHz frequency range and (b) same data on zoomed horizontal and vertical scales.

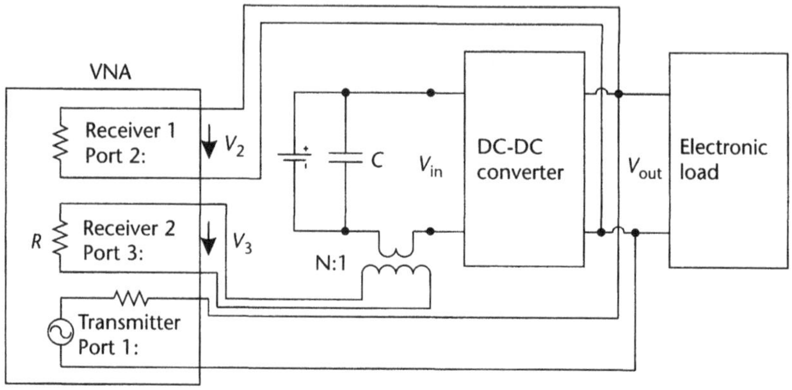

Figure 7.27 Setup to measure output-to-input transfer admittance with current-sensing transformer.

attenuations to the two inputs are similar, we do not need further calibration (since we are measuring a ratio). The output-to-input transfer admittance is:

$$Y_{transfer} = \frac{V_3}{V_2}\frac{N}{R} \tag{7.4}$$

The nonideal response of the step-down transformer can be characterized separately, and its effect can be removed from the measured data.

We can eliminate the N:1 current-sensing transformer from the setup by using a small-valued current-sensing resistor to measure the current on the input side. The setup is shown in Figure 7.28.

The R_s current-sensing resistor has to be of sufficiently small value so that the input of the converter still encounters low source impedance. Moreover, unless the instrument has floating inputs, we need an isolation transformer or amplifier to break the cable-braid loop between port 1 and port 2. With the instrument, we need to measure the ratio of V_2 and V_3. If the two inputs track and the cable attenuations to the two inputs are similar, since we measure a ratio, we do not need calibration. The output-to-input transfer admittance is:

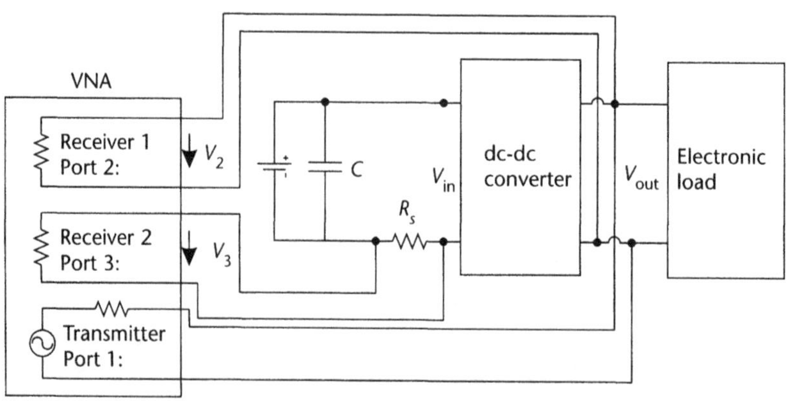

Figure 7.28 Setup to measure output-to-input transfer admittance with current-sensing resistor.

$$Y_{transfer} = \frac{V_3}{V_2}\frac{1}{R_s} \quad (7.5)$$

As an illustration, we show output-to-input transfer admittance curves measured with the setup shown in Figure 7.28. The measured data is shown in Figure 7.29 for the same converter that was used for Figure 7.26. The R_s sense element was a precision 10-mΩ shunt resistor. The transfer admittance was measured with 0-A, 3-A, and 5-A dc load currents. The traces for the different load currents run close; but, in contrast to the input-to-output voltage transfer ratio shown in Figure 7.26, the three curves show a small but definite difference. The difference increases near the switching frequency.

7.1.2.5 Measuring Loop Stability of DC-DC Converters

The traditional way to evaluate the performance of small-signal control-loop performance is to examine the open-loop and closed-loop transfer functions of the control circuitry. On the loop-transfer functions, the gain margin and phase margin are defined and measured. While gain and phase margins provide intuitive insight of the loop's small-signal health, there are two practical limitations for the end users. First and foremost, the direct way of measuring loop stability requires the injection of a test signal in series to the loop. This requires the opening of the loop; with analog off-the-shelf converters this is not easy. Second, due to the nonlinear relationship between the loop transfer function and the small-signal output impedance (which is the primary target parameter for the end user), there is no simple method of extracting the output impedance from the loop transfer function without knowing the internal parameters of the converter.

Therefore, we do not cover the classic loop-stability measurements. The interested reader is referred to the widely available literature (see, for instance, [3, 4]).

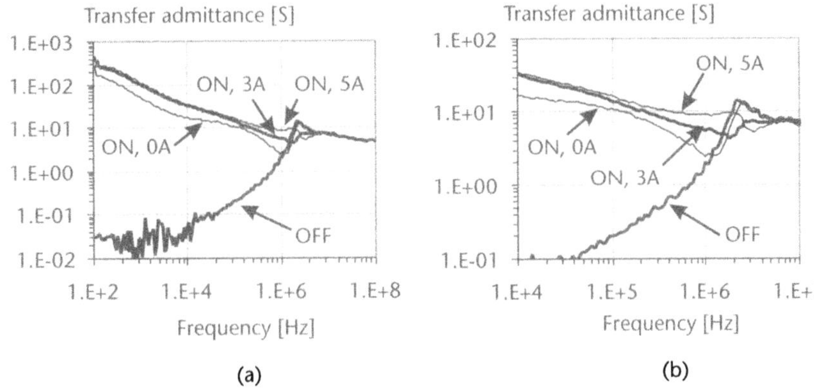

Figure 7.29 Output-to-input transfer admittance measured on a nonisolated dc-dc converter with three different load currents and without input power: (a) full scale and (b) an enlarged view of horizontal and vertical scales.

7.1.2.6 An Exception: Time-Domain Check of DC-DC Converters

There are a few exceptions when time-domain measurements on a PDN are more useful than frequency-domain measurements. One such exception is the time-domain noise of dc-dc converters. High-frequency ringing on the switching edges tends to be low in duty cycle and may not be fully periodic. For these reasons, identifying the ringing in the frequency domain is difficult. The resulting short bursts with changing frequency create smeared spectral peaks which may be hard to distinguish from the background noise. Moreover, in such cases, the peak-to-peak time-domain noise value is probably of higher interest.

An illustration is shown in Figure 7.30. The noise was measured with an oscilloscope and a spectrum analyzer. The DUT was a 25-A nonisolated point-of-load dc-dc converter, with 12-V input and 2.5-V output voltages, which is common to feed DDR-I memory. The ripple at the 350 kHz switching frequency was 30 mVpp. On the switching edges, however, a burst ringing with multiple frequencies appears. Figure 7.30(a) shows the close-up of the ringing, which reaches 174 mVpp. The fundamental ringing is 50 MHz with additional ringing components at 170 MHz and 220 MHz. The spectrum of the output voltage shows peaks at these frequencies, but without seeing the time-domain transient waveform, the low levels and smeared nature of these peaks would not necessarily be alarming.

7.1.3 Measuring Low Impedances at High Frequencies

As it was shown earlier, cable performance limits the two-port VNA measurements at low frequencies due to the cable-braid ground loop. At high frequencies, in addition to the noise floor of the instrument itself, the finite surface transfer impedance and resonances in the cable braids set the error floor.

As PDN impedances are seldom close to the 50Ω instrument impedance, there is usually a large mismatch and reflection at the connection to the DUT. Through the finite surface transfer impedance of the cable braids, the reflections and leakage show up as erroneous peaks in the response.

Figure 7.31(a) shows the equivalent impedance reading from two coaxial cables of type RG178 connected to a VNA port with their far ends shorted with SMA caps. The two cables were about 5 cm (2 inches) apart. The same cables in the same posi-

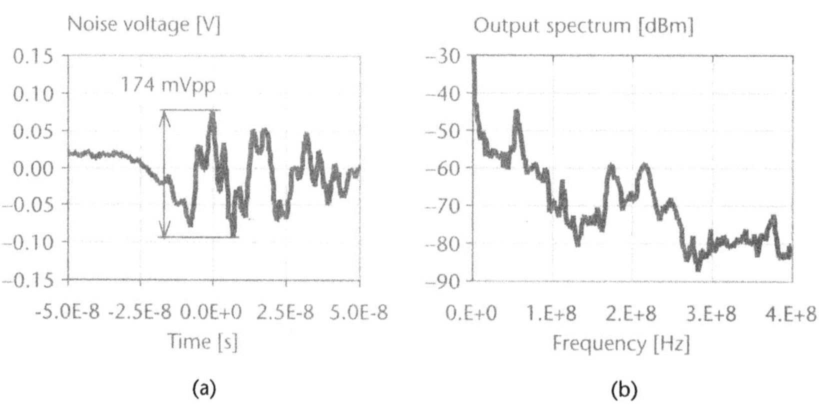

Figure 7.30 (a) Time-domain and (b) frequency-domain high-frequency noise at the output of a dc-dc converter.

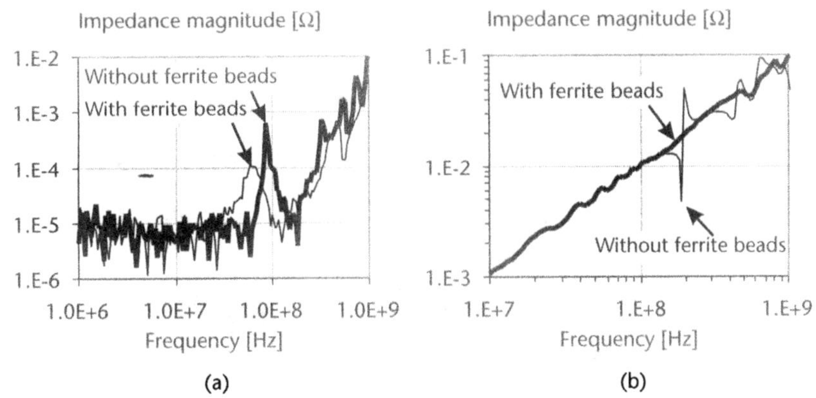

Figure 7.31 Illustration of cable performance at high frequencies. (a) Equivalent impedance magnitude with two-port VNA measurements, the two cables are shorted at their ends, 5 cm (2 inches) apart, with and without ferrite beads. (b) Residual impedance magnitude measured on plane short with semirigid coax probes, with and without ferrite beads on the coaxial cables (shown in Figure 7.4).

tion were also measured with a different number of small ferrite clamps placed around them. Figure 7.31(b) was measured across a plane short in a medium-size PCB, with cables and probes shown in Figure 7.32. The slope corresponds to the impedance magnitude of a 1.5-pH inductance. The few pH equivalent inductance of the resonance-free error floor is achievable only if we cover the entire length of the cable braid with ferrite absorber sleeves, place ferrite sleeves around the semirigid probe, and isolate the probe-holder metal parts from the cable.

There are coaxial cables available with good surface transfer impedances, which would reduce this error. Unfortunately, cables with better quality braids and shields tend to be more expensive as well as heavier, bulkier, and stiffer. When we have to reach with two probes inside a larger system or to connect to a premounted fixture, flexible cables are essential.

Figure 7.33 shows a large probe station with high-performance coaxial cables. When measuring low-inductance PDN components, this setup still requires a few large ferrite clamps along the cable to further suppress cable resonances.

Figure 7.32 Ferrite-covered flexible coaxial cable and semirigid probe with ferrite sleeve. For lowest residual error, the probe and probe holder metals are isolated in the probe-holder head with plastic foam. The probe holder is marked by an arrow.

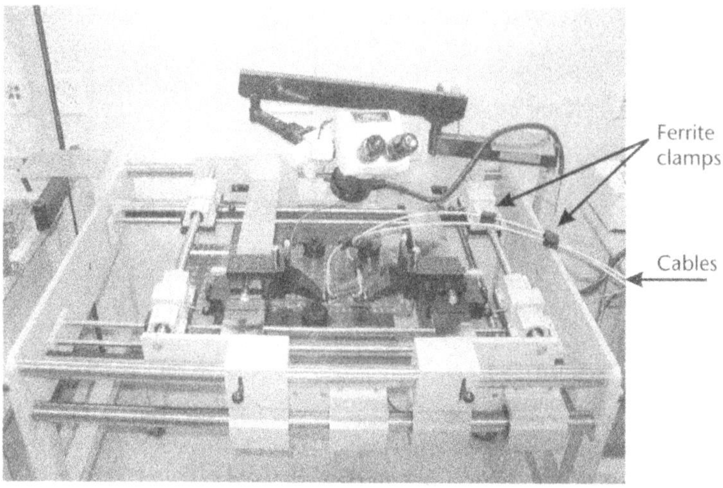

Figure 7.33 Large probe station with high-performance coaxial cables with a few ferrite clamps. Note the ferrite clamps at all locations where the cables otherwise would touch either each other or the metal frame of the probe station.

7.2 Making the Proper Measurements

In PDN measurements, especially if we have probes instead of connectors in the setup, we need to be especially cautious to ensure that that the measured data is valid and correct (within acceptable error limits). The first (and best) line of defense is to have reasonable expectations; when the result we get is unexpected, we should check and double-check the result. We can measure the DUT several times, possibly with different instruments and settings, and separately check the noise and error floor of the measurement setup.

7.2.1 Multiple Measurements, Multiple Instruments

If possible, it is good practice to take multiple measurements on the same DUT to reduce the chance of errors in calibrations, connections, and setups. This practice is not recommended to enhance the absolute accuracy of the measured data but rather to ensure the integrity of the measured data. Without coaxial connections on power distribution systems, our test results are not traceable; therefore we need other means to verify our results.

Multiple measurements are forced when the target frequency range is too wide to be covered with one instrument. In such cases we need two (or sometimes three) independent instruments with their own calibrations to get all the data. In such cases, we can check the integrity of the data at the frequencies where the different datasets are joined or overlap. If there is overlap in the frequency ranges from the different instruments, the overlapping regions should exhibit a smooth continuation and good agreement.

Figure 7.34 is an illustration of joining two sets of measurement data on the same DUT, a small capacitor test site with shorted pads. Low-frequency measurements were done in the 100-Hz–10-MHz frequency range. A high-frequency measurement was taken in the 1-MHz–1.8-GHz frequency range. Note that all three sets

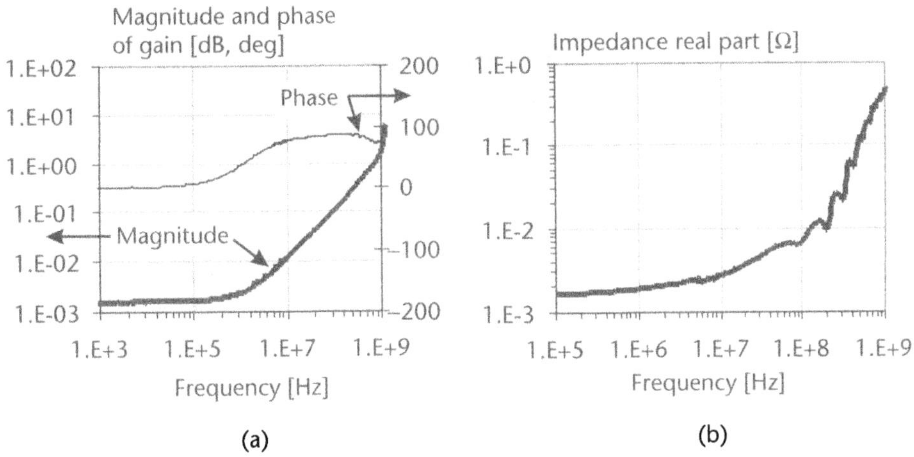

Figure 7.34 Joining two good sets of measurement data on a shorted capacitor test site: (a) impedance magnitude and phase and (b) impedance real part.

of traces (magnitude, phase, and real part) continue smoothly without a significant jump or offset.

As an illustration of joining datasets with insufficient accuracy, Figure 7.35 shows the measurement data on a similar shorted capacitor test site: one of the two measurements was taken with insufficiently clean probe pins; this compromised the accuracy. The impedance real part data has a jump around 5 MHz where the low-frequency dataset ends and the high-frequency dataset starts. Note, however, that even though the impedance real part shows a noticeable jump, the discontinuity on the impedance magnitude and phase traces is harder to notice.

7.2.2 Averaging, Smoothing, and Bandwidth

VNAs detect the test signal synchronously; this makes it possible to suppress independent, random noise by averaging. Averaging is useful when we have very low or very high impedance values to measure with the two-port shunt-through connec-

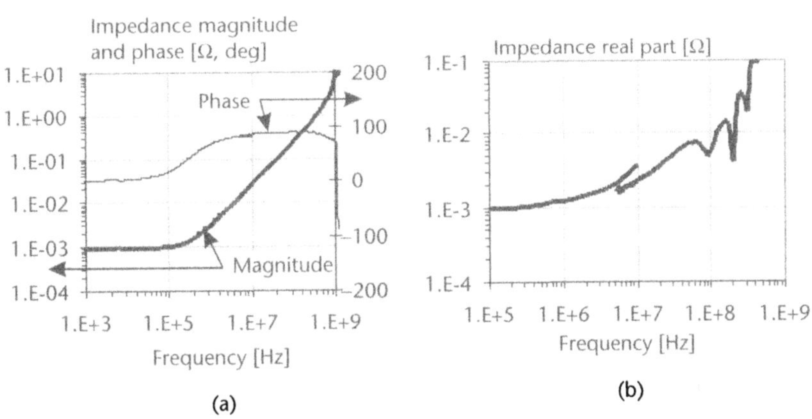

Figure 7.35 Joining two compromised sets of measurement data on a shorted capacitor test site: (a) impedance magnitude and phase and (b) impedance real part.

tion. Figure 7.36 shows the output impedance of a powered dc-dc converter with instantaneous reading and with 128-times averaging. The nonaveraged impedance magnitude shows the sign of random noise below 10 kHz where the impedance magnitude falls below 2 mΩ. The graph with 128-times averaging has smooth traces at all frequencies, confirming that the noise was in fact random and independent of the test signal.

Averaging is effective when the sweep is fast. If one sweep takes several seconds or possibly minutes, averaging may take very long. With the setting used to take data for Figure 7.36, one sweep lasted 47 seconds. In such cases, we should consider applying a moving average (also called smoothing) to the raw data; in a moving average each data point is replaced with the average of the data point itself and a certain number of nearby data points. Figure 7.37 shows the same data we had in Figure 7.36(a) with a 3-point and 11-point moving average. The 11-point smoothing suppresses the noise almost as well as the 128-times averaging. Note, however, that smoothing works only if we do dot have sharp peaks or dips in the data (because those would be smoothed out as well).

The measurement bandwidth also influences the noise. In Section 6.4.2, we showed that in a well-maintained instrument we may not need to recalibrate if we

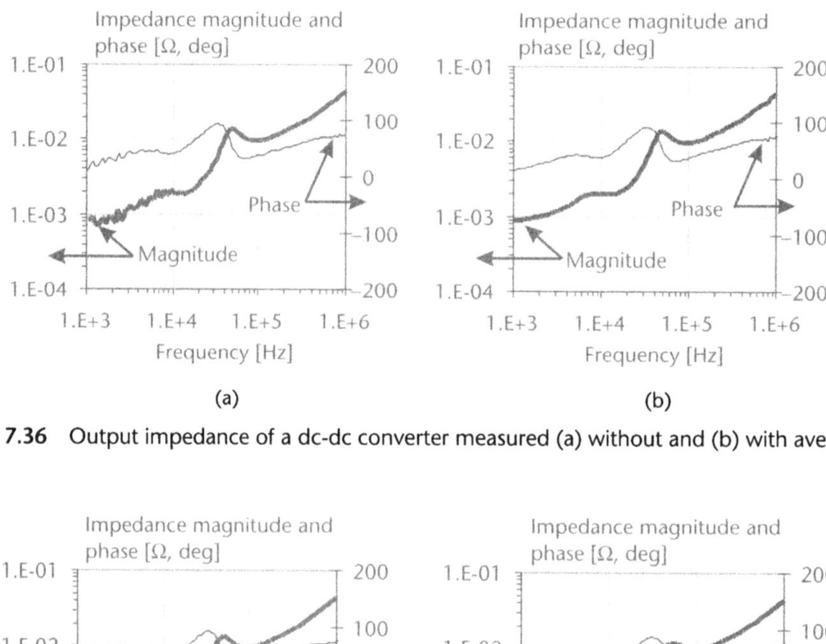

Figure 7.36 Output impedance of a dc-dc converter measured (a) without and (b) with averaging.

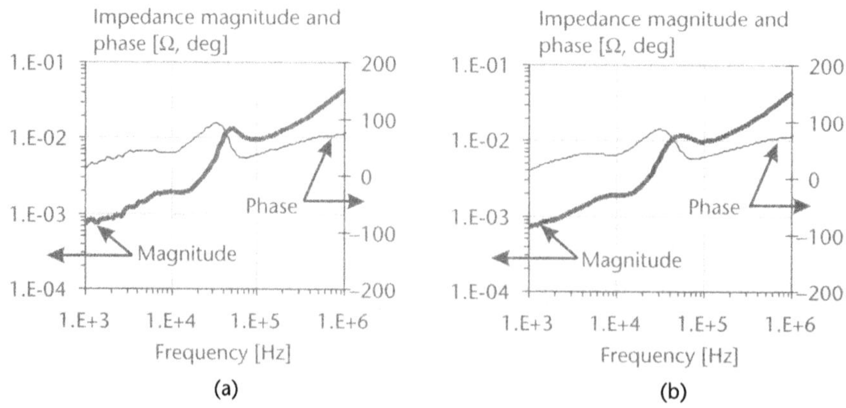

Figure 7.37 Effect of smoothing. Data from Figure 7.36(a) with (a) 3-point and (b) 11-point moving averages.

change the IF bandwidth. To speed up the sweep, we may increase the IF bandwidth during measurement. With very low impedance values to measure and with increased bandwidth, the instrument noise will become the limitation. Figure 7.38 shows the impedance of a metal shorting bar in the 10-kHz–100-kHz frequency range and with different measurement bandwidths, taken with an Ultimetrix P4800 VNA. Note that data taken with 10-Hz intermediate frequency bandwidth (IF BW) follows a smooth curve on the logarithmic frequency scale; but, as the IF BW increases, the trace noise introduces ripples in the measured impedance magnitude. Note also that in this entire frequency range the measured impedance stays below 0.5 mΩ.

In dedicated instruments, such as impedance analyzers, the bandwidth of the measuring hardware may be fixed, so the user may have only a limited ability to change it. In general-purpose instruments, such as VNAs and spectrum analyzers, the measurement bandwidth is user selectable. VNAs usually have a choice for the bandwidth of the intermediate frequency circuits (IF BW). Some spectrum analyzers also have a settable bandwidth for the video postprocessing of the data. When selecting an IF bandwidth, we are often choosing between measurement speed and noise floor. In general, narrower bandwidth results in lower noise floor but longer sweep time. Narrow measurement bandwidth also helps to suppress spurious deterministic components. When a live system is measured, the clock signals and the switching ripple of dc-dc converters may appear at fixed frequencies. The IF bandwidth and sweep parameters (number of points, sweep type, start and stop frequencies) determine whether these constant-frequency signals are captured in the measured data. Figure 7.39 shows two sets of data taken on the same PDN where a dc-dc converter was powered and running. The dataset in Figure 7.39(a, b) was taken with a bandwidth of 30 Hz and 10 Hz, respectively.

The peak seen in Figure 7.39(a) can be eliminated either by reducing the IF bandwidth or by changing the sweep parameters so that there is no measurement frequency point at the switching frequency.

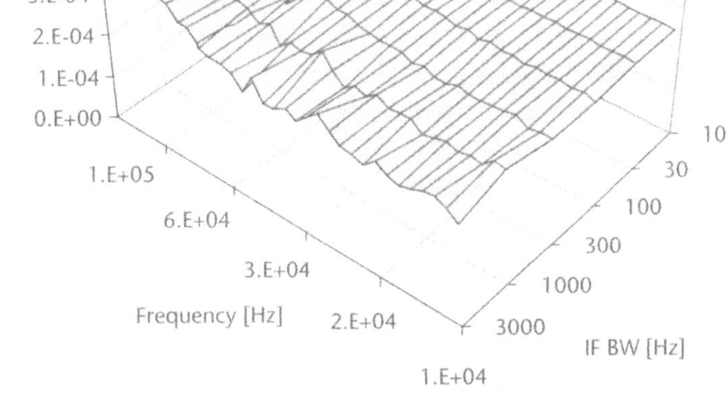

Figure 7.38 Measured impedance versus bandwidth on a metal shorting bar.

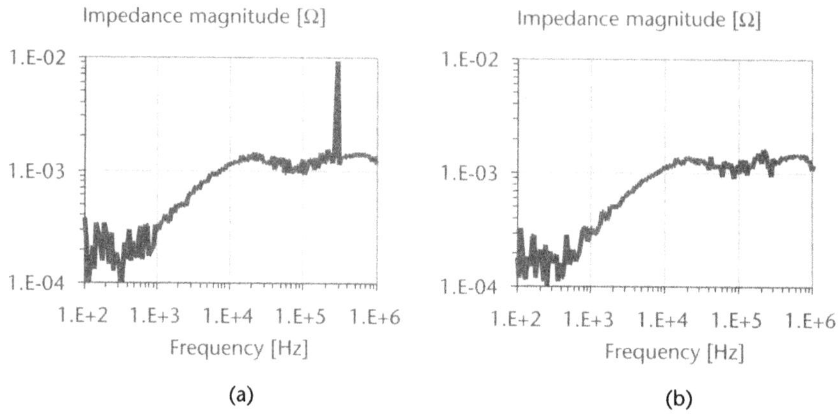

Figure 7.39 Switching ripple of dc-dc converter showing up in the output impedance data. IF bandwidth is (a) 30 Hz and (b) 10 Hz.

Note that interfering peaks originated from discrete-frequency signals show up as sharp peaks; we should not confuse this peak with broader peaks associated with antiresonances or loop instability.

7.2.3 Background Noise, Noise Floor

As the DUT impedance decreases, eventually we hit the noise floor of the instrument. We can obtain the noise floor of the setup by taking the reading after calibration with port 2 shorted and otherwise unconnected. Figure 7.40 shows the noise floor converted to self-impedance on two different low-frequency VNAs. Note that in generic VNAs the impedance noise floor may strongly depend on the instrument settings (among others, IF bandwidth, port attenuation, and power level).

7.2.4 Repeatability of Data

In making measurements on low-impedance structures, the repeatability of the connections has to be examined. There is a valid concern that the finite contact resis-

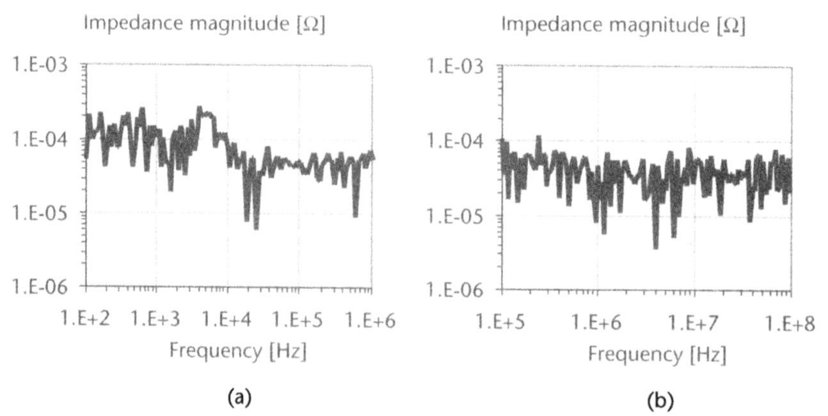

Figure 7.40 Noise floor, converted to self-impedance magnitude, on (a) a low-frequency VNA and (b) mid-frequency VNAs. Note the different horizontal scales.

tance between the probe tips and DUT connection points may alter the data. To illustrate the effect, a low-ESR bulk capacitor was measured repeatedly in the same small PCB fixture without soldering. The sample was inserted into the through-holes of the fixture; then it was held in place by slightly twisting the capacitor body by hand so that the leads made connections to the through-hole walls. Figures 7.41 and 7.42 show the setup and the result.

Note that the extracted impedance minimum (ESR) shows a variability of about 1 mΩ, but the inductance reading is very consistent. This suggests that to measure low-ESR bulk capacitors or active VRMs, soldered connections may be needed. But at higher frequencies, where inductance dominates the path, the repeatability may be sufficient without soldered connections.

7.3 System Measurements

Having looked through the fundamentals of PDN measurements, we can apply the rules and tools to take measurements on more complex DUTs. In this section, we show measurement results of full boards, unpowered and powered. Additional measurements of packages, and different combinations of boards, packages, and active devices will be shown in Chapter 9.

To gain confidence in the data and to provide cross correlation of the measured information, it is useful to measure as many of the building blocks separately as possible and then to measure them combined. On a full board, this means we can measure the bare board first and separately one of each type of components on the selected supply rail: the dc-dc converter, bypass capacitors, and packaged chips. Bypass capacitors are the smallest entities for these measurements. With dc-dc converters and packaged chips, however, we may repeat the procedure: measure the building blocks first, then measure the full module. If we use third-party dc-dc converter modules, we can treat it as a single component (since we may not want to do measurements inside the module). In embedded dc-dc converters, or if we design and build our own dc-dc converter modules, we can measure the properties of the

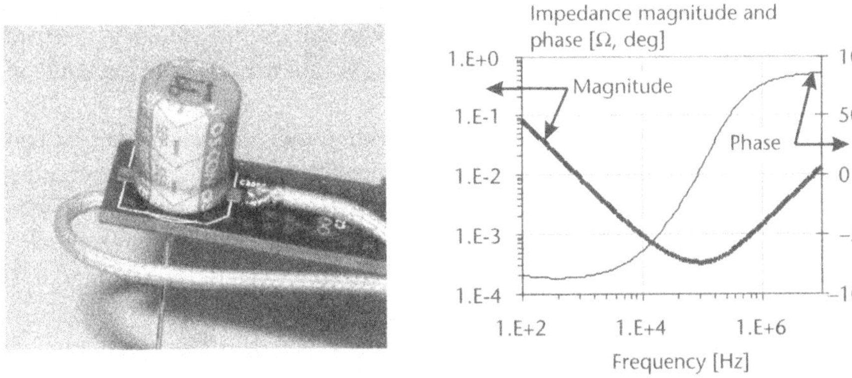

(a) (b)

Figure 7.41 Setup to check repeatability of data at low frequencies in fixture without soldering: (a) photo of the fixture with one of the samples inserted and (b) representative impedance magnitude and phase versus frequency.

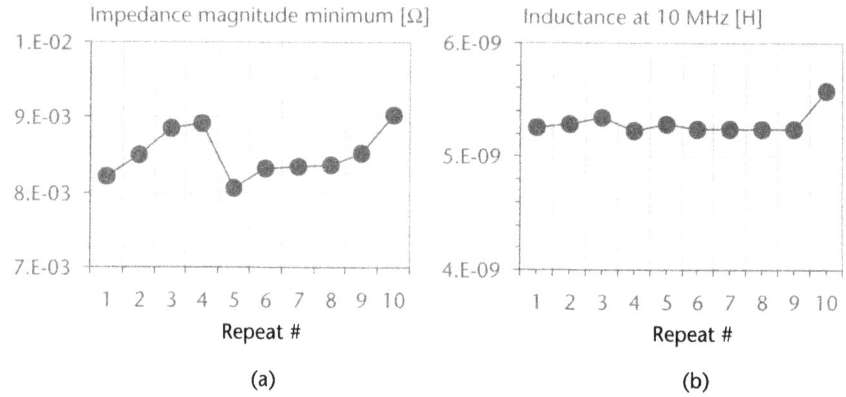

Figure 7.42 (a) Minimum of impedance magnitude, and (b) extracted inductance at 10 MHz. The same sample piece was measured ten times in the same fixture, without soldering.

bare module board and components before we measure the finished module. Similarly for packaged chips, we can measure the bare package followed by the package-silicon combination. Each component category has its own special measurement requirements for operating conditions and frequency range.

After the components are measured, we measure the populated module or populated board. We can measure the dc voltage drop on high-current rails and then measure the impedance profiles at selected points. We can measure the populated boards without power applied, so that dc-dc converters and packaged devices show their "cold" impedance, and also with input power applied. As shown earlier, except for low-frequency measurements (below 1 MHz), we need dedicated test points to make accurate measurements. This limits us to measuring the supply rail at locations where test points had been placed or can be created. Eventually we need to repeat the measurement process for each supply rail we want to characterize. Dependent on the system design, we may also need to measure the interaction between and among different supply rails.

The board can be measured with and without major active devices attached. If major chips should be included in the measurements, the same step-by-step approach can be applied: first, we measure the packaged chips and populated board separately; then we measure the board with the chip(s) attached. The same approach can be followed for packaged chips as well: first measure the package without active device then with active device.

When active devices are involved in powered measurements, care should be taken to ensure that the active devices get the proper initialization, input/output terminations, power-up and power-down sequencing, so that the devices do not get damaged even if functionality is not necessarily required for the PDN measurements. Unless we measure a fully functional system, sometimes it is difficult to ensure all of these conditions.

For a complex system, the frequency range of interest may cover several decades, from (nearly) dc to several gigahertz. The low end may be in the kilohertz region for unpowered systems and tens or hundreds of hertz for powered systems. The upper end depends on the size and functions of the structure: larger size boards with conventional parallel buses may require a few hundred megahertz upper end.

Smaller boards, packages, memory modules (especially if they have to handle Gbps signaling) may require an upper end of several gigahertz. At present, this very wide band cannot be covered with one instrument. Instead, we need to use different instruments to cover the full range; this allows us to cross correlate data from the independent instruments in overlapping frequency bands.

Typically, the purpose of full-system measurements is to validate a design. At the same time, we can create and validate models for the building blocks and for the entire PDN system. In validation and characterization, we need to keep in mind the three major functions of PDNs: providing clean power to devices, (optionally) providing sufficient return path for signals, and ensure that the PDN does not contribute to EMI problems. To validate the clean-power performance, we need an impedance target profile. Since we do not cover the PDN design processes in this book, we assume that the target to verify is known and available. Even if we do not have an impedance target, we want to check for unnecessary peaks or resonances in the PDN impedance. By doing so, we automatically eliminate the most likely EMI offenders. The measurement of return-path functions of high-speed signals requires measurements on PDN together with signals traces. It usually involves the measurement of the integrity of the power-ground structure and the measurement of stray coupling to noisy or resonating cavities.

7.3.1 Measurements of Powered Boards

In Figure 7.43 we show the measurement setup of a populated and working board. The board was measured with semirigid probes and with two VNAs. For low-frequency measurements, a 4395A VNA was used with an isolation input amplifier working in the 100-Hz–1-MHz range. The 1-MHz–1,800-MHz frequencies were covered with a 4396A VNA and 85046 S-parameter test kit. The low-frequency VNA was calibrated with a simple through calibration, the high-frequency VNA used a full two-port calibration. The same probes and cables were used for both VNAs. This simple measurement was a quick check without separate measurements of components and building blocks.

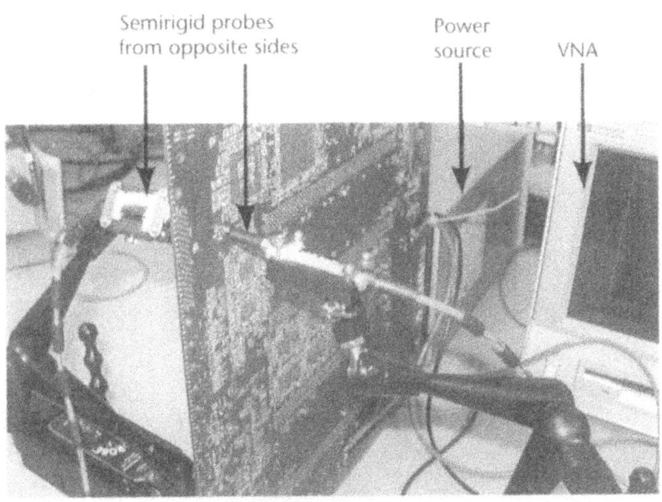

Figure 7.43 Setup for full-board PDN measurements.

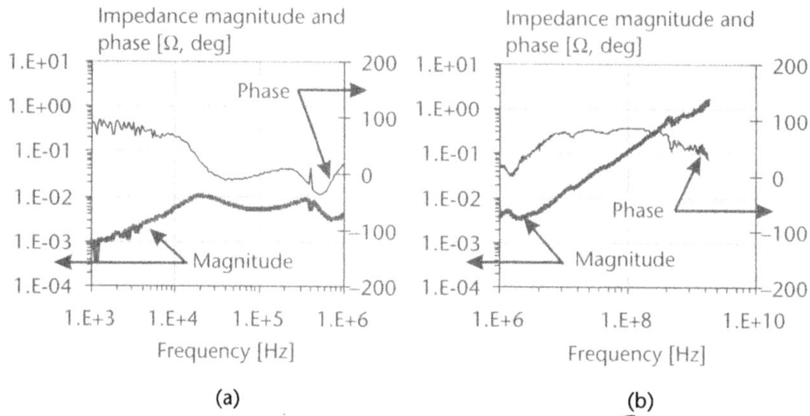

Figure 7.44 Impedance magnitude and phase of a 30-A supply rail from the board shown in Figure 7.43: (a) low-frequency plot and (b) high-frequency plot.

The measured impedance profiles collected with the two instruments are shown in Figure 7.44. Both plots show the self-impedance of a medium-current supply rail, powered up and running.

References

[1] Venable, D., "Source-Load Interactions in Multi-Unit Power Systems," *Proc. of the 29th Intersociety Conf. on Energy Conversion Engineering,* Monterey, CA, August 1994, pp. 351–358.

[2] Burns, S., "Distributed Systems—What Causes These Systems to Oscillate?" *PCIM Power Electronics Systems,* May 1999, pp. 14–33.

[3] Ridley, R., "Loop Injection," http://www.ridleyengineering.com, 2005.

[4] Agilent Technologies, "Switching Power Supply Evaluation with Agilent 4395A," Product Note 4395-2, 5968-7274E, 2000.

CHAPTER 8

Characterization and Modeling of Bypass Capacitors

In many electronic circuits, capacitors are used either in power distribution bypassing or in high-speed signal termination and dc-blocking applications. For modeling purposes, there is a distinct difference between the two applications. In high-speed signal applications, the capacitor is usually embedded in 50Ω single-ended environment or a 100Ω sometimes 150Ω differential environment; this makes the parasitic capacitance to ground from the capacitor body and pads an important element of the model. On the other hand, bypass capacitors are connected to power distribution networks which typically have low impedance compared to the usual 50Ω trace impedance. Also, one terminal of bypass capacitors is usually connected to the return (ground) network. This application difference is highlighted in Figure 8.1.

Figure 8.1(a) shows a high-speed signal application where the capacitor is in series to a signal line. The C_p capacitance at either end represents the capacitance of solder pads to ground. Each C_p capacitance also represents half of the lumped body capacitance to ground. The pad capacitance is a few tenths of a picoFarad, (possibly a few picoFarads for large parts). In a 50Ω or 100Ω impedance environment, it is important to model this capacitive reactance. The connection of a bypass capacitor, as shown in Figure 8.1(b), effectively eliminates one pad capacitance and connects the remaining one in parallel to the bypass capacitor. Except for specialized cases, the capacitance of a bypass capacitor is orders of magnitude greater than the pad and body capacitance; therefore, it can be ignored. This is shown in Figure 8.1(c). The box around capacitor C indicates that it could be replaced with any of the more detailed models shown later in this chapter.

8.1 Simple *C-R-L* Models and Spreadsheet Correlations

Figure 8.2 shows three of the simplest capacitor models. The parameters of the part and the frequency range of an application determine which one is acceptable. For the moment, we will assume that all elements in these equivalent circuits are frequency independent. Figure 8.2(a) is the simplest equivalent circuit: an ideal capacitor without any parasitics. Though it may look overly simplistic, there are practical situations for which this simple approximation still may be sufficient. For instance, measured data of Figure 8.3(a) shows the impedance of a small fixture in the 1-MHz –1-GHz frequency range; at first look, over three decades of frequency the impedance appears sufficiently close to that of a capacitance. We can use the ideal capacitor model because we stay below the resonance frequencies of the part. However,

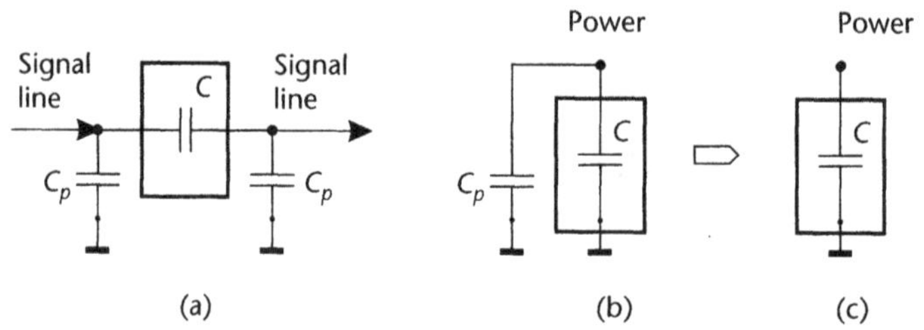

Figure 8.1 Equivalent circuit of a capacitor (a) in high-speed signal circuits and (b) in PDNs. The simplified equivalent circuit is shown in (c).

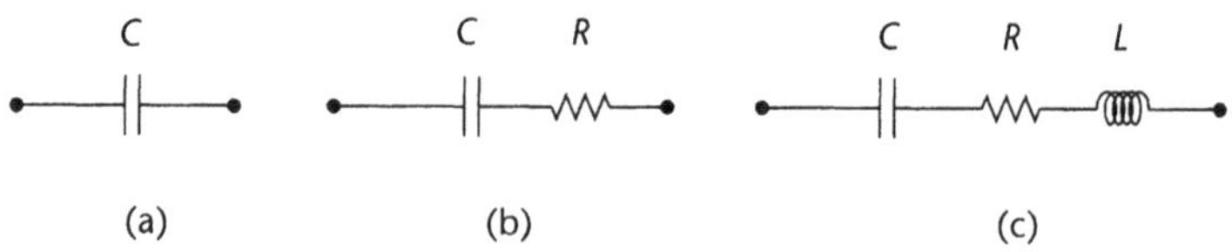

Figure 8.2 Simple C-R-L capacitor models: (a) capacitor only, (b) *C-R*, and (c) *C-R-L*.

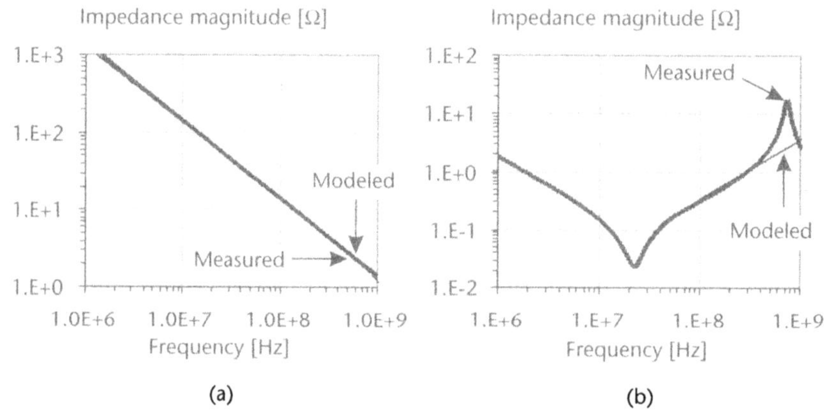

Figure 8.3 (a) Measured and modeled impedance magnitude of a small fixture and an ideal capacitor and (b) measured and modeled impedance magnitude of the same fixture with a 0.1-μF ceramic capacitor mounted.

when the frequency range extends both below and above the series resonance frequency, we need a minimum of three components in the equivalent circuit: a capacitance, resistance, and inductance, as shown in Figure 8.2(c).

Spreadsheets are convenient to calculate and display the impedance of simple capacitor models. Figure 8.3 shows the measured and modeled impedance magnitudes of a small PCB fixture, with and without a mounted capacitor. Figure 8.3(a) shows data on the bare FR-4 PCB fixture; the extracted capacitance was 112 pF at 100 MHz. Even though we know that the capacitance of the FR-4 material drops a few percentage points with every decade increase of frequency, for the purposes of modeling impedance magnitude, the difference between the measured and modeled data traces is negligible. The chart in Figure 8.3(b) is the impedance magnitude plot of the same fixture with a 0402-size 0.1-μF MLCC part. The extracted parameters for the MLCC parts were $C = 85.5$ nF (at 10 MHz), $R = 0.022\ \Omega$, $L = 0.57$ nH (at 30

MHz). Using these three constant values and ignoring the static capacitance of the fixture, we get a fairly good approximation of the impedance magnitude of the mounted capacitor: below 300 MHz, the measured and modeled traces completely overlap. The resonance peak at 700 MHz in the measured data is due to the fixture capacitance that we neglected.

C-R models, such as shown in Figure 8.2(b), are common for low-Q high-ESR capacitors at low frequencies and the output capacitors of dc-dc converters are often modeled this way.

Figure 8.4 illustrates further possibilities of the spreadsheets and also the limitations of these simple models. We can easily calculate the parallel equivalent of complex impedances with a spreadsheet, thus modeling bypass capacitors (assuming ideal lumped connections among them). However, when we combine different models that alone may appear to be sufficiently accurate, the limitations of the simple frequency-independent models may be revealed. The graph in Figure 8.4(a) combines the fixture capacitance and capacitor *C-R-L* model that were used in Figure 8.3. The correlation is still fairly good below 100 MHz, but there is a substantial difference between the measured and modeled parallel resonance frequencies. The difference results from the frequency dependent inductance and capacitance of the mounted capacitor, both of which decrease as frequency increases. We do not notice this effect in Figure 8.2, because the vertical scale spans a range of 1:10,000. The horizontal displacement of the sharp peak, however, is very noticeable. We can adjust the inductance of the mounted capacitor to match the parallel resonance frequency. By adjusting the inductance to 280 pH, the measured and modeled parallel resonances line up closely, as shown in Figure 8.4(b). Unfortunately, this adjustment means the series resonance frequency is no longer captured properly. To capture both the series and parallel resonances accurately, we need to include the frequency dependencies of component parameters. This will be shown later in this chapter.

Figure 8.5 illustrates two additional aspects of simple spreadsheet modeling. Through macros, spreadsheets can implement tolerance analysis also. The frequency independent *C-R-L* parameters for the three capacitors are listed in Table 8.1. Capacitors C_1 and C_2 may represent ceramic capacitors with typical ESR and

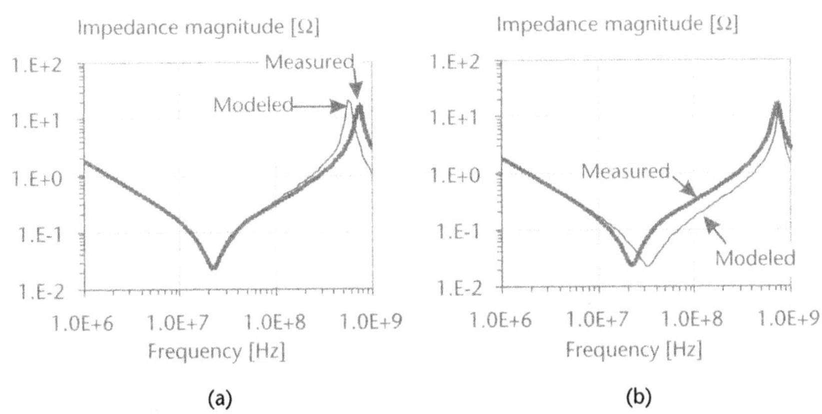

Figure 8.4 Measured and modeled impedance magnitude of the small and 0.1-μF ceramic capacitor, with the 112-pF parallel static capacitance of the fixture included. (a) $C = 85.5$ nF, $R = 0.022\Omega$, $L = 0.57$ nH, and (b) $C = 85.5$ nF, $R = 0.022\Omega$, $L = 0.28$ nH.

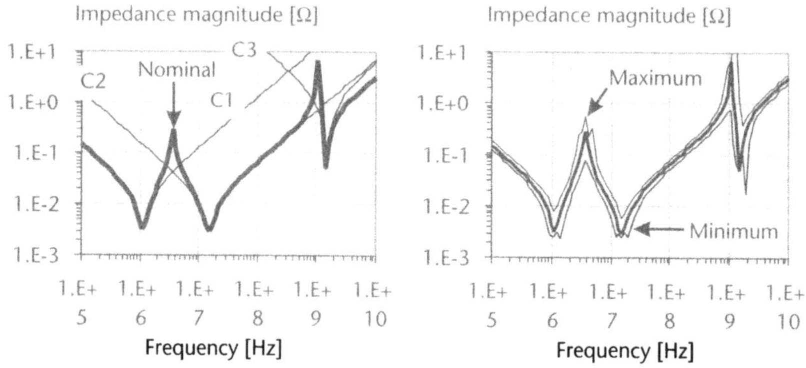

(a) (b)

Figure 8.5 (a) Impedance magnitude of three parallel-connected capacitors and (b) tolerance analysis of the three paralleled capacitors.

Table 8.1 *C-R-L* Parameters of the Three Capacitors for Figures 8.5 and 8.6

	C_1	C_2	C_3
C [μF]	10	0.1	0.0001
R [Ω]	0.003	0.03	0.01
L [nH]	2	1.2	0.1

ESL values. Capacitor C_3 may be considered as the first-order equivalent circuit of a small plane puddle.

The chart in Figure 8.5(a) shows the impedance magnitudes of three individual capacitors, C_1, C_2, and C_3, as well as the impedance magnitude of the three capacitors connected in parallel. Figure 8.5(b) shows the result of the tolerance analysis assuming ±20% tolerance on all nine parameters: accumulated minimum, maximum, and nominal impedance curves of the three paralleled capacitors.

The two charts in Figure 8.6 are illustrations of a potential pitfall: the same three capacitors, if we erroneously sum the impedance magnitudes (instead of the complex impedances), will not show the antiresonance peaks properly.

8.2 Wideband Characterization

For frequency-dependent characterization and modeling, we usually need wideband measurements, which can sufficiently capture the change of capacitance at low frequencies, as well as the change of inductance at high frequencies. The first example uses a bulk capacitor. Figure 8.7 shows the results of a three-terminal facedown organic capacitor measured with two separate VNAs on a small fixture in the 100-Hz–1,800-MHz frequency range.

Figure 8.8 shows the measured impedance magnitudes of two multilayer ceramic capacitors. Both capacitors had nominally 1-μF capacitance as well as 0612 reverse-geometry form factors. The first capacitor was a regular low-ESR type; the

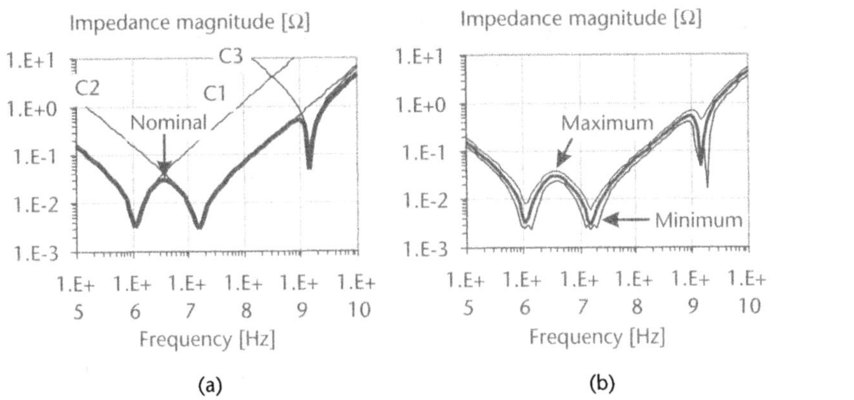

Figure 8.6 The same three capacitors as in Figure 8.5, except summed up in the wrong way: by their magnitudes instead of complex impedances: (a) contributing impedance magnitudes and sum and (b) tolerance analysis.

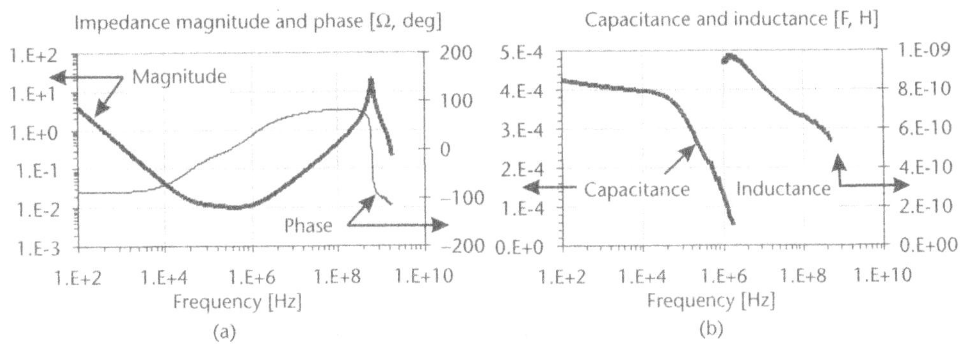

Figure 8.7 (a) Wideband impedance profile and (b) extracted capacitance and inductance (b) of a three-terminal facedown bulk capacitor.

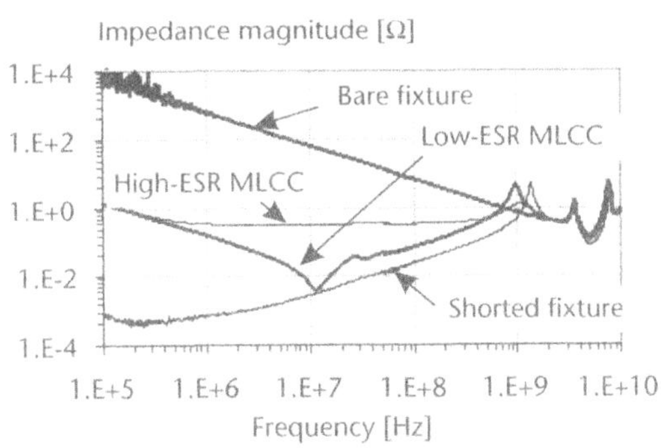

Figure 8.8 Wideband characterization results of 0612-size 1-μF multilayer ceramic capacitors.

second was a low-Q high-ESR capacitor, reported in [1]. The capacitors were soldered on a small test-fixture PCB, with approximately 2.54 × 2.54 cm (1 × 1 inch) size and a thin laminate close to the surface. To allow the measurement with one-sided probe stations, there were two pairs of vias, one on either side of the capacitor body. The distance between these via pairs was approximately 2.5 mm (100 mils). The transfer impedances between the two via pairs were measured under different conditions. First, the fixture was characterized. The impedance profile of the bare test fixture follows the static capacitance up to 1 GHz, followed by modal resonance peaks at 3.6-GHz and 8-GHz frequencies. Second, the test fixture was measured with the capacitor site shorted. Lastly, the capacitor sample was soldered on and measured.

Data was taken with two independent instruments: HP4396A VNA in the 0.1–100 MHz frequency range and HP8720D VNA in the 50-MHz–10-GHz range. Wafer probes of 450-μm pitch were used with full two-port calibration on both VNAs. The data in the 50–100-MHz range was taken with both VNAs; the smooth continuation of traces indicates good repeatability and accuracy of the independently recorded datasets. Note the shallow minimum in the shorted fixture's impedance magnitude at around 250 kHz. One would expect a flat impedance curve at the low end; instead, the response slightly increases as we approach 100 kHz. The reason for this increase is twofold. The first reason, as pointed out in Section 7.1, is that the cable's ground loop creates a low-frequency error that has to be reduced or eliminated with ferrite rings on the cable or with an isolation amplifier. For this figure, data was taken with ferrite-covered cables without an isolation amplifier. The ferrite cover pushed the frequency of minimum reading to 250 kHz, but below that minimum frequency the cable-braid loop error gradually shows up. The second reason is that the particular *S*-parameter test kit is specified from 300 kHz; below the specified minimum frequency the system shows an increasing error.

8.3 Impact of Geometry on Electrical Parameters

As several illustrations in this section will show, the actual parameters of a bypass capacitor (capacitance, resistance, and inductance) all may depend not only on the internal geometry of the capacitor, but also on the geometry of user application. The most difficult and most controversial of these parameters is the parameter defined as the inductance of bypass capacitor. It is commonly called the effective series inductance (ESL). First, we look at some of the possible definitions of ESL, followed by examples of resistance and inductance manipulated by various internal and external connection geometries.

8.3.1 How to Define ESL

The ESL of a part has several possible definitions. Inductance in real circuits is always realized by current loops. While the concept of partial self- and mutual inductances allows us to break down the loop inductance into components, measuring these inductance components directly and separately is not easy. In all of the examples in this book, unless otherwise noted, the measured or extracted inductance refers to some type of loop inductance.

For PDN design and validation, however, the loop inductance may not be the most convenient parameter to characterize the high-frequency behavior of a bypass capacitor. This will be explained using Figure 8.9.

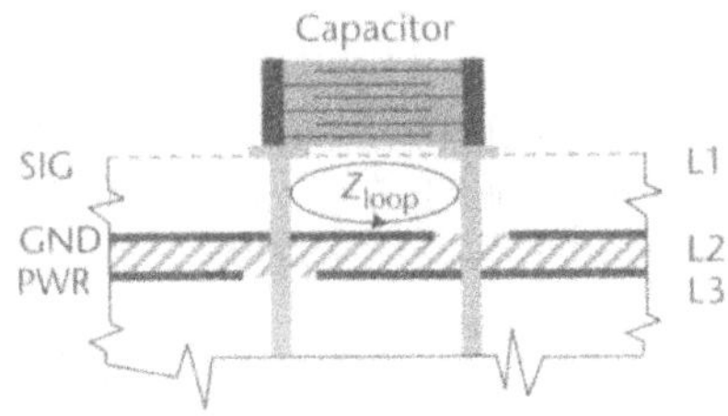

Figure 8.9 Sketch of a bypass capacitor mounted on a PCB and connected to an internal plane pair through vias.

The figure shows one bypass capacitor connected to a pair of planes through vias. The full construction creates a Z_{loop} impedance with its corresponding inductance, which in general is frequency dependent. The loop inductance can be expressed in terms of partial self- and mutual inductances:

$$L_{loop} = L_{capacitor-self} + L_{plane-self} - 2L_{capacitor-plane-mutual} + 2L_{via-self} - 2L_{via-mutual} \quad (8.1)$$

In (8.1), the $L_{capacitor\text{-}self}$ is the partial self-inductance of the capacitor body. We assume that the inductances of both via barrels from the surface to the upper plane are identical. The mutual inductances are assumed to be between the opposite parallel surfaces: the capacitor body and planes as well as between the vias. The sketch of Figure 8.9 shows one via at each terminal and no horizontal escape traces from the capacitor pads to the vias. For low-inductance mounting, multiple vias per pad are common. Also, with multiterminal capacitors, there may be multiple vias per capacitor side. With multiple vias, the $L_{via\text{-}self}$ and $L_{via\text{-}mutual}$ values should be adjusted accordingly in (8.1). If horizontal escape traces are present in the user's geometry, $L_{via\text{-}self}$ should be replaced with $L_{via\text{-}self}$ + L_{escape}, and $L_{capacitor\text{-}plane\text{-}mutual}$ should be replaced with $L_{capacitor\text{-}plane\text{-}mutual}$ + $L_{escape\text{-}plane\text{-}mutual}$. The two vias are assumed to have the same geometry; otherwise their self- and mutual inductances could be accounted for separately.

In this model the only neglected contribution is the possible 3D interaction among the orthogonal vias, plane, and capacitor body (and escape traces, if present).

Based on the above inductance picture, we can define the inductance of the capacitor in several different ways [2]. The simplest and most obvious option is to call the partial self-inductance, $L_{capacitor\text{-}self}$, the inductance of the capacitor. While this definition gives a unique value that depends only on the capacitor itself, this value does not capture the influence of the user geometry in the remainder of the loop. Furthermore, the coupling between the capacitor body and the underlying planes as well as the cover thickness of the capacitor body are completely ignored. As we will see later in the chapter, the cover thickness is a crucial factor in the high-frequency performance of the capacitors.

The term *mounted inductance* (introduced in [3]) captures the full loop of the capacitor body and the user geometry. Unfortunately, there is an item that makes the mounted inductance nonunique to the capacitor: the $L_{plane\text{-}self}$ term. This term captures not only the horizontal inductance of the upper plane but also the inductance associated with the portion of the current loop that passes through the dielectric layer between the upper and lower planes. The $L_{plane\text{-}self}$ term is dependent on the horizontal location, as well as frequency dependent. As a result, for the same capacitor body and same mounting geometry, the mounted inductance (or simply the loop inductance) will depend on the location of the capacitor on the planes. This is illustrated in the following figures.

The test board that was shown in Figure 4.38 had a grid of capacitor locations to attach eight-terminal 1206-size capacitors. We used this test board to illustrate the correlation between measured and simulated bare-board and shorted-board parameters. Figure 4.41 already indicated that the low-frequency equivalent inductance of the (shorted) plane is location dependent. Here we expand the investigation to include capacitors. Figure 8.10 shows the cross-section of two different eight-terminal capacitors mounted on one of the test sites. This figure shows the upper three layers in the PCB stackup together with some of the lower plates of the mounted capacitor. The dark capacitor body contains the horizontal plates. The brighter strip between the capacitor body and the top metal layer of the PCB is the cover layer of the MLCC part.

The figure illustrates several important things. First, the cover thickness can be significant compared to the vertical height in the PCB between the top layer and the first plane layer. In this example, the first plane was 75 μm (3 mils) below the surface. Together with the thickness of the top metal layer, the bottom surface of the capacitor body was approximately 125 μm (5 mils) above the upper plane layer. Second, the thickness of the cover layer may vary from part to part and from manufacturer to manufacturer. The capacitor on the left has an approximately 150-μm (6-mil) cover layer. The capacitor on the right, which was nominally the same part but from a different vendor, has the capacitor plates starting even higher; the cover thickness is approximately 300 μm (12 mils). The third point implicitly follows from the figure: there is no reason to assume, and there is no specification to guarantee, that the cover thicknesses on the top and bottom sides are the same. Besides, the top and bottom is not differentiated by any marking on today's ceramic capacitors.

In the cross-section photos, only a corner of each capacitor and the top three layers of the board are shown. Note the blind via connection under the capacitor terminal connecting the pad to layer 2. During the experiment, the same piece of capacitor was sequentially soldered on multiple test sites; the loop inductance was extracted at the nearest test point. The distance and relative orientation between the capacitor site and the test point was maintained. The same side of the capacitor

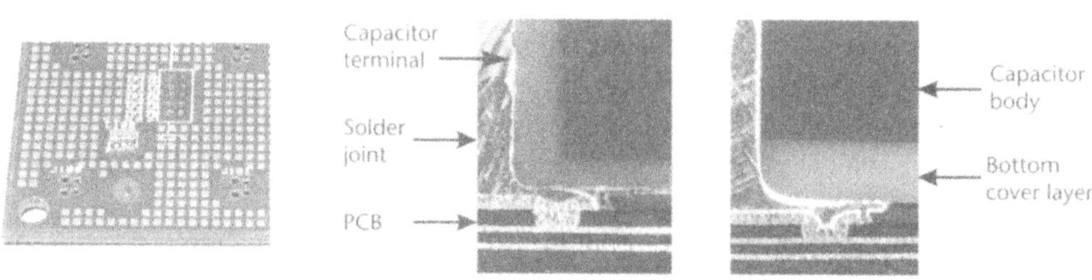

(a) (b)

Figure 8.10 Eight-terminal capacitor mounted on multilayer PCB: (a) corner of the test board with a mounted capacitor and (b) cross-sections of the top three metal layers of the test board, with two different capacitors.

always faced the pads. The capacitor was pushed all the way down to the pads manually during soldering to eliminate the solder layer between the pads and the capacitor. Figure 8.11 shows some of the measured and correlation results.

The graph of Figure 8.11(a) shows the impedance magnitude measured at three different capacitor locations. Notice that at low frequencies all traces coincide because the low-frequency response is dominated by the capacitance of the MLCC part. There is a peak around 60 MHz, which originates from the static capacitance of the board and the mounted inductance of the capacitor. The measured traces above the 10-MHz series resonance frequency are clearly different dependent on the location: the frequencies and values of peaks are different. The graphs on the figure have linear horizontal scales to clarify the differences around the impedance peak.

Similar to the shorted-pad case, we can establish a best-fit capacitor model that gives a location-independent characterization of the capacitor on the pads. Note that the model will describe the capacitor and the via/pad geometry together. Since the ESR of the capacitor was much greater than the via resistance, the skin-resistance portion of the via model does not appear separately. However, the ESR and attached inductance of the capacitor both exhibit noticeable frequency dependence.

Curve-fit models were created separately at the series resonance frequency (around 8 MHz) and around the parallel resonance frequency (around 60 MHz). The best-fit model was 6 mΩ and 160 pH at 8 MHz [from (8.2)], and 11 mΩ with 120 pH at 60 MHz [from (8.3)]:

$$Z_{attached\ at\ 8MHz} = 6E^{-3} + j2\pi f 160E^{-12}\ [\Omega] \tag{8.2}$$

$$Z_{attached\ at\ 60MHz} = 11E^{-3} + j2\pi f 120E^{-12}\ [\Omega] \tag{8.3}$$

The correlation at site 1 is shown in Figure 8.11(b). The same model provided similarly good correlation at all test sites involved in the experiment because this model captures only the current path above the planes and inside the capacitor; but this model does not include the location-dependent plane impedance.

In the exercise above, we realized that the current loop does not close on the upper plane only: it has to go through the dielectrics between the two planes and through the antipad taking the via connection to the lower plane. Therefore the

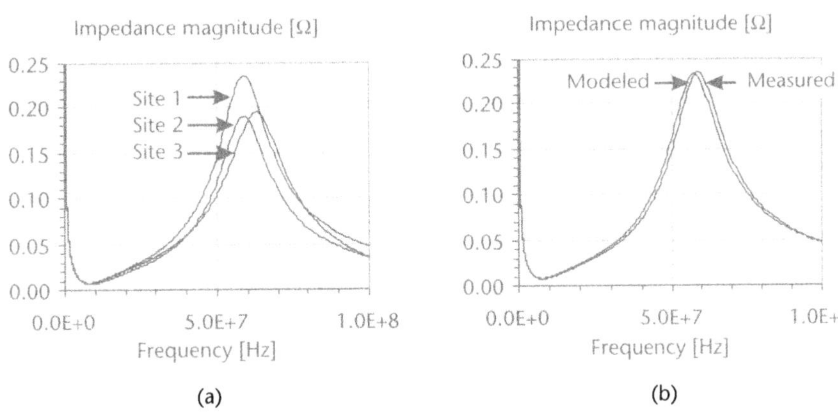

Figure 8.11 (a) Measured impedance of test board with the same capacitor mounted on three different test sites. (b) Correlation results at site 1.

$L_{plane\text{-}self}$ component in (8.1) does depend on the plane separation, antipad, and via geometries. Moreover, the inductance of a plane pair is frequency and location dependent. This means that the mounted inductance also depends on where the capacitor is on the plane. So, for simulation purposes, we would need to generate a different simulation model for the same capacitor depending on the location of its connection.

We can generate a location and plane-independent simulation model for the bypass capacitor if we break down the loop impedance differently, as shown in Figure 8.12. Equation (8.4) defines the attached resistance and attached inductance as part of the $Z_{loop} = R_{loop} + j\omega L_{loop}$ impedance:

$$R_{loop}(f) = R_{attached}(f) + R_{plane}(f) \quad L_{loop}(f) = L_{attached}(f) + L_{plane}(f) \tag{8.4}$$

In (8.4), the $L_{attached}$ component captures the partial self-inductance of the capacitor, plus the self- and mutual inductances of the vias, and the mutual inductance between the planes and the capacitor body. Depending upon the coupling strength between the capacitor body and the underlying planes, the attached inductance can be either larger or smaller than the partial self-inductance of the capacitor body itself: with aggressive mounting, the mutual inductance between the capacitor and planes will create a low attached inductance.

Note that if the attached inductance is much larger than the plane inductance, the attached inductance and the mounted inductance will closely match. As we use more aggressive mounting techniques, however, the attached inductance can become low compared to the plane inductance, and the difference becomes negligible.

The attached inductance still depends on the mounting geometry, so we have to generate a separate simulation model for each of the capacitors/pads/via geometries we want to use. However, the resulting $L_{attached}$ will not depend on the location on the planes, nor will it carry the frequency dependency of the plane inductance.

As shown above, the attached inductance can be determined separately for shorted capacitor pads and for the mounted capacitor. The advantage of the attached impedance/inductance concept is that it can provide a simulation model for a given capacitor and pad/via geometry that is location and plane independent.

Note also that while the attached impedance of a given pad/via combination may show consistent results all over the board (because the geometry can be tightly controlled), the attached impedance or inductance of a capacitor will depend on the height of the lowest capacitor plates above the PCB planes. This height is determined

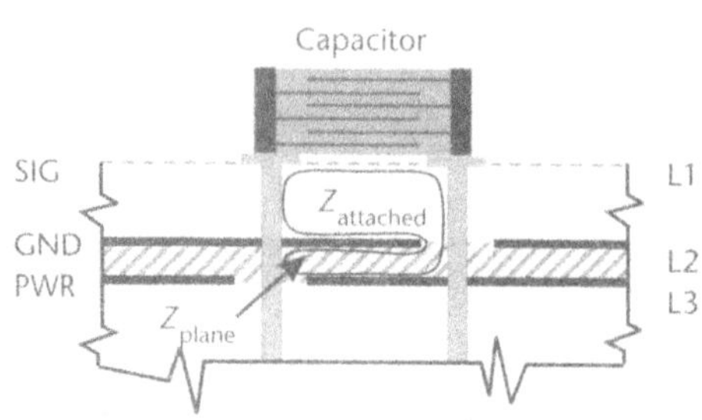

Figure 8.12 Breakdown of the loop impedance to attached impedance and plane impedance.

primarily by the solder thickness between the pad and capacitor bottom and the bottom cover thickness of the capacitor. The cross-section photos of Figure 8.10 show that the cover thickness of a capacitor can be greater than the dielectric layer height separating the first plane from the surface; in those cases, the variation of the cover thickness can be significant. Therefore, the good correlation shown in Figure 8.11 is simply an illustration of the consistency of the attached impedance/inductance concept: if different pieces of capacitors are used at the various locations, the statistical variations of their geometry should be considered.

The attached inductance is convenient for simulations, since by adding it to the frequency-dependent plane model, the two models together capture both the location and frequency dependence of inductance.

The way to generate the attached inductance is through curve fitting (which is a simulation exercise). Obtaining the attached inductance through measurement, however, is more difficult, because it requires the generation of a detailed plane model for the fixture's planes (as in the exercise above). When the characterization rests primarily on measurements instead of attached inductance, we can either use the loop inductance (mounted inductance) or the added inductance. Added inductance will be described in Section 8.3.6.

8.3.2 Impact of Body Geometry on ESL of MLCC

The major geometry contributors of a mounted part in the cross section, which determine the loop inductance and attached inductance of the MLCC part, are shown in Figure 8.13. In this model, there are two horizontal and four vertical geometry parameters:

- H_1: body length of capacitor;
- H_2: spacing of vias connecting the pads to planes;
- V_1: stack height of capacitor plates;
- V_2: bottom-cover thickness of capacitor;
- V_3: mounting height; the distance between the upper surface of upper plane to the bottom of capacitor body. It is the sum of the dielectric thickness from

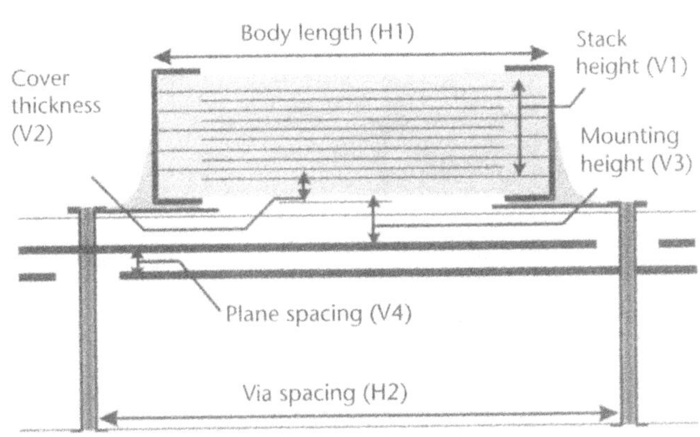

Figure 8.13 Sketch defining the major contributors of loop inductance of a mounted capacitor.

the board surface to the upper plane and the thickness of pads and solder layer.

- V_4: spacing of power-ground planes.

We notice that the extracted equivalent inductance trace in Figure 8.7 exhibits a negative slope. This indicates that the inductance decreases with increasing frequency, which in turn indicates that the current loop shrinks. The approximate current loop sizes and cross-section areas are shown in Figure 8.14. At dc we can assume that the current distribution vertically is uniform. The center of gravity of the current flow is at the midpoint of the capacitor height. The high-frequency sketch shows the reduced current loop, because the proximity effect forces the current to close on the bottom plates of the capacitor.

For bypass capacitors, this brings up the question, how does the inductance depend on the geometry of the capacitor itself and its mounting? Most specifically, for a given capacitor body size, how does the inductance depend on the height of the cover thickness and the total height of the capacitor body? To answer these questions, several exercises were performed with measurements and simulations on small-size plane pairs with capacitor pads.

One series of tests used up to five capacitors (first measuring one; then later adding stacked capacitors one at a time). The small plane pair was characterized with open and shorted with a sheet metal. One capacitor was soldered on the pads and on the nearby via pair, the self-impedance of the DUT was measured. From the self-impedance, the equivalent inductance of the entire loop of capacitor-pad-vias-planes was extracted. One by one additional capacitors were soldered (piggybacked), on top of the previous capacitors. This ensured that the geometry between the fixture's planes and closest capacitor plates were left unchanged (only the total capacitor height was increased). During soldering, special attention was paid not to melt the solder under the parts that were already on the fixture to make sure that any solder layer between the parts stayed the same as parts were added.

The fixture and the mounted capacitors are shown in Figure 8.15, with one and five capacitors on the pads. The capacitors were reverse-geometry 0508 parts with 4.7-μF nominal capacitance. Each capacitor was separately characterized to make sure that its capacitance, resistance, and inductance were identical within the resolution and repeatability of the measuring system. The impedance profile was taken with an HP4396A VNA after full two-port calibration up to the semirigid probes. Figure 8.16 shows the measured impedance magnitudes. The traces show the imped-

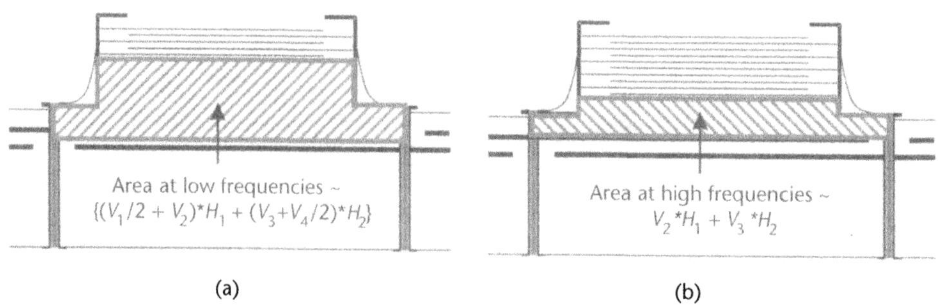

Figure 8.14 Sketches of current loops at extreme frequencies: (a) dc and (b) very high frequencies.

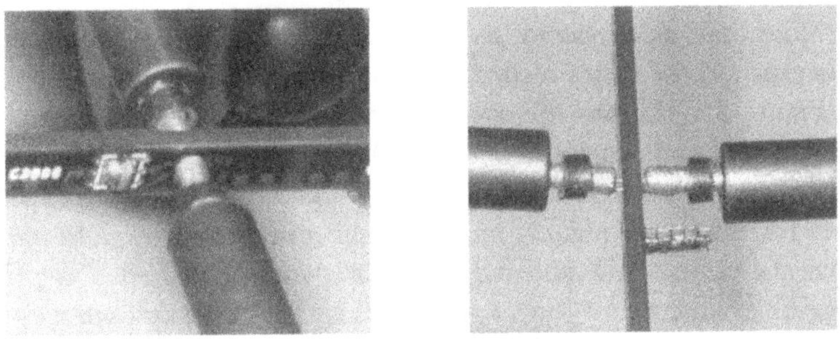

(a) (b)

Figure 8.15 Small fixture with (a) one and (b) five stacked capacitors on the pads.

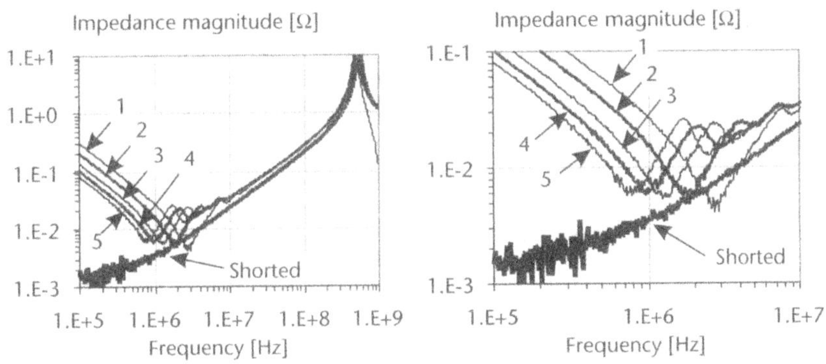

(a) (b)

Figure 8.16 Measured impedance magnitude of the small fixture with shorted pads and one, two, three, four, and five stacked capacitors on the same pads. Same data is shown with (a) full frequency scale and (b) enlarged to a 0.1–10-MHz frequency scale.

ance magnitude of the full DUT (small plane plus capacitors) with a reference trace, where the capacitor pads were shorted. The trace of the shorted fixture runs somewhat below the other traces, indicating lower inductance.

As capacitors were added, the series resonance frequency was shifted down. Table 8.2 lists the measured series resonance frequencies and the equivalent loop inductance values calculated from the series resonance frequency and total nominal capacitance.

Table 8.2 Extracted Capacitance, Series Resonance Frequency, and Total Loop Inductance of Vertically Stacked Capacitors Mounted on Fixture

	Capacitance [μF]	*SRF [MHz]*	*Loop inductance [nH]*
$1 \times 4.7\,\mu F$	4.7	2.8	0.7
$2 \times 4.7\,\mu F$	9.4	1.7	0.93
$3 \times 4.7\,\mu F$	14.1	1.2	1.25
$4 \times 4.7\,\mu F$	18.8	0.98	1.4
$5 \times 4.7\,\mu F$	23.5	0.75	1.92

Note that, as expected, as the total height of the capacitor stack increases, the inductance at the series resonance frequency also increases [4].

Figure 8.17 shows the extracted inductance versus frequency. The inductance with shorted pads varies between 500 pH at 1 MHz and 350 pH at 100 MHz. The inductance with different numbers of stacked capacitors varies strongly close to the series resonance frequency, but all traces quickly converge to values about 100 pH above the trace of the shorted fixture's inductance. This 100-pH difference comes from the extra current-loop size created by the cover thickness of the lowest capacitor in the stack. Above 20 MHz there is no noticeable difference among the inductance values with the various numbers of stacked capacitors. In contrast, the low-frequency value of inductance varies approximately in proportion to the number of stacked capacitors: it is 700 pH with one capacitor, and 2 nH with five stacked capacitors.

We can conclude that the high-frequency inductance of a mounted bypass capacitor primarily depends on the cover-layer thickness, not on its total height. The inductance around and below SRF, however, is proportional to the total height of the mounted part.

8.3.3 ESR and ESL of Very Tall Capacitors

A second set of tests used up to 10 stacked capacitors; it focused not only on inductance, but also on resistance as a function of total component height. The arrangement is shown in Figure 8.18. A SPICE-grid simulation study accompanied the measurements. The simulations are shown in Section 8.7, and it visualized the internal current distribution of MLCC parts. The results confirmed that the current distribution in the dielectrics and capacitor plates starts to deviate from uniform distribution already well below SRF.

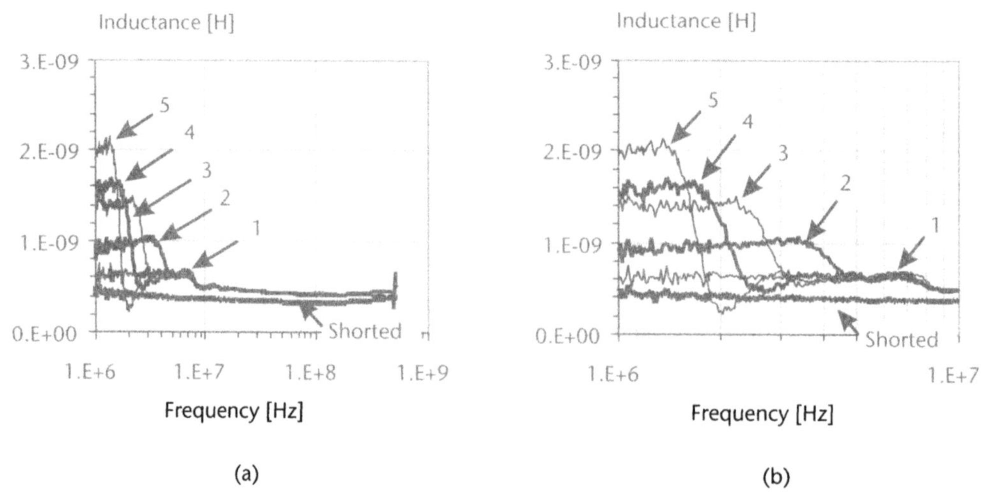

Figure 8.17 Extracted inductance of the small fixture with shorted pads and one, two, three, four, and five stacked capacitors on the same pads. The labels indicate the number of capacitors in the stack. Same data is shown with (a) full frequency scale and (b) enlarged to a 0.1–10-MHz frequency scale.

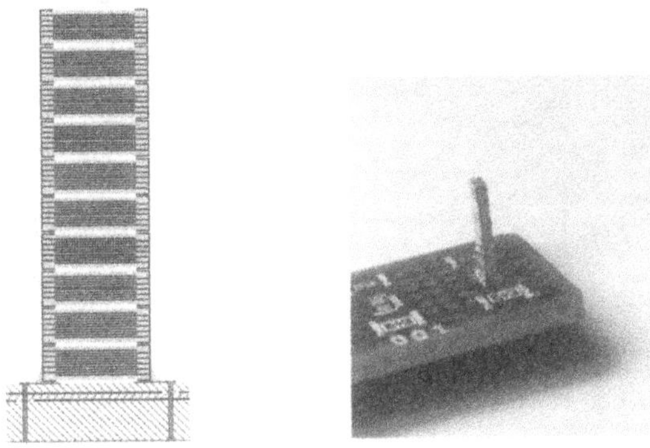

(a) (b)

Figure 8.18 (a) Side-view sketch and (b) photo of stacked 10-μF 0508 reverse-geometry capacitors mounted on a small test fixture. The plane pair was 10 × 15 mm (400 × 600 mils) size. The 50-μm (2-mil) dielectric laminate was 0.1 mm (4 mils) below the surface.

The measurement study was conducted with 10-μF 0508 reverse-geometry MLCC parts. Ten parts were selected. The parts were marked individually and their two sides (top and bottom) were also marked. All 10 parts were measured in the same fixture, in both configurations: first bottom facing the fixture and then top facing the fixture. The parts were hand soldered to the fixture by pushing them in the melted solder all the way to the fixture pad. This method was employed to eliminate most of the uncertainty of unknown solder height between the fixture pads and capacitor terminals. Though eliminating the solder fillet in this manner cannot be done in volume manufacturing, in this study we were interested in capacitor characteristics originated from their internal geometry and construction, which justifies the exclusion of external variables. The part-to-part and top-versus-bottom variability was found to be almost nonmeasurable among the selected 10 parts.

With this consistent set of 10 parts, different horizontal and vertical combinations were built. Horizontal configurations with a different part have been reported in [5]. Here we show one of the vertical configuration studies, where the parts were stacked on top of each other (starting with one and gradually adding parts one-by-one up to a stack height of 10).

Each time a capacitor was added to the stack, the impedance of the fixture was measured and the parameters extracted. The measurements were done with two separate VNA setups; one covering the 100-Hz–10-MHz range and the other covering the 100-kHz–1.8-GHz range. The two independent sets of curves were merged, and the impedance magnitudes and impedance real parts were plotted.

Figure 8.19 shows the impedance magnitude with 1, 2, 3, 4, 6, and 10 parts stacked. Traces for 5, 7, 8, and 9 stacked parts followed the same trend and those traces were left out to increase the clarity of graphs. First, we can verify that as we add more capacitors, the impedance magnitude below SRF moves to the left in accordance with the increased total capacitance. We can also notice on the plot that above 10 MHz all traces coincide. In other words, this result confirms once again the findings shown in Figure 8.17, that the capacitor height does not degrade (nor does it improve) the inductance associated with the part at high frequencies.

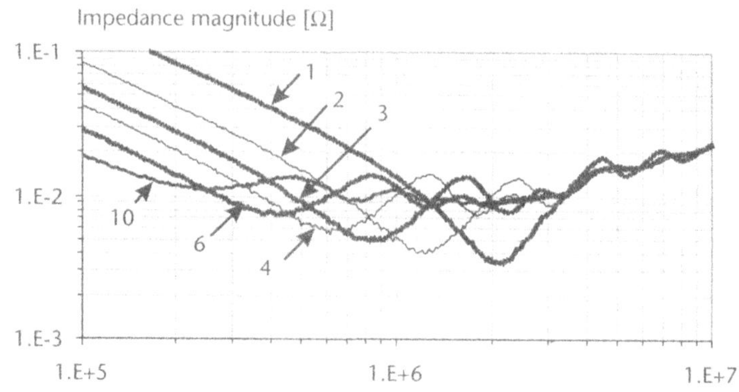

Figure 8.19 Impedance magnitude of 1–10 stacked 10-μF 0508 MLCC parts. The labels show the number of parts stacked.

On the plots around SRF, one would expect that as we add capacitors on top of the stack, the impedance minimum would continue to decrease (perhaps saturating at a certain component height and number of parts stacked). Note, however, that the measured data shows something strikingly different: as we add parts, the impedance minimum continues to increase monotonically. This increase occurs in spite of the fact that we only add parts to the top of the stack, without changing the geometry of the parts already on the fixture. The impedance minimum with one part on the fixture is 3.5 mΩ. It increases to 4.3 mΩ and then to 5.1 mΩ as soon as we stack a second and third part on top of the first, respectively. With four parts stacked, the impedance minimum is 5.9 mΩ. By the time we have a stack height of 6 and 10 parts, the impedance minimum rises to 7.6 mΩ and then 11 mΩ, respectively.

We can capture the same trend more clearly when we plot the real part of measured impedance. Figure 8.20 shows the real part of the impedance traces for the same configurations: 1, 2, 3, 4, 6, and 10 stacked parts. Though it was not noted, the data in Figure 8.16 already showed that ESR increases as we add components vertically; this is equivalent to increasing the height of a single component. Also, both

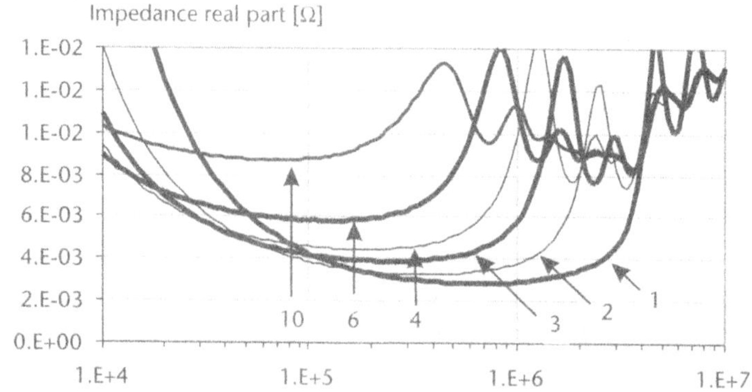

Figure 8.20 Impedance real part of 1–10 stacked 10-μF 0508 MLCC parts. The labels show the number of parts stacked.

Figures 8.19 and 8.20 suggest that increasing the capacitor height not only increases ESR, but it also creates a wider frequency band over which the higher ESR is maintained with less fluctuation. The impedance-magnitude and ESR numbers are summarized in Table 8.3. These figures underline the possibility of creating controlled-ESR lossy MLCC parts by careful construction of their internal geometry.

Figure 8.21 shows the extracted inductance. Note that with 10 parts stacked, the inductance stabilizes at about 1.5 MHz, from which point it rises with increasing frequency; this pattern is, again, counterintuitive. In the 1.5–5-MHz range, the inductance rises by a factor of 2. At 5 MHz, it joins the slowly decreasing inductance curves of shorter capacitor stacks. The temporary rise of inductance is attributed to the fact that taller capacitor stacks show more pronounced resonances; with stronger resonances, when the standing-wave pattern rolls towards the lower plates, it pushes more current over a wider frequency band towards the bottom. In the frequency range of stronger resonances, the overall inductance is lower than the inductance at higher frequencies. As initial resonances die out, inductance rises in an interim frequency range. This will be further illustrated and explained in Section 8.7.

Table 8.3 Summary of Impedance Minima and Frequency of Impedance-Minimum Values from Figure 8.19, ESR Minimum (Minimum of Real Part of Impedance), and Frequency of ESR Minimum Values from Figure 8.20 as a Function of Parts Stacked

Parts	*Imp. min. [mΩ]*	*Fr. of imp. min. [MHz]*	*ESR min. [mΩ]*	*Fr. of min. ESR [MHz]*
1	3.5	2.1	2.8	0.8
2	4.3	1.2	3.3	0.36
3	5.1	0.85	3.9	0.24
4	5.9	0.63	4.4	0.19
6	7.6	0.4	5.8	0.12
10	11	0.2	8.7	0.07

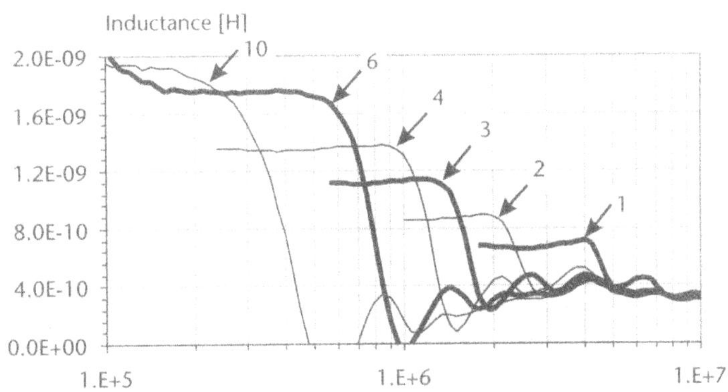

Figure 8.21 Extracted inductance of 1–10 stacked 10-μF 0508 MLCC parts. The labels show the number of parts stacked.

Finally, it was verified empirically that this vertical resonance behavior is primarily the property of the capacitor itself; the external connecting geometry has little (but clearly more than zero) influence on it. The stack of 10 capacitors was lifted carefully from the fixture pads, so that the solder holding the 10 capacitors together did not melt. Two solid copper spacers with rectangular cross sections were soldered on the fixture pads, then the stack of 10 capacitors was soldered back on top of these spacers. The height of the spacers was approximately the height of a single capacitor in the stack. Figure 8.22 shows the photo of stacked capacitors on the riser and the impedance magnitude curves for the 10 stacked parts, with and without the riser. Note that with and without the risers there is no difference in the impedance magnitude below 100 kHz. In the 100-kHz–1-MHz range, the impedance magnitude is slightly higher without the spacers. Above 1 MHz, the extra inductance caused by the spacers gradually manifest itself. Note that both with and without the spacers the low-ripple bottom portion of the impedance magnitude curves is more than a decade wide.

The above experiments show that the parameters of the bypass-capacitor equivalent circuit are not only frequency dependent, but also that they are not unique: the inductance and resistance are functions of the application geometry as well.

This section showed data on very tall parts, and the building block (a single 10-μF 0508-size capacitor) was tall enough to exhibit significant vertical resonances. For this reason, the impedance minimum continued to increase monotonically while stacking these capacitors. We can assume that if we start with a thin capacitor with no pronounced vertical resonance, its impedance minimum will first decrease as we increase its height. Eventually we would reach a height where vertical resonances become pronounced; from that point on, increasing the height further results in an increase of impedance minimum.

8.3.4 Impact of Vertical MLCC Mounting on ESL and ESR

The impedance magnitude plots of Figures 8.16 and 8.19 show a series of secondary resonances starting an octave above the series resonance frequency. The resonances are more pronounced as the height of the capacitor body increases with respect to the mounting height; this is also true as ESR of the part gets lower with respect to the

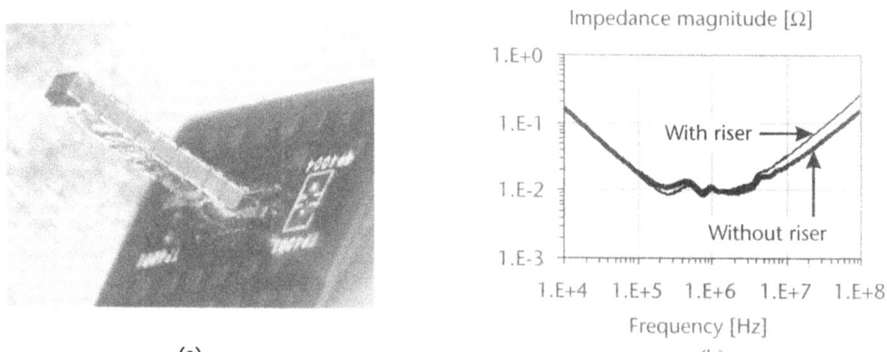

(a) (b)

Figure 8.22 Effect of vertical spacer on tall capacitors. (a) Photo of stacked capacitors on riser bars. (b) Impedance magnitude versus frequency curves of 10 stacked capacitors with and without riser bars.

resistance of the rest of the loop. Short capacitors, capacitors high above the planes, and high-ESR capacitors all exhibit this phenomenon to a much less significant degree or not at all.

Secondary resonances in multilayer capacitors have long been known in the microwave industry [6]. It was shown that this phenomenon can be observed when the capacitor plates are parallel to the PCB planes. With a horizontal orientation of the main PCB, this means horizontal capacitor plates. It was also observed that the same capacitor exhibits minimal or no secondary resonances when the capacitor plates are perpendicular to the underlying PCB planes (which means vertical capacitor-plate orientation with horizontal PCB). Traditionally, multilayer ceramic capacitors had plates arranged horizontally because this arrangement is easier and cheaper to manufacture with capacitor bodies of greater width than height. Vertical capacitor plates have also been available, but only in specialty parts [7]. With the constant push for higher capacitance densities, a larger percentage of contemporary MLCCs have close to a square cross-section in the side view. At present most bypass capacitors do not carry markings, and therefore with a square side view there is very little to differentiate (visually) between horizontal and vertical plate mounting. Knowing the benefits of vertical plate orientation, it is reasonable to assume that eventually capacitors will be marked so that users can make a conscious choice how they want to mount them.

A capacitor with square aspect ratio appeared in Figure 8.15(b). One of those capacitors, a 0508-size 4.7-μF reverse-geometry MLCC was measured in the same fixture with horizontal and vertical plate orientations. The comparison is shown in Figure 8.23. The impedance magnitudes are shown in the 10-kHz–100-MHz frequency range. The extracted loop inductance is shown from 1–500-MHz (500 MHz was the parallel resonance frequency between the fixture capacitance and loop inductance).

The impedance magnitude of a single part with horizontal mounting shows a mild secondary resonance around 10 MHz. The secondary resonance is accompanied by a sharp drop of inductance at the same frequency. With vertical plate mounting, the same part exhibits almost no secondary resonance. The drop of inductance is nearly constant from 1 MHz to 500 MHz (there is a very minor

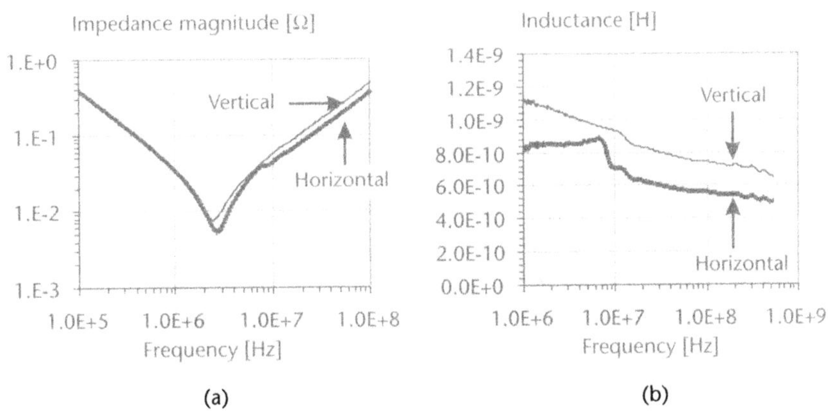

Figure 8.23 (a) Impedance magnitude and (b) inductance of the same 0508-size 4.7-μF capacitor with horizontal and vertical mounting.

increase of the slope around 15 MHz). Note that a constant offset of 190 pH occurs between the inductance curves above 20 MHz. This is due to the different cover-layer thicknesses of the capacitor on the bottom versus on the sides. With the same cover-layer thicknesses, the asymptotical high-frequency inductance would match closely. This assumption was verified with a second set of experiments.

The second experiment used a regular-geometry 1210-size 220-μF MLCC part measured on a small PCB fixture. Figure 8.24 shows the test fixture and the relevant cross-section geometry. Ten samples were measured, each in horizontal and vertical orientations. Instead of soldering, the parts were attached to the fixture with uncured silver-filled epoxy. The capacitor was pushed onto the pads with a clamp, providing approximately the same contact force from sample to sample. Figure 8.25 compares the impedance magnitudes and extracted loop inductance of one sample with horizontal and vertical orientations.

Below 100 kHz and above 1 MHz, the orientation has little impact. In the 100-kHz to 1-MHz range, the vertical mounting provides a smoother profile. The particular sample yielded approximately the same high-frequency inductance, as shown in Figure 8.25(b), because the bottom and side cover thicknesses were similar.

8.3.5 Impact of Special Geometries on ESL and ESR

As measured data in the previous sections clearly indicates, the inductance and resistance of a mounted capacitor can be influenced by changing the geometry either inside or around the capacitor. The previous two sections examined the impact of capacitor height. The examples with multiple stacked parts were shown, with the understanding that the extreme hand-built geometries and aspect ratios cannot be applied to volume manufacturing. In this section, we consider other possible geometry manipulations that have a direct influence on the inductance and resistance of the part [8].

In MLCC parts, a large part of the series resistance comes from the capacitor plates. For conventional two-terminal parts, the plate construction is shown in Figure 8.26. The cumulative resistance of the plates is:

$$ESR = \frac{L}{\sigma TWN} \tag{8.5}$$

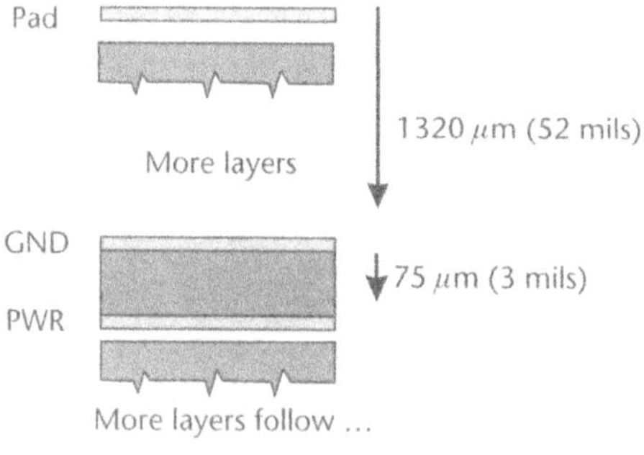

(a) (b)

Figure 8.24 Measurement of a 1210-size 220-μF MLCC sample. (a) Photo of the sample on the fixture and (b) relevant cross-section of the fixture.

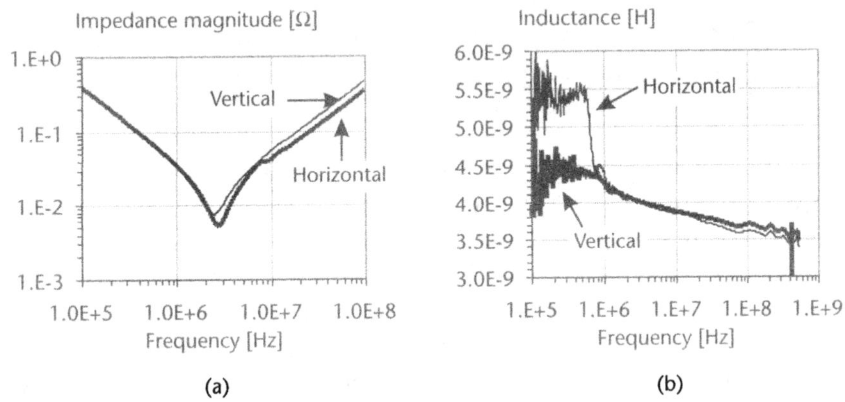

Figure 8.25 (a) Impedance magnitude and (b) inductance of the same 1210-size 220-μF capacitor with horizontal and vertical mounting.

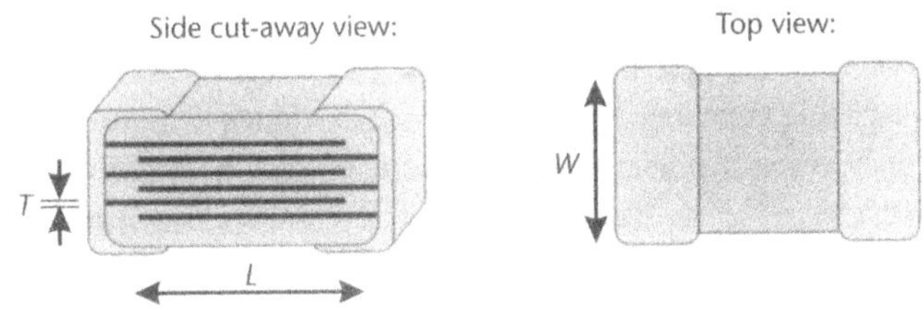

Figure 8.26 Capacitor construction and geometry definition for an MLCC part.

where σ is the conductivity of the capacitor plates, L, W, and T are the length, width, and thickness of the plates, respectively, and N is the number of metal plates. Figure 8.26 and (8.5) refer to regular-geometry MLCCs. For reverse-geometry, parts L and W should be swapped. For a given style and dielectric material, higher capacitance requires a greater number of thinner dielectric layers, which increases N. If N increases more quickly than T decreases, ESR goes down. Note that (8.5) ignores the resistance of the terminals that connect the capacitor plates. The expression also assumes that the current is uniformly distributed over the plates, which is true for frequencies only well below the series resonance frequency.

Without changing the numbers in the above ESR expression, resistance is increased if we force current through narrowed channels on the plates. This is called *patterning*. Patterning can be accomplished by creating narrow voids on the conducting plates, because these voids block the current flow in certain directions. If we pattern only the plates outside the high-frequency current loop, we can increase ESR without increasing the high-frequency inductance. However, to accomplish this we have to assume that the orientation of the capacitor mounting is known. The part of the capacitor through which the high-frequency current loop passes can be declared a keep-out zone. Patterning the capacitor plates outside the keep-out zone will influence ESR with minimal impact on high-frequency inductance. Some of the options are:

- Use vertical capacitor plates and pattern them away from the PCB planes, above the keep-out zone.

- Pattern horizontal plates above the keep-out zone.
- Pattern horizontal plates outside the keep-out zone, horizontally displaced from the high-frequency current loop.

The keep-out zone is illustrated on the sketch of Figure 8.27(a). The current decays gradually with height within the capacitor body; therefore, the boundary of the keep-out zone depends on frequency and on the permissible error. An additional possibility is shown in Figure 8.27(b) where the capacitor is viewed from the terminal end; therefore, the two vias line up, and we see only the one in front. On this sketch, we assume that the capacitor body has an overhang portion that extends beyond the footprint of the high-frequency current loop defined by the terminals and pads. In the construction shown on the sketch the overhang occurs perpendicular to the plane defined by the two terminals. Alternately, the overhang can also occur in line with the terminals, as an extension beyond the pads and vias.

With the overhang construction, the inductance and resistance of the part can be manipulated even without patterning the plates. This is illustrated with striplike, multilayer film capacitors. The plastic film capacitors have low volumetric capacitance, and their primary application is low-distortion audio circuits. For low-inductance applications, plastic film capacitors can be made in reverse geometry form factor with significant length. Figure 8.28 shows the simplified top view of the test board together with the geometry of the tested capacitors. The width of the capaci-

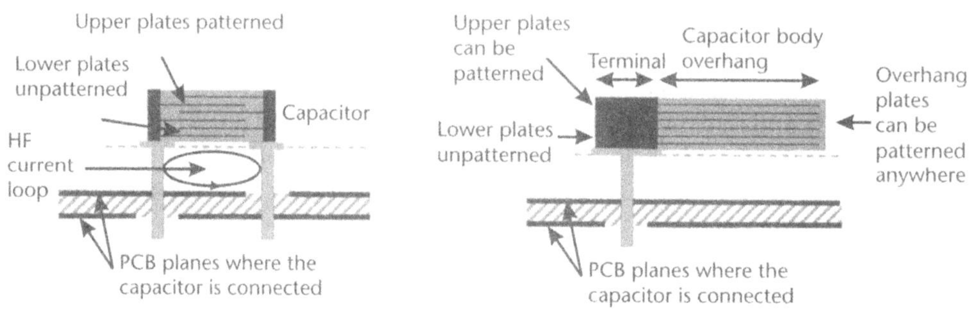

Figure 8.27 Patterning options with horizontal plates. Definition of (a) keep-out zone and (b) overhang construction.

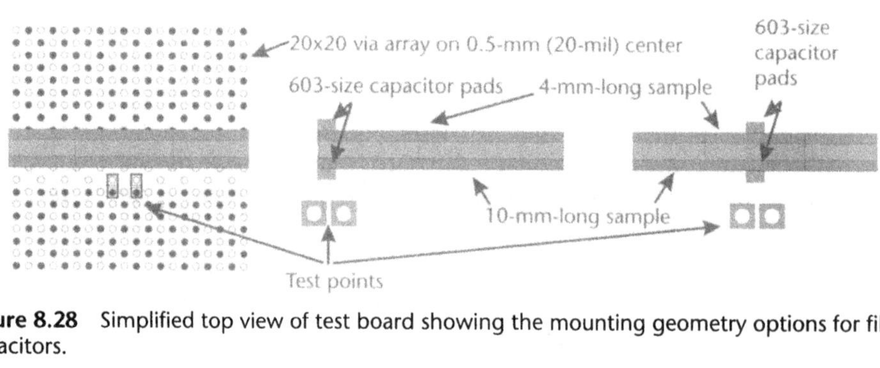

Figure 8.28 Simplified top view of test board showing the mounting geometry options for film capacitors.

tor strip corresponded to the length of a regular 0603-size capacitor: 1.5 mm (60 mils). The lengths of the measured samples were 0.75 mm, 2 mm, 4 mm, 6 mm, 8 mm, and 10 mm. The test site on the left had an array of power-ground vias and pads connecting to the planes below. The sample was always centered, regardless of its length. The measurement was taken at the same via pair next to the middle of the capacitor sample. As the length of the capacitor varied, a different number of vias connected each terminal to the planes. The longest, 10-mm sample used 10 power and 10 ground vias along the length of the part. The test sites in the center and the right had regular-size pads for 0603-size components. The site in the center was used with capacitor bodies always lined up at their ends. The site on the right was used with center-aligned samples.

Figure 8.29 shows the measured impedance of the 4-mm and 10-mm samples in a via array and on regular 603-size pads mounted in the middle and at the end, as defined in Figure 8.28. The impedance profiles show that the electrical properties of long reverse geometry capacitors depend significantly on the connection geometry to the terminals. ESR of the 10-mm-long part can be manipulated in a 10:1 range merely by changing the mounting and connection geometry. As a reference, the impedance magnitude produced by a 0603-length size film capacitor is also shown on the graphs.

Patterning of vertically oriented plates or geometry variations along the vertical axes also have an impact on the inductance and resistance of the mounted parts.

Sketches in Figure 8.30 show vertical capacitor plates. Terminals on the left and right of the plates are omitted from the drawings. The sketch in Figure 8.30(a) shows the possibility of not connecting the entire height of plates to the terminals. The vertical position of entry and exit tabs influence the inductance, whereas the width of the tabs influence the resistance. The sketch in Figure 8.30(b) shows the optional patterning of the vertical plates, which directs the current according to the voids, thus influencing both the resistance and inductance. The two solutions can also be combined.

The default connection of plates in multiterminal capacitors is to connect all of the plates to all of the corresponding terminals, thus minimizing resistance and inductance. We can, however, purposely skip connections to certain terminals. The connection pattern will eventually determine the resistance and inductance of the

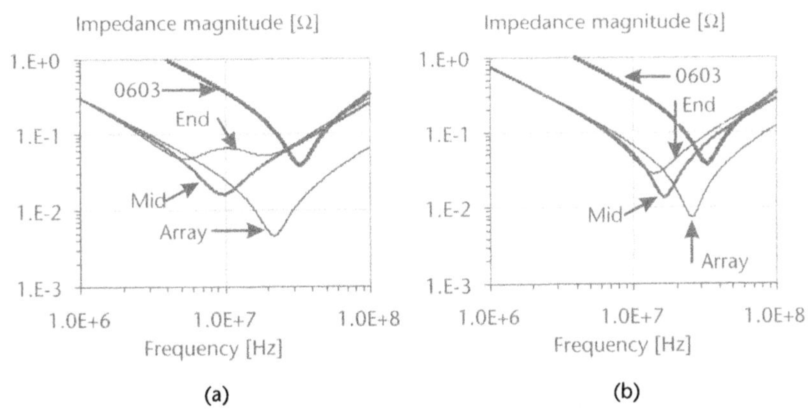

Figure 8.29 Impedance magnitude of (a) a 10-mm-long and (b) a 4-mm-long film capacitor in three different mounting configurations, as defined in Figure 8.28.

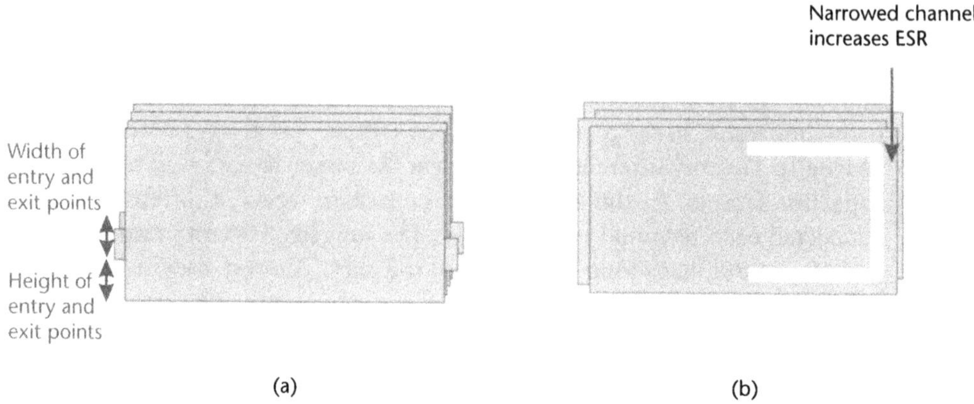

Figure 8.30 Geometry variations influencing the inductance and resistance of MLCCs with vertical plate orientation. (a) Varying the height and width of connection tabs and (b) changing current flow with cutouts on the planes.

part. Figure 8.31 shows the conventional connection scheme in an eight-terminal capacitor, where each capacitor plate has four tabs, connecting either to terminals A-C-E-G or B-D-F-H. In contrast to conventional solutions, we can connect each capacitor plate only to one terminal in a rotating pattern. Due to the thin capacitor plates, all terminals are still connected to plates at a very low height, close to the PCB return path; thus inductance is hardly impacted.

Figure 8.32 compares the impedance real parts and inductances of eight-terminal capacitors from two different manufacturers but with the same case size and same nominal capacitance. Part from vendor A used the conventional geometry, connecting each capacitor plate to all four corresponding terminals. The part from vendor B used a single-terminal plate connection. Note that with the single-terminal plate connection, high-frequency inductance increases only by a few percentage points, whereas resistance increases by a factor of at least three.

8.3.6 Uniqueness of Parameters

We showed in Sections 8.3.2 through 8.3.5 that both inductance and resistance of a mounted capacitor depend on the geometry not only internal, but also external to the capacitor. In Section 8.3.1 we looked at three possible inductance definitions: partial inductance, mounted inductance, and attached inductance. We showed that

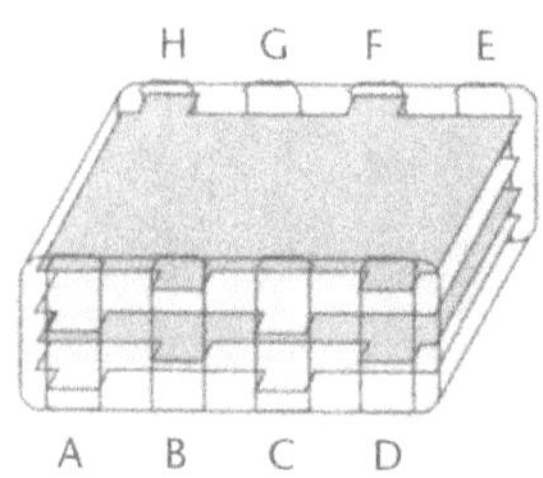

Figure 8.31 Construction of multiterminal capacitor with each plate connecting to four terminals.

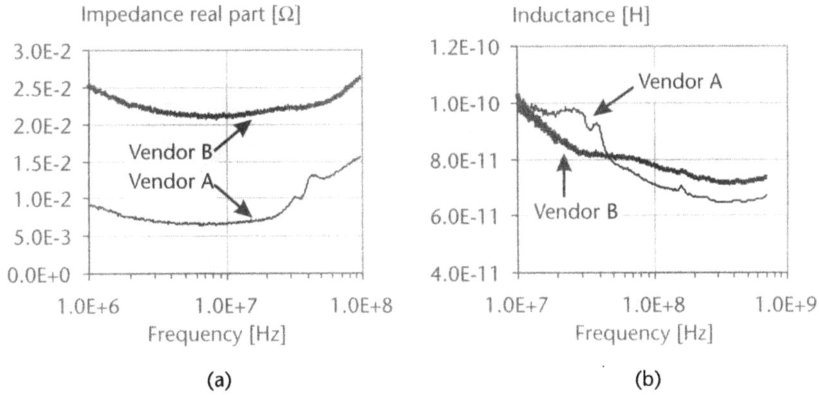

Figure 8.32 Comparison of (a) impedance real part and (b) extracted loop inductance of two eight-terminal capacitors with the same nominal capacitance.

among these three, partial self-inductance, though unique to the part, does not capture the effect of cover thickness. The mounted inductance for the same hookup geometry depends on the location on the planes, and attached inductance alone describes the capacitor and its environment uniquely. The attached inductance, however, requires a detailed model of the fixture and planes. Also, attached inductance is dependent on the surrounding geometry; therefore, we have to determine the attached inductance for the same part for each application geometry. Another possibility is to define an inductance that is the difference between the loop inductances of the fixture with the DUT and the same fixture with a shorting bar. We call this difference the *added inductance* (L_{added}). In a similar manner we can define the *added resistance* (R_{added}) as the difference between the impedance real parts of the fixture's loop impedance with a shorting reference versus DUT.

$$L_{added} = L_{shorted-fixture} - L_{fixture-with-DUT} \tag{8.6}$$

$$R_{added} = R_{shorted-fixture} - R_{fixture-with-DUT} \tag{8.7}$$

Note that some measurement setups provide L_{added} and R_{added} because the subtractions in (8.6) and (8.7) are performed through fixture compensation, or they are built into the calibration process. In this section we compare the mounted inductance and added inductance under various circumstances to see if they uniquely describe the capacitor's performance.

The first set of tests was done with three-terminal face-down capacitors on three different test sites. Figure 8.33 shows the relevant stackup and the test-site layout.

Sites A and B were on the top side of the board. Site A had blind via connections, and sites B and C connected to the planes with through-holes. The layouts of sites B and C were identical; site B was on the top side, and site C was on the bottom, connecting to the planes with longer via barrels. Ten capacitors were selected for the study. The parts were numbered and all parts were measured on all three fixtures. The parts were attached to the fixtures without soldering: silver-filled uncured epoxy paste was put on the terminals, and the parts were pressed onto the pads with a clip, which provided repeatable contact force during the measurements. Data was

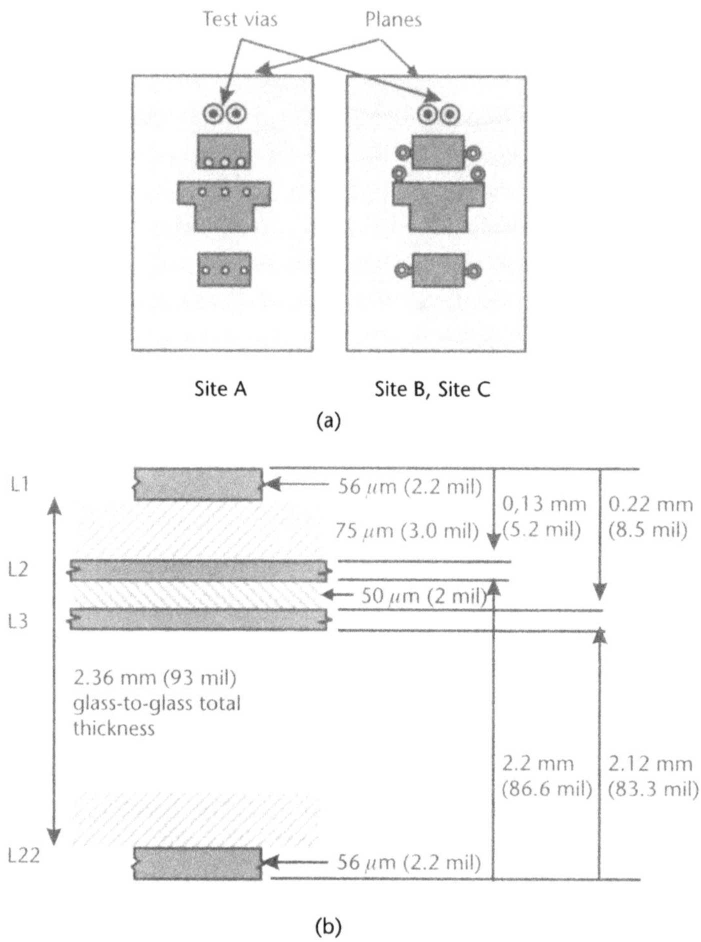

Figure 8.33 (a) Layout and (b) stackup of test sites for three-terminal face-down capacitors.

taken with two different VNA setups, covering the 100-Hz–10-MHz and the 100-kHz–1,800-MHz ranges.

The measured loop inductances were extracted as a function of frequency. Figure 8.34 shows the extracted inductance at 100 MHz for each sample, separately in the three test sites. The shorts were made with the same shorting bar on all three sites. At 100 MHz, the loop inductances of the shorted sites were: 176 pH for site A, 310 pH for site B, and 1.19 nH for site C.

Figures 8.35(a) and 8.35(b) show the impedance magnitude statistics in site A and the inductance versus frequency plots of the three shorted sites, respectively. Note the low inductance of site A and site B. Site C has a much higher inductance because of the longer via path. Note also the consistency of the data: though the data points in Figure 8.34 do exhibit a spread, the inductance values in the three test sites track very closely. The conclusion is that the same capacitor creates a larger absolute inductance increase (more added inductance) on a low-inductance site and smaller absolute inductance increase (less added inductance) on a high-inductance site. This phenomenon causes the ranking of the sites to reverse between the two graphs of Figure 8.34.

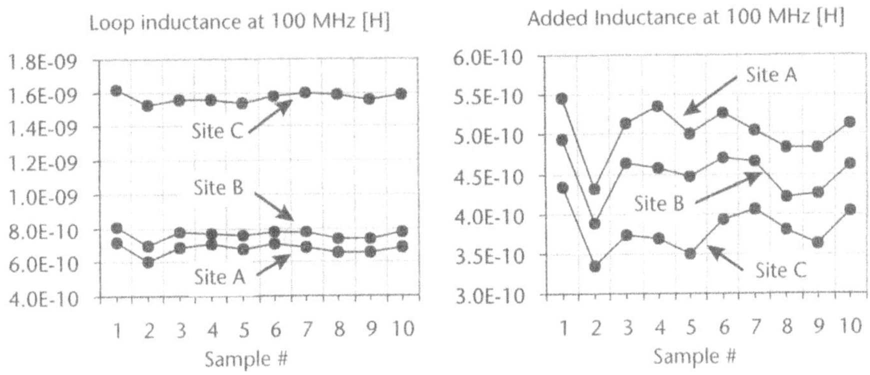

(a) (b)

Figure 8.34 (a) Loop inductance and (b) added inductance of 10 pieces of 7343-size 330-μF face-down capacitor samples in the test sites shown in Figure 8.33.

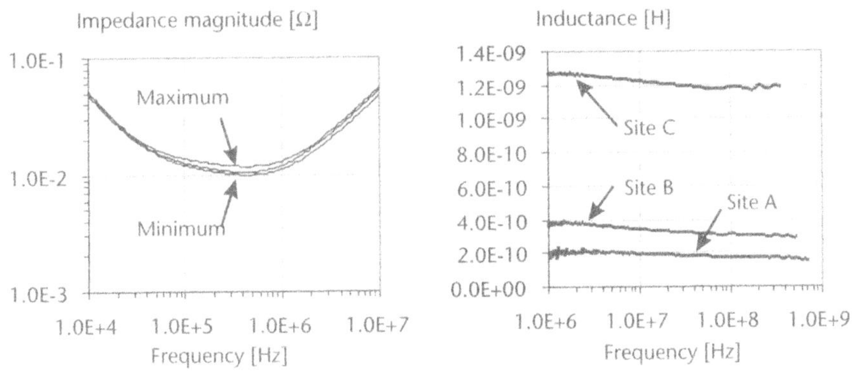

(a) (b)

Figure 8.35 (a) Impedance magnitude statistics of the 10 measured parts in site A and (b) extracted loop inductance of the three shorted test sites.

A similar set of tests were conducted with reverse-geometry 0306-size 1-μF MLCC samples on two different test sites: the first with blind vias, the second with through-holes. The stackup and layout are shown in Figure 8.36. The samples were soldered on the test sites by pushing the parts down all the way to the pads during soldering. The parts were individually identified, and the same side of the part was always facing the pads.

The loop inductance and added inductance for the 10 parts are shown in Figure 8.37. We see the same trend that we saw in Figure 8.34: the same parts consistently show higher added inductance on a test site with lower shorted inductance.

From the illustrations in this section, we conclude that the inductance of a given piece of capacitor depends on both the capacitor itself and the characteristics of the test fixture. Therefore, when components are compared, it is important to characterize them in a test environment that closely resembles the geometry details of the intended application.

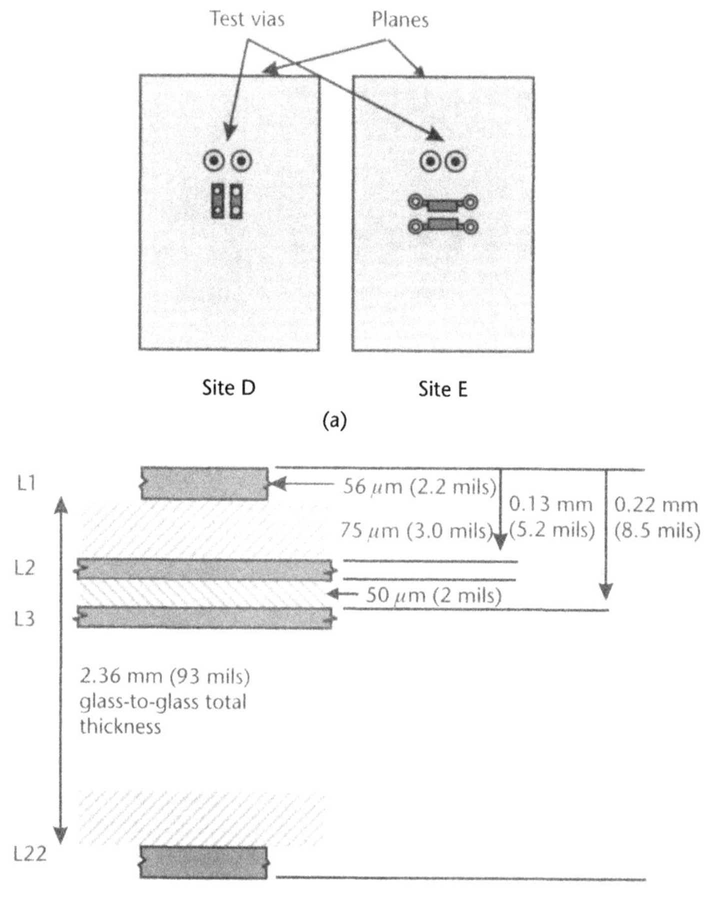

Figure 8.36 (a) Layout and (b) stackup of test sites for 0306-size ceramic capacitors.

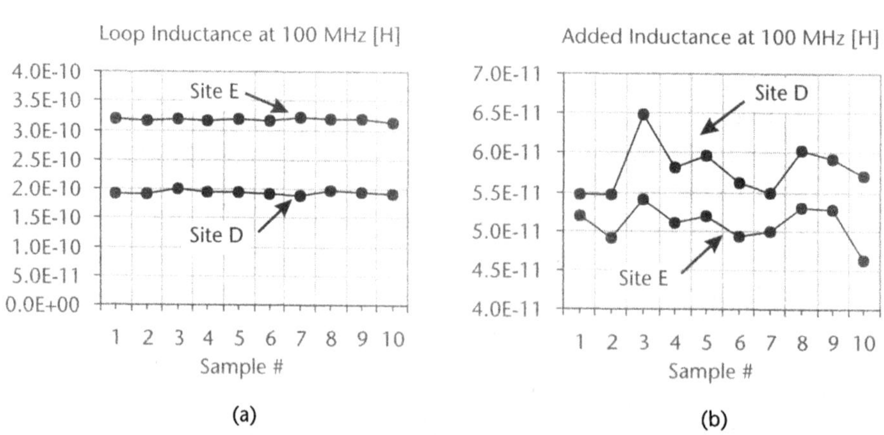

Figure 8.37 (a) Loop inductance and (b) added inductance of 10 pieces of 0306-size 1-μF capacitor samples in test sites shown in Figure 8.36.

8.4 Effect of Other Variables on Capacitor Parameters

In addition to material properties and the internal and external geometry, there are several other variables that may have an impact on the electrical parameters of capacitors [9]. Some of these parameters are expected to have a quasi-static effect, such as dc bias, temperature, and aging.

8.4.1 Effect of DC and AC Bias Voltage and Piezo Effect

Ceramic materials with a high-dielectric constant tend to have significant sensitivity to the imposed electric field, both static and changing. The dc bias voltage, ac bias voltage, and a combination of them may change capacitance. With a given voltage across the part, thinner ceramic layers will result in a higher electrical field, and hence more change.

8.4.1.1 Effect of DC Bias Voltage

Figure 8.38 shows the extracted capacitance versus dc bias voltage as a function of frequency for an organic polymer capacitor. The part was a 7343-size 470-μF capacitor with a 2.5-V rating. The impedance of the sample was measured in a small fixture with a different dc bias and ac source voltages. It was found that the ac source voltage had no measurable effect on the extracted capacitance. When the dc bias was varied with the proper polarity, the capacitance-frequency curve showed just a slight decay. At 100 Hz, the capacitance was 531 μF with a 0-V bias. It dropped to 524 μF with a 2.5-V bias. The capacitance stayed relatively stable up to 1V of reverse bias. With larger negative bias voltages, the extracted capacitance (and also the leakage current) increased sharply at low frequencies.

Figure 8.39 is an illustration of significant sensitivity to dc bias, exhibited by a 100-μF 1210-size ceramic capacitor. The part was 6.3V rated, and it was stressed up to ±10-V dc bias. The nonpolarized ceramic capacitor showed the same bias dependence for both positive and negative bias. The measurement was done with an Agilent 4294A impedance analyzer. Ceramic capacitors with thin dielectric layers are sensitive to both dc and ac bias. To remove the effect of ac bias, the oscillator

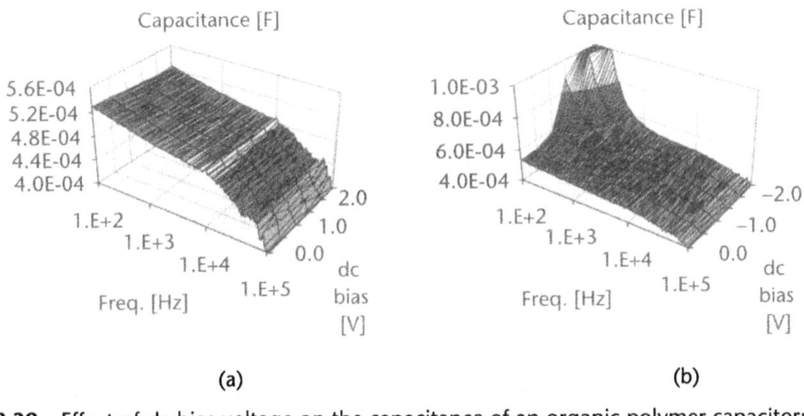

Figure 8.38 Effect of dc bias voltage on the capacitance of an organic polymer capacitor: (a) forward bias and (b) reverse bias.

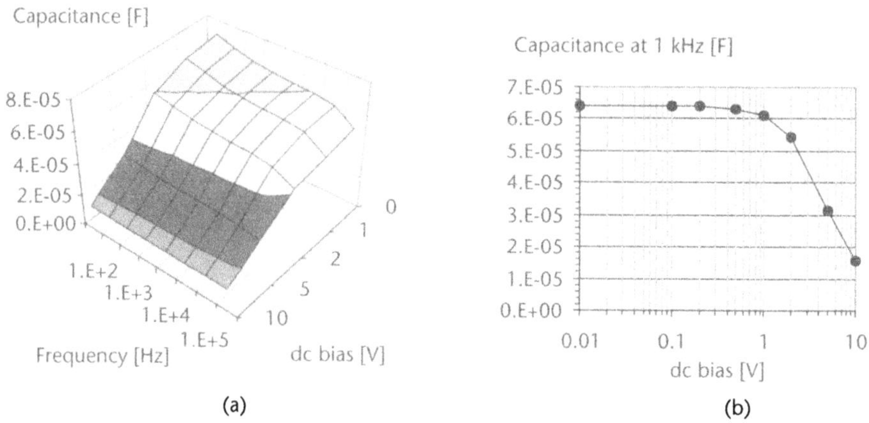

Figure 8.39 Capacitance as a function of bias voltage and frequency measured on a 1210-size 100-μF 6.3V X5R part: (a) surface plot as a function of frequency and dc bias and (b) 2D plot at 1 kHz as a function of dc bias.

voltage was set to 0.1 Vrms. It was found that reducing the oscillator voltage further did not change the capacitance plot noticeably. The horizontal axes are logarithmic on both plots: capacitance hardly changes up to about 1V. Beyond 1-V bias, the capacitance drops sharply.

On Figure 8.40 we show the variation of capacitance with dc bias and frequency on a 4.7-μF 0508-size ceramic capacitor. The measurement was done with an Agilent 4396A VNA, with –3-dBm output power. The logarithmic frequency scale starts at 100 kHz, where the capacitive reactance is already sufficiently small that the ac bias across the DUT was less than 10 mVrms. Note that the two peaks in the extracted capacitance above 8 MHz are the residual artifacts of capacitance extraction. The series resonance frequency was around 5 MHz. This part had relatively smooth and uniform voltage dependence.

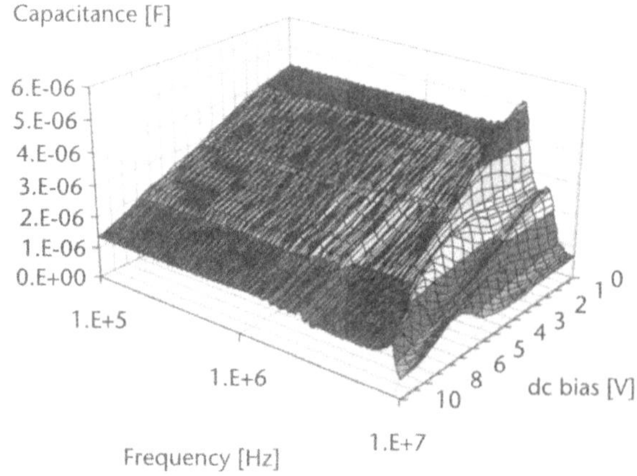

Figure 8.40 Capacitance as a function of bias voltage and frequency, measured on a 0508-size 4.7-μF 6.3V X5R part.

8.4.1.2 Effect of AC Bias Voltage

The first example uses the capacitance versus frequency and ac bias of the same 7343-size 470-μF capacitor that was shown in Figure 8.38. The measurement was taken with an Agilent 4294A impedance analyzer with 0 dc bias and with oscillator voltages ranging from 5 mVrms to 500 mVrms. The change of capacitance in Figure 8.41 is negligible, as the ac bias changes.

Some dielectrics, most notably ceramics with high dielectric-constant, show strong variation of capacitance as a function of ac bias. In Figure 8.42 we show the variation of capacitance as a function of ac and dc bias voltages at 100-MHz frequency. The capacitor was the same 1210-size 100-μF X5R piece that we used for Figure 8.39. The ac oscillator voltage of the Agilent 4294A impedance analyzer was set to 0.1 Vrms, 0.2 Vrms, and 0.5 Vrms.

Note that the capacitance increases significantly as the ac oscillator voltage is increased from 0.1 Vrms to 0.5 Vrms. In bypass-capacitor applications, 0.5-Vrms noise is seldom acceptable. With lower overall noise, the capacitor will exhibit lower capacitance. With 1-V dc bias and at 100 Hz, the capacitance increases from 67.4 μF to 94.4 μF (an increase of 40%), as the ac oscillator voltage is increased from 0.1 Vrms to 0.5 Vrms. Capacitor data sheets usually specify the capacitance with 0.5-Vrms oscillator voltage, so in a bypass-capacitor application, we most likely will obtain less capacitance from the parts. Also, as it was shown in Section 6.4.2, in most instrumentation the ac bias voltage across the DUT changes with frequency and setup.

8.4.1.3 Piezoelectric Effect

Ceramic capacitors are ferroelectric devices; as such, they show a more or less piezoelectric effect [10]. *Piezoelectric effect* means that the part behaves like an electroacoustic transducer: it translates electrical energy to mechanical and vice versa. For a dc voltage bias the capacitor tends to bend; if we bend the capacitor, it produces a voltage. When excitation varies with time, the part may show a series of resonances as a function of frequency. Tantalum, niobium, electrolytic and Class-I ceramics (COG) capacitors do not exhibit piezoelectric effects.

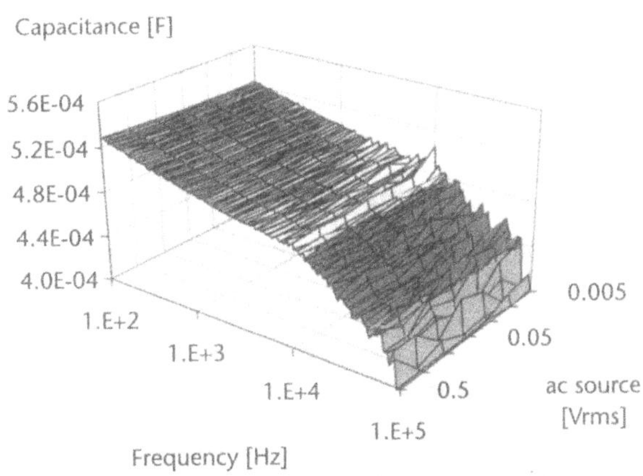

Figure 8.41 Effect of ac bias voltage on the capacitance of an organic polymer capacitor.

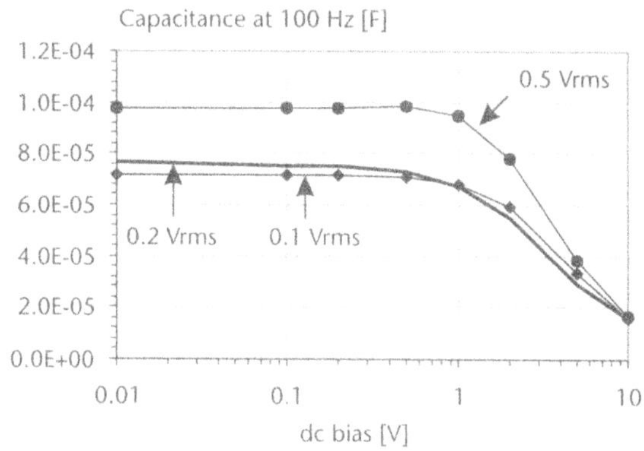

Figure 8.42 Capacitance versus ac and dc bias of a 1210-size 100-μF X5R ceramic capacitor.

Typically, in bypass applications of high-power digital circuits, the conversion of mechanical vibrations to electrical signal is not a problem. However, it may be a problem in bypassing of low-power or low-noise analog circuits. The more disturbing effect occurs when the noise voltage across the capacitor is translated into acoustical noise. The piezoelectric effect is more pronounced with a higher relative dielectric constant of the ceramic material (or as the dc and ac excitation field strength increases). The field strength in the ceramic material depends linearly on the voltage applied across the device and inversely on the thickness of individual ceramic layers. As a result, the piezoelectric effect is a concern primarily in ceramic capacitors with the highest available capacitance in the given case size.

The piezoelectric effect also shows up in impedance profiles. Figure 8.43 illustrates this with the impedance magnitude and impedance real part plots of the 4.7-μF 0508-size capacitor that we used in Figure 8.40. With 0-V dc bias, the impedance magnitude plot follows the smooth V curve around the series resonance frequency. As the dc bias increases, the impedance magnitude shifts upwards, the series resonance moves towards higher frequencies indicating loss of capacitance, and ridges appear at frequencies below the series resonance. Similar ridges are also present above the series resonance, but those are caused by secondary resonances in the capacitor body, so they are present at all dc bias levels. The piezoelectric resonances can be seen even more clearly on the surface plot of the impedance real part. With 0-V dc bias, the plot is smooth up to 5 MHz, where the secondary resonances of the capacitor construction begin to emerge. With increasing dc bias, ridges and peaks appear at lower frequencies as well.

8.4.2 Effect of Environmental Variables

Environmental variables, such as temperature, humidity, and time, may also have an impact on electrical parameters.

8.4.2.1 Effect of Temperature

Temperature has an impact on both the capacitance and resistance of bypass capacitors. Since inductance is primarily determined by the geometry of the current flow, it shows minimal variation with environmental effects.

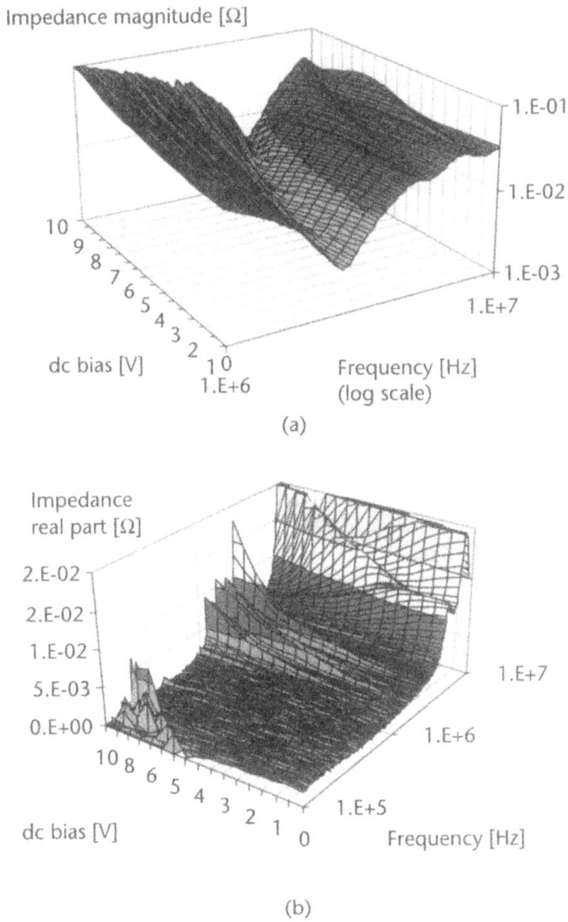

Figure 8.43 (a) Impedance magnitude and (b) impedance real part of the ceramic capacitor from Figure 8.40, showing the signs of piezo-electric resonances.

Figure 8.44 shows the temperature dependence of electrolytic, tantalum, and ceramic capacitors. Figure 8.45 shows the aging of some ceramic capacitors. Note that both figures are for illustration purposes only; for actual data on a particular device, the data sheet has to be consulted.

8.4.2.2 Aging

The capacitance drops with time in capacitors. The capacitance drops approximately with a constant percentage over each decade of time. This results in a straight line on a linear-logarithmic plot. A static adjustment of the capacitance in the electrical model is sufficient.

Resistance is also temperature and time dependent in some types of capacitors. Most sensitive are aluminum electrolytic capacitors. In ceramic capacitors, the resistance is determined by the metallization, which has its own temperature coefficient.

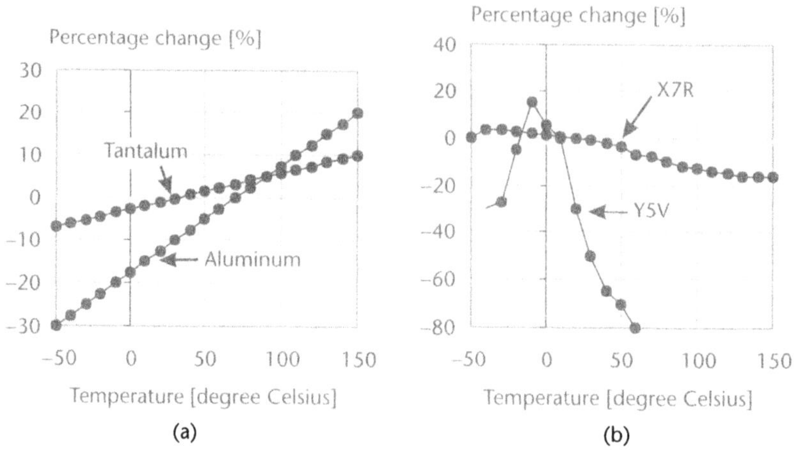

Figure 8.44 Temperature dependence of various capacitors: (a) aluminum electrolytic and tantalum capacitors and (b) X7R and Y5V ceramic capacitors.

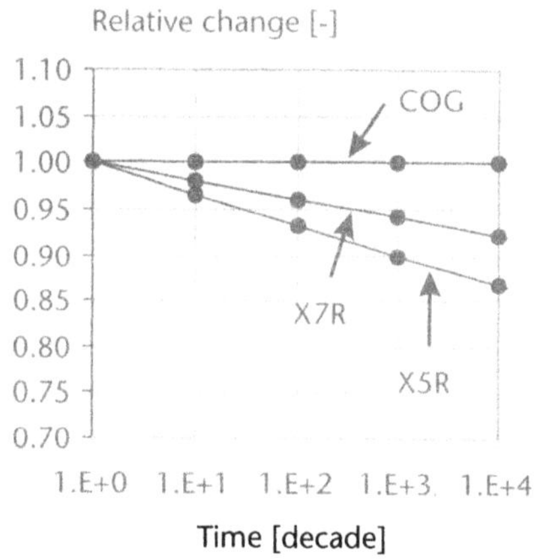

Figure 8.45 Capacitance versus time of various ceramic capacitors.

8.5 Multicomponent *C-R-L* Models

In Section 8.1 we examined simple *C-R-L* models and concluded that those models can capture the first-order behavior of the capacitors: the capacitive region, series resonance, and the inductive region. In Sections 8.3 and 8.4 we saw that capacitance, resistance, and inductance are functions of frequency and that they depend on internal and external geometry, as well as on environmental factors. For frequency-domain characterization, the primary concern is the frequency dependency of the *C-R-L* elements in the equivalent circuit, due to material characteristics and internal/external geometry. Environmental effects usually vary slowly, so they can be taken into account by statically adjusting the parameters in the model.

Multicomponent models are built of frequency-independent *C-R-L* elements, so they are compatible with both frequency-domain and time-domain simulators.

8.5.1 Multicomponent Models for Bulk Capacitors

In bulk capacitors, the important frequency-dependent parameters are capacitance and resistance. Though inductance is also frequency dependent, it is usually a weak function of frequency and a model with a constant inductance is suitable for many applications.

In bulk capacitors, the capacitance (and also the real part of impedance) may be strongly frequency dependent. This is due to the combined effect of large capacitance, (optionally) low ESR, and internal geometry. Simple spreadsheet approximations reveal that a three-element *C-R-L* equivalent circuit with frequency-independent values may have a noticeable error of impedance prediction close to the series resonance frequency. As an example, we apply a six-element model to the capacitor from Figure 5.20. Figure 8.46 shows the model with the values used for the correlation together with the impedance-magnitude correlation. The impedance plot of the measured data and the six-element *C-R-L* impedance approximations show good correlation.

To check the correlation further, we examine the imaginary and real parts of the impedance. The correlation of phase and real part on the same set of data is shown in Figure 8.47.

The correlation on the phase plot appears to be good, even though the modeled phase plot has a steplike approximation that weaves around the measured data. The modeled real part of impedance, however, shows significant deviation from measured values: it flattens below 2 kHz and above 200 kHz, whereas the measured data increases monotonically in these regions.

To achieve a better approximation, eventually we will use frequency-dependent *C*, *L*, and *R* values. However, this approach will limit the application of the equivalent circuit to ac simulations. Alternately, more elements can be added to the model. Figures 8.48 and 8.49 show a 10-element model for an organic polymer capacitor together with the correlation of impedance magnitudes and impedance real parts [11].

The above examples show that bulk capacitors can be modeled with reasonable accuracy with a set of frequency-independent *C-R-L* components. The difficulty is that each model may require a different topology to achieve a reasonable accuracy; this makes the creation of generic, parameterized models difficult.

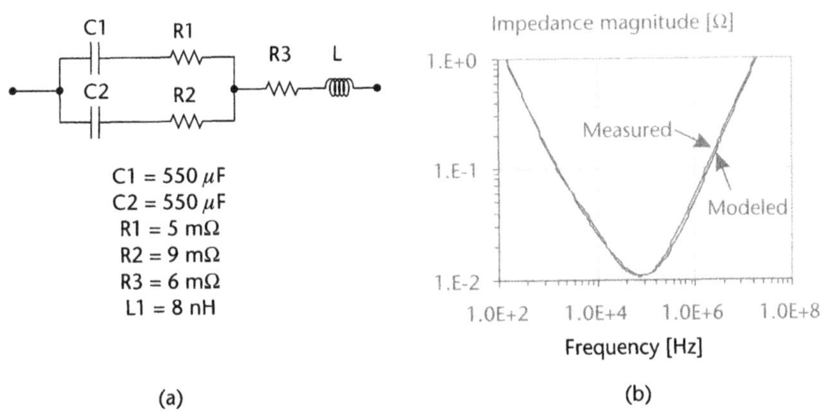

Figure 8.46 (a) Six-element equivalent circuit and (b) its correlation for the 1,200-μF polymer capacitor.

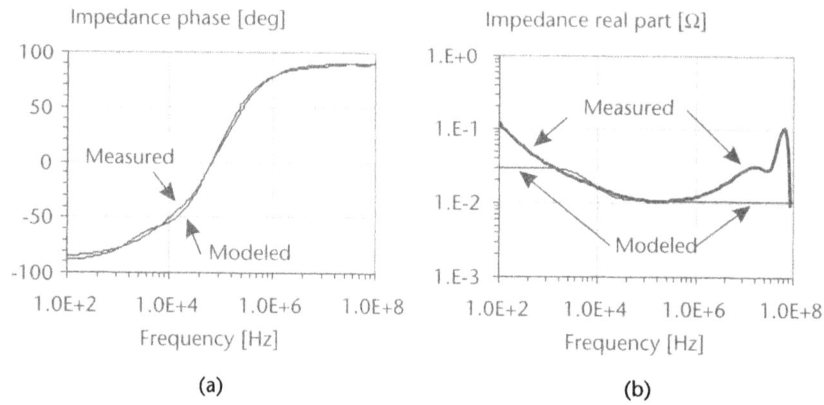

Figure 8.47 Phase of impedance (a) and impedance real part (b) correlation for the 1,200-μF polymer capacitor with the five-element model from Figure 8.46.

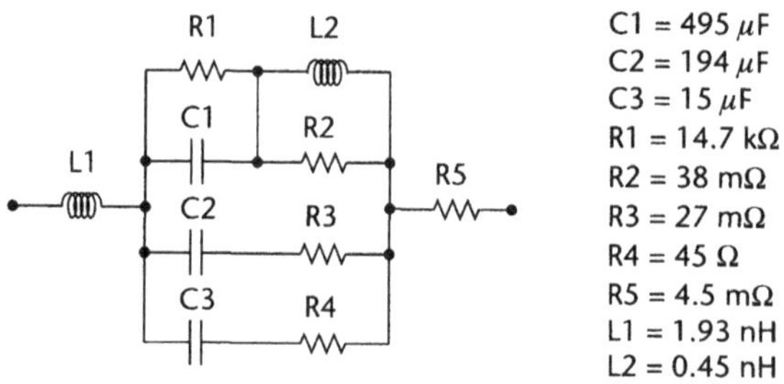

Figure 8.48 The 10-element equivalent model for an organic polymer capacitor. (Data courtesy of Sanyo.)

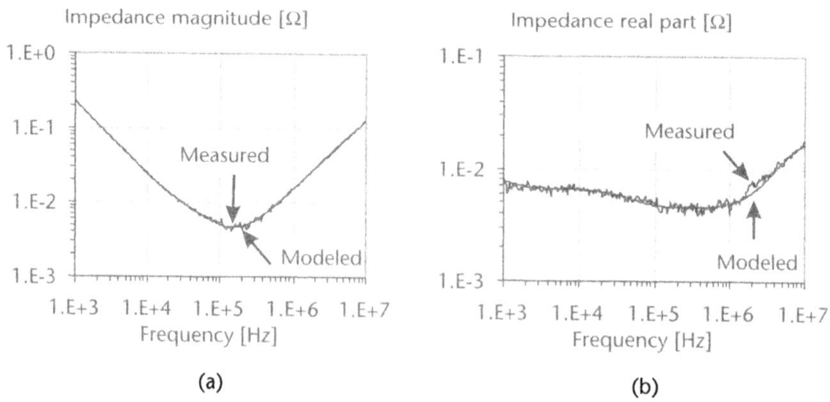

Figure 8.49 Correlation of 10-component model for 2R5TPF680M6L organic polymer capacitor: (a) impedance magnitude and (b) real part of impedance. (Data courtesy of Sanyo.)

8.5.2 Multicomponent Models for Ceramic Capacitors

In ceramic capacitors, the frequency-dependent inductance may be equally important to capture. The frequency dependency of inductance is more pronounced with aggressive low-inductance mounting or in capacitors with more complex internal geometry. As the first illustration, we look at the modeling of an eight-terminal capacitor [12]. Figure 8.50 shows the photo of the test card that was used in the characterization. The upper two photos show the bare test board in top view and bottom view. The two photos in the bottom row show the test site shorted and with DUT mounted. The test site has a combination footprint, which can take 8-terminal and 10-terminal parts, sharing four pads. The test board had three metal layers and an approximate thickness of 250 μm. The top pad layer was 50 μm above the first plane. The second plane on the bottom layer had the opening around the through-holes connecting to the capacitor pads and to the first plane. Connections to the wafer probes were on the bottom.

The measurements were conducted with an Agilent 8358A VNA in the 0.3-MHz–9-GHz-frequency range with 500-μm picoprobes. The graph in Figure 8.51(a) shows the measured impedance magnitude and the correlation with a regular three-component model. The simple *C-R-L* model captures the impedance magnitude with sufficient accuracy up to about 20 MHz. Above 20 MHz the measured impedance is lower than it was predicted to be, suggesting that the actual inductance is lower at higher frequencies than what the lumped model predicts. More complex models can be created in several different ways. The graph in Figure 8.51(b) shows the correlation with a nine-element model, where the frequency dependent inductance is achieved with three branches of R-L circuits, while the capacitance is represented with a single element.

The two equivalent circuits are shown in Figure 8.52. The nine-element model reproduces the performance of the capacitor very well up to about 6 GHz. Good correlation was achieved in this case with the nine-element model, because the

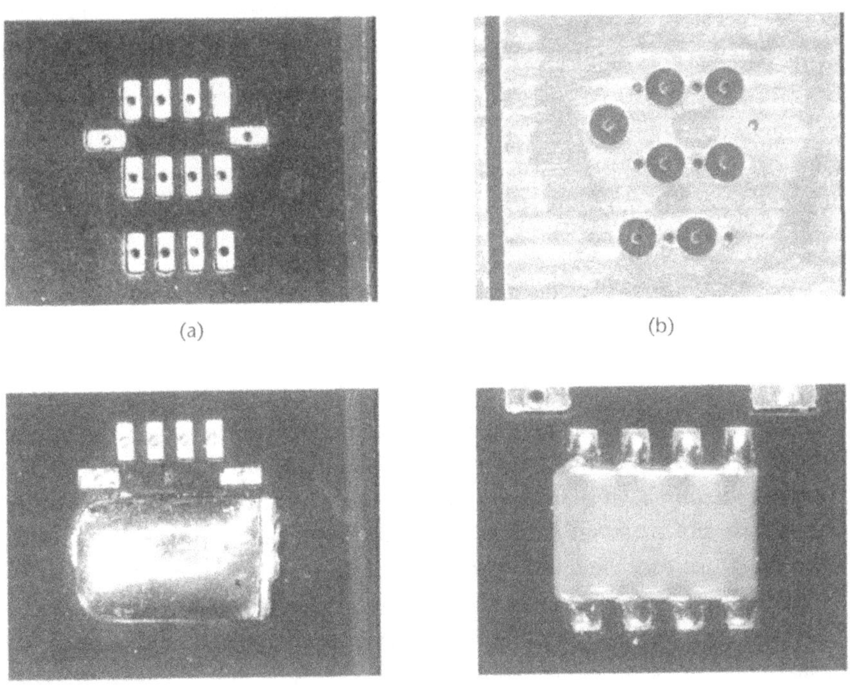

(a) (b) (c) (d)

Figure 8.50 Test board for the characterization and modeling of an 0508-size eight-terminal capacitor. (a) Top view of board with a combination of 8- and 10-terminal footprints. (b) Bottom view. (c) Top view with shorting bar. (d) Mounted capacitor. (Photographs courtesy of AVX.)

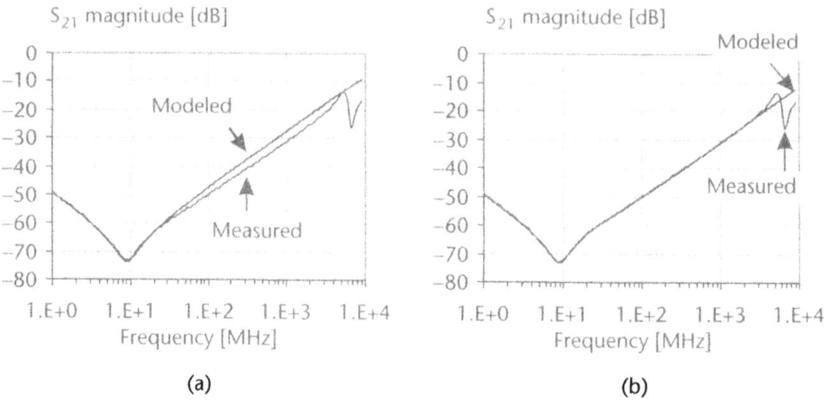

Figure 8.51 Measured and modeled S_{21}magnitude of a 0508-size eight-terminal capacitor in the test board shown in Figure 8.50: (a) correlation with a three-element lumped model and (b) correlation with a nine-element branched model. (Data courtesy of AVX.)

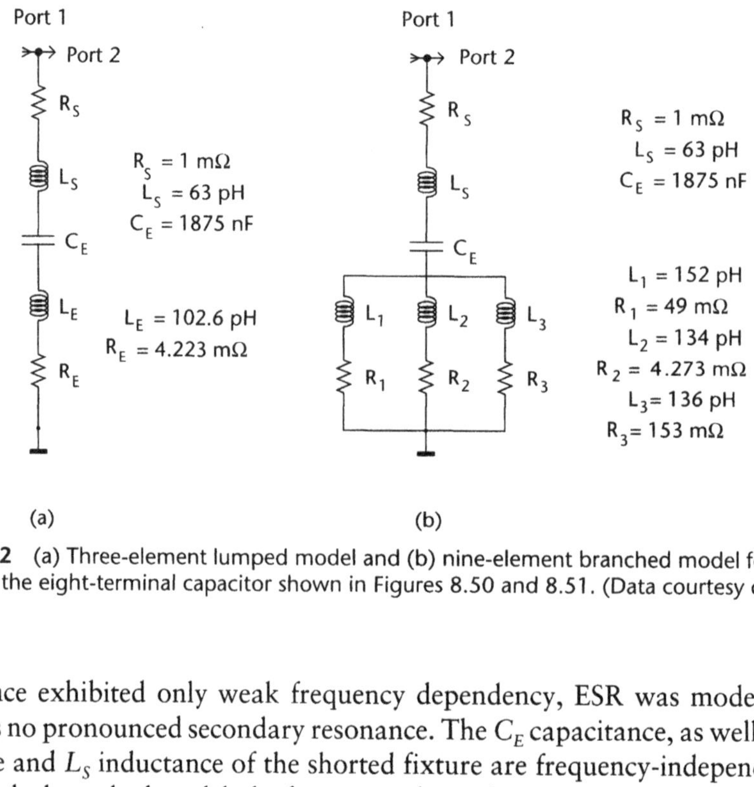

Figure 8.52 (a) Three-element lumped model and (b) nine-element branched model for the correlation of the eight-terminal capacitor shown in Figures 8.50 and 8.51. (Data courtesy of AVX.)

capacitance exhibited only weak frequency dependency, ESR was moderate, and there was no pronounced secondary resonance. The C_E capacitance, as well as the R_S resistance and L_S inductance of the shorted fixture are frequency-independent constants. In the branched model, the frequency-dependent resistance and inductance of the part are modeled with the parallel, connected three branches of *R-L* circuits. Note the small resonance at 5.6 GHz in the measured data caused by the parallel resonance between the fixture's 10-pF capacitance and the inductance of the mounted part.

Our next example is a low-ESR low-ESL 0508-size X5R ceramic capacitor. The capacitor was mounted with multiple vias to the planes of a small test fixture close to the surface. This aggressive mounting, together with the low ESR of the part, creates pronounced secondary resonances that cannot be captured by the previous models. The measured impedance profile of the part with the extracted capacitance and inductance are shown in Figure 8.53.

The secondary resonance is created by the interaction of the inductance of the vertical terminals and the distributed *C-R-L* impedance of the horizontal plates. First, we apply a ladder model from [4] to this measured data. Next, in Sections 8.7 and 8.8, we will use the same capacitor data for different models. Figure 8.54 shows the 10-leg ladder model with the component values. Figure 8.55 shows the correlation.

This 44-element model is suitable for either time-domain or frequency-domain simulations. It predicts the frequency of the secondary resonances with reasonable accuracy.

8.6 Black-Box Model

Instead of multielement lumped models, we can use the simple equivalent circuit of Figure 8.2(c). The complex frequency dependency of C, ESR, and ESL can, in general, be taken into account by turning the three elements frequency dependent [13]. This will prevent us from using the models in some time-domain simulators, but the model fits well to frequency-domain simulators and spreadsheetlike solutions. This

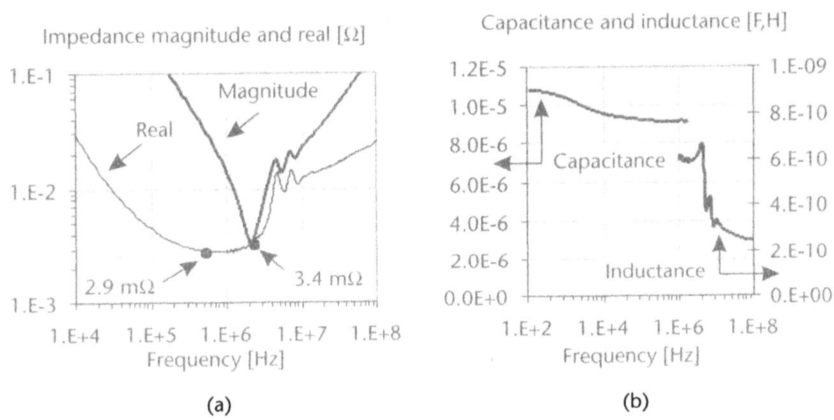

Figure 8.53 (a) Measured impedance magnitude of a 10-μF 0508 MLCC with the real part of the impedance and (b) extracted capacitance and inductance versus frequency.

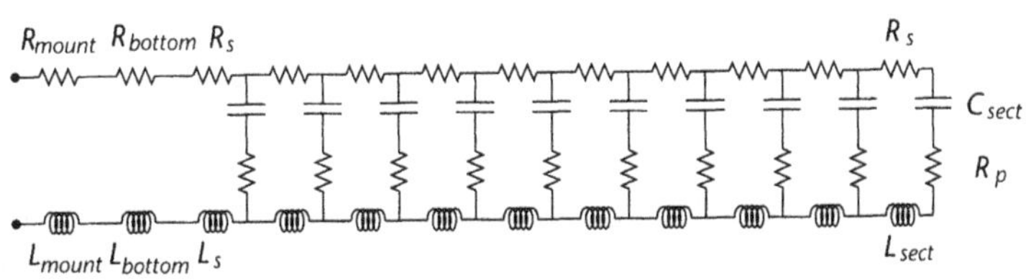

Figure 8.54 Ten-leg ladder model of ceramic capacitor. In the ladder portion of the schematics all legs are identical with R_s = 0.5 mΩ, L_{sect} = 70 pH, C_{sect} = 1 μF, and R_p = 1 mΩ. The connection components are L_{bottom} = 150 pH, R_{bottom} = 0.5 mΩ, L_{mount} = 150 pH, and R_{mount} = 0.5 mΩ.

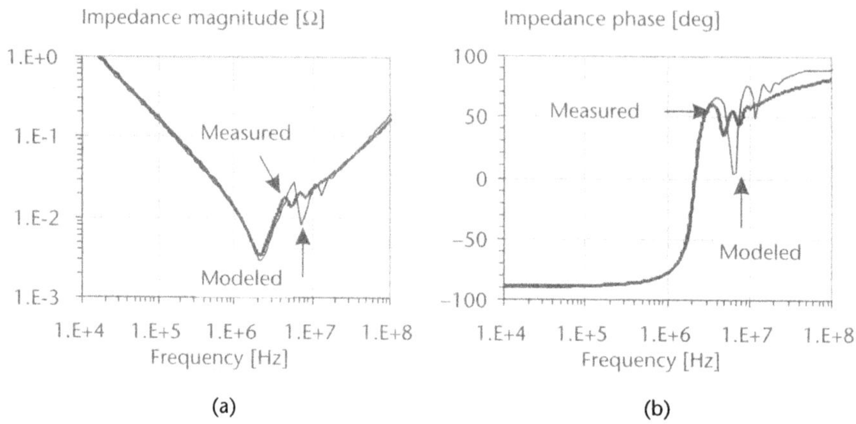

Figure 8.55 Correlation with 10-leg ladder model of capacitor: (a) correlation of impedance magnitude and (b) correlation of phase of impedance.

approach follows the widely used standard solution to describe the complex propagation of high-speed interconnects, traces, and transmission lines with frequency-dependent *R-L-C* parameters.

8.6.1 The Building Blocks

In black-box modeling we do not necessarily need to understand the physical cause of the behavior to be modeled. However, we do need to know how the signatures vary over the range of parameters to be modeled so that we can find an effective way to describe them. With a small set of measured data, we could find fairly accurate curve fitted functions under any arbitrary circumstances. However, small datasets may lead to different approximation functions for different cases, while knowledge of the general trends would enable us to find common functions. If we can find simple building blocks so that each block is just a weak function of frequency, the actual response can be constructed as the linear superposition or product of the building blocks.

In frequency-domain characterization of electronic circuits many response functions have a simpler expression if plotted on logarithmic frequency scale. In Chapter 4, we saw that the dielectric constant and loss tangent of causal models change with a fixed percentage over logarithmic frequency changes. This behavior can be captured with exponential functions, for instance the following multiplier:

$$\frac{1}{\left(1+\left(\frac{f}{f_0}\right)^2\right)^m} \tag{8.8}$$

In (8.8), f is frequency, f_0 is the corner frequency (or the start of the slope), and m sets the slope. Figure 8.56(a) illustrates the exponential term with a logarithmic frequency scale, normalized to f_0. Below the corner frequency, all traces saturate at one. At higher frequencies, all traces approach zero. With small slope, the traces approxi-

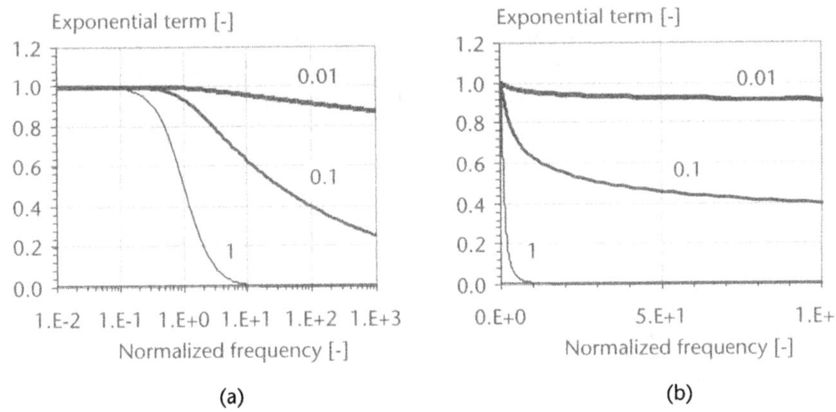

Figure 8.56 Illustration of the exponential term in the black-box model: (a) logarithmic frequency scale and (b) linear frequency scale.

mate a straight line on the linear-logarithmic scale. We will use this building block to describe the frequency-dependency of capacitance and inductance.

Steplike changes with different transitioning slopes can be approximated with sigmoidal functions [14]. The building block of (8.9) has an f_0 corner frequency and a slope parameter, m.

$$\frac{1}{1+e^{\frac{\log\left(\frac{f}{f_0}\right)}{m}}} \tag{8.9}$$

The sigmoidal building block is illustrated in Figure 8.57 with three different m values. The frequency scale is normalized to the corner frequency. The function always starts from one at low frequencies and approaches zero at high frequencies. A multiplier in the nominator can be used to set the step magnitude to values other than one. Similar to the exponential building block, plotting the sigmoidal expression on linear frequency scale yields a more complex shape. This sigmoidal building block will be used to capture the sudden change of inductance and resistance near the secondary resonances.

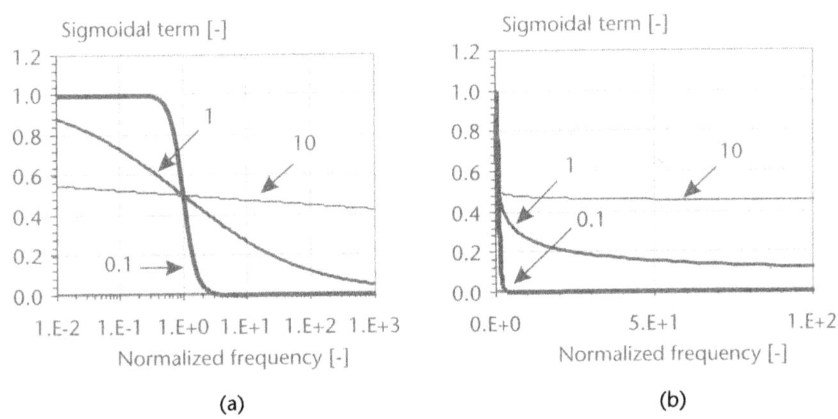

Figure 8.57 Sigmoidal function: (a) logarithmic frequency scale and (b) linear frequency scale.

There are parameters, for instance the real part of impedance, which follow a more-or-less even function around an f_0 center frequency on the logarithmic scale. A slope parameter, m, can be included in the power term to adjust the steepness of wings:

$$\left|\log\left(\frac{f}{f_0}\right)\right|^m \tag{8.10}$$

The power block is illustrated in Figure 8.58. The linear frequency scale shows a shape that would be more difficult to capture with a simple function.

This building block will be used to model the real part of impedance. In the next section we will use these three building blocks to approximate the measured impedance of bypass capacitors.

8.6.2 Modeling of Capacitance Versus Frequency

To find the proper model, first we examine the measured data to see how the extracted capacitance of different capacitors varies from low frequencies up to the series resonance frequencies [13].

Figure 8.59 shows the extracted capacitance of various bulk and ceramic capacitors measured from 100 Hz upwards. Four capacitors in Figure 8.59(a) are radial type organic polymer parts in cans of different diameters and heights. The 2R5TPL330M capacitor is a three-terminal face-down low-height part. Figure 8.59(b) shows three MLCC parts, exhibiting various degrees of capacitance drop as frequency increases. The curve for each capacitor is truncated at SRF.

The measured data suggests that the capacitance follows a constant-percentage drop model segment-by-segment with varying slopes. As shown in Chapter 4, this behavior is characteristic of having constant loss tangent for each segment. Each segment with constant slope can be modeled with an exponential term. It was found that three exponential terms with an additional C_0 multiplier can model the fre-

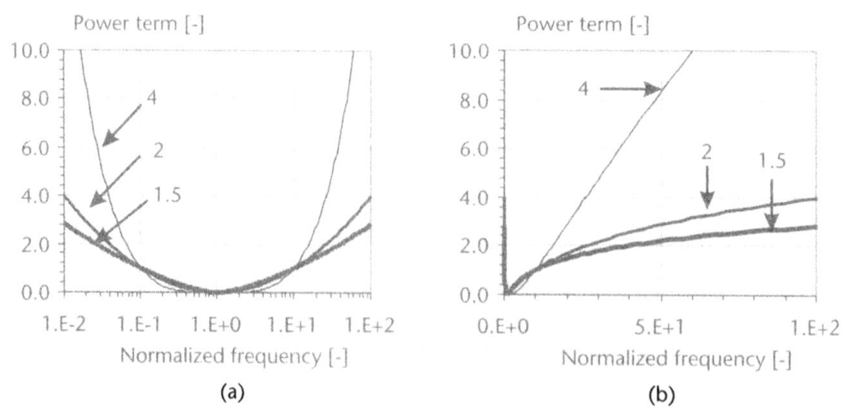

Figure 8.58 Illustration of the power block: (a) logarithmic frequency scale and (b) linear frequency scale.

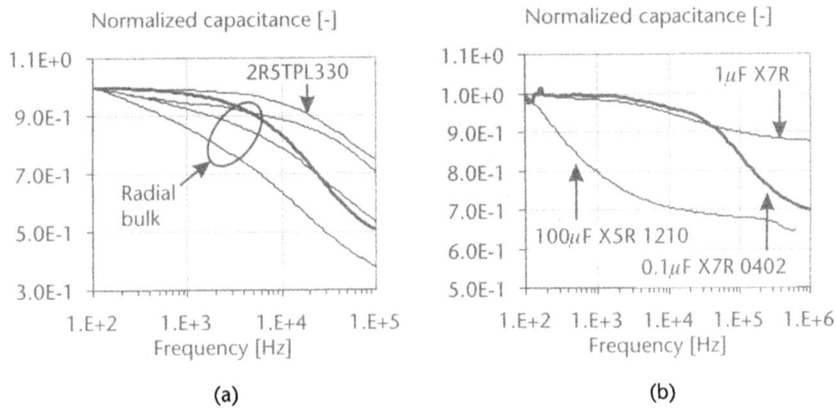

Figure 8.59 Measured capacitance versus frequency curves: (a) bulk capacitors and (b) multilayer ceramic capacitors.

quency dependent capacitance of most parts. In the formula, C in the indices refers to capacitance:

$$C_a = \frac{C_0}{\left(1+\left(\frac{f}{f_{C1}}\right)^2\right)^{m_{C1}} \left(1+\left(\frac{f}{f_{C2}}\right)^2\right)^{m_{C2}} \left(1+\left(\frac{f}{f_{C3}}\right)^2\right)^{m_{C3}}} \tag{8.11}$$

The process is illustrated in Figure 8.60, and the curve-fit parameters are shown in Table 8.4. The complex impedance of a 1,500-μF organic polymer radial-can bulk capacitor was used for the example, since this capacitor requires all three segments in the approximation. The extracted capacitance was automatically curve fitted by varying the seven parameters. The squares of the relative differences between measured and modeled capacitances were summed over the frequency range of interest from 100 Hz up to SRF (which for this part was around 100 kHz). The solver function of spreadsheet software was used to perform the curve fitting automatically, by minimizing the error sum. Alternately, a simple macro was written to step each parameter systematically and find the best fit.

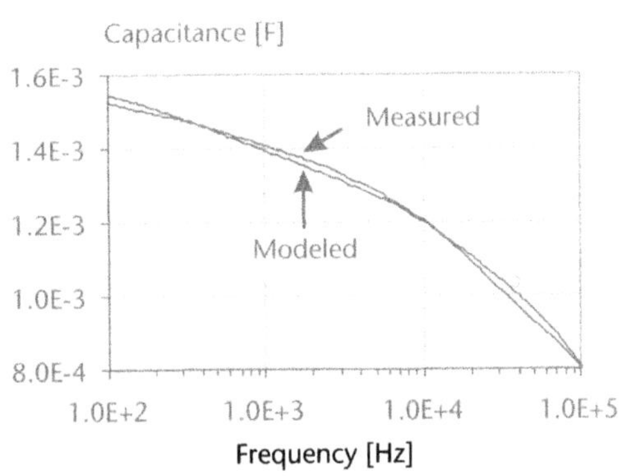

Figure 8.60 Curve fit of the extracted capacitance of a 1,500-μF organic polymer capacitor with (8.11). The curve-fitted parameters are shown in Table 8.4.

Table 8.4 Curve-Fit Parameters for Figure 8.60

Parameter	*Value*
C_0 [μF]	1,580
m_{C1} [—]	0.0254899
f_{C1} [Hz]	90
m_{C2} [—]	0.045951
f_{C2} [kHz]	10
m_{C3} [—]	1.54853
f_{C3} [kHz]	400

8.6.3 Modeling of Inductance Versus Frequency

The inductance of capacitors may vary in a complex way with frequency. In bulk capacitors, the variation of inductance with frequency is not as pronounced or strong because (usually) it follows a one-pole exponential rolloff between the dc and high-frequency values. In tall MLCCs, secondary resonances create a sudden drop of inductance at frequencies two to three times above SRF. Beyond the secondary resonance, the inductance drops smoothly and slowly so it can be approximated with a one-pole exponential roll off, similar to modeling the capacitance drop. The sudden drop of inductance is not monotonic; as we saw in Section 8.3.3, inductance rises before it starts to drop sharply. We model this drop of inductance with a sigmoidal function and ignore the nonmonotonic ringing. The slow decay of inductance at higher frequencies is a good match for an additive exponential term. An L_{inf} high-frequency asymptotical inductance is added to complete the approximation:

$$L_a = L_{\text{inf}} + \frac{L_1}{\left(1+\left(\frac{f}{f_{L1}}\right)^2\right)^{m_{L1}}} + \frac{L_2}{1+e^{\frac{\log\left(\frac{f}{f_{L2}}\right)}{m_{L2}}}} \tag{8.12}$$

Figure 8.61 shows the extracted and approximated inductance versus frequency of a reverse-geometry MLCC. The exponential and sigmoidal terms in the approxi-

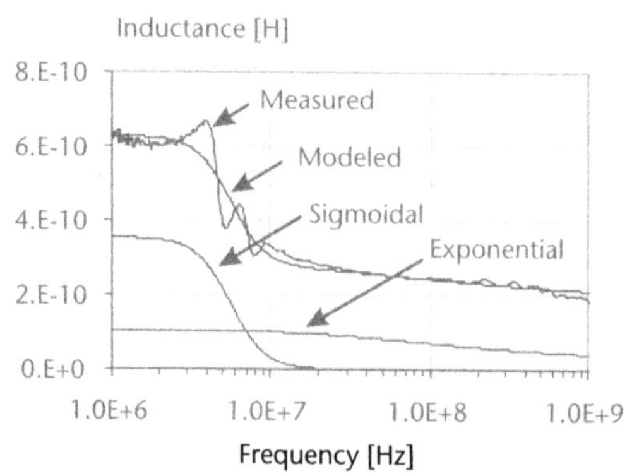

Figure 8.61 Inductance modeling. The extracted inductance from the measured impedance of a 10-μF 0508 reverse-geometry MLCC part, together with the approximation using (8.12). For the approximation, we used the values from the first column of Table 8.5.

mation formula are also plotted separately. Similar to fitting measured capacitance, the curve fitting of inductance can be performed manually or semiautomatically. In manual fitting, the parameters can be changed in the spreadsheet, then the result can be iterated quickly, based on the visual feedback. Table 8.5 shows the minimum and maximum limit values that were used by the macro to minimize the error function. Note that for many bulk capacitors, the sigmoidal component may not be necessary. Without changing the expression, we can remove the sigmoidal function from the approximation by setting $L_2 = 0$.

The only nonautomated step of the curve-fitting process is the selection of reasonable minimum and maximum values to guide and bound the fitting procedure. Note that without any user-supplied limits, the solver function of spreadsheets may fail to find the correct solution: one of the reasons is the high sensitivity of the sigmoidal function to the m_{L2} parameter.

8.6.4 Modeling of Resistance (ESR) Versus Frequency

When capacitor losses are transformed into a single series equivalent resistance value, the frequency dependency tends to follow a bathtub curve with a minimum near the series resonance frequency. At low frequencies, the parallel dielectric losses increase the series resistance. At high frequencies, similar to skin and proximity effects in conductors, the current loop reassigns itself to minimize inductance, filling only a portion of the capacitor body; hence, resistance also increases at high frequencies. This behavior can be captured with the power term of (8.10).

In tall MLCCs at frequencies where inductance drops steeply, measured and simulated data shows that ESR increases in a similar sharp manner. As with inductance, this sudden increase of ESR is not seen in most bulk capacitors. The sharp increase, except the ringing of the secondary resonance, can be captured with a sigmoidal function. Together with the constant R_{min}, seven parameters are used to approximate resistance:

$$R_a = R_{min} + \frac{R_1}{\left|\log\left(\frac{f}{f_{R1}}\right)\right|^{m_{R1}}} + \frac{R_2}{1 + e^{\frac{\log\left(\frac{f_{R2}}{f}\right)}{m_{R2}}}} \tag{8.13}$$

Table 8.5 Curve-Fitted Parameters for Figure 8.61

Parameter	*Value*	*Lower Limit*	*Upper Limit*
L_{inf} [pH]	172	100	250
L_1 [pH]	60	30	150
f_{L1} [MHz]	21.8	1	100
m_{L1} [—]	0.42	0	1
L_2 [pH]	87.8	30	200
f_{L2} [MHz]	22	5	50
m_{L2} [—]	0.02	0.01	1

The lower and upper limits show the range limits used in curve fitting.

The illustration for ESR curve fitting uses the same MLCC part that was used for Figure 8.61. The fitting of ESR data can also be automated. Either the solver function of spreadsheet software or a macro can be used to perform the curve fitting. Figure 8.62 shows the impedance real part and the curve-fit result. The first column in Table 8.6 shows the results of the curve-fit procedure. The second and third columns show the selected minimum and maximum values that were used in a macro to produce the semiautomated curve that fit. In the semiautomated curve fit solution, the min/max bounds are the only required user inputs; the rest is automated.

8.7 Bedspring Capacitor Model

Figure 8.53 showed the measured impedance and the extracted *C*(*f*), ESR(*f*), and ESL(*f*) values for a 10-μF 0508 reverse-geometry MLCC part mounted on a small fixture. In contrast to usual expectations, the minimum of the impedance real part is not at the SRF. At 600 kHz ESR is 2.9 mΩ, whereas at the 2.1 MHz SRF, ESR is 3.4 mΩ. Moreover, the extracted ESL(*f*) value is 600 pH at SRF, but the inductance first increases with frequency reaching a 660-pH peak at 4.2 MHz before it starts to drop

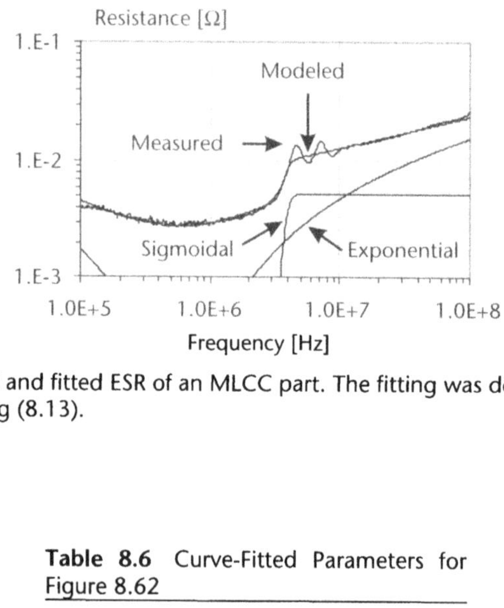

Figure 8.62 Measured and fitted ESR of an MLCC part. The fitting was done with the parameters shown in Table 8.6 using (8.13).

Table 8.6 Curve-Fitted Parameters for Figure 8.62

Parameter	*Value*	*Lower Limit*	*Upper Limit*
R_{min} [mΩ]	3.32	2	5
R_1 [mΩ]	5	1	10
f_{R1} [MHz]	3.02	2	5
m_{R1} [—]	1.2	1	5
R_2 [mΩ]	13.5	5	30
f_{R2} [MHz]	18	10	20
m_{R2} [—]	0.0272	0.01	0.05

The lower and upper limits show the range limits used in curve fitting.

sharply. Are these unexpected signatures due to measurement errors or a deficiency in the extraction procedure? Do ESR and ESL actually behave contrary to common assumptions?

To examine the frequency dependency of the parameters of an MLCC part, a simplified 2D bedspring model was constructed with 10 horizontal capacitor plates and 10 sections each of capacitor plates and dielectrics. Figure 8.63 shows the partial schematics of the bedspring model. For a physical design with N capacitor plates, each horizontal path of the model represents $N/10$ plates. The conductive plates are modeled by their series resistances and inductances. To capture the resonances both vertically and horizontally, each model of a capacitor-plate group is further divided horizontally into 10 segments. Each segment is represented by an R_p, plate resistance, and L_p, plate inductance. The R_p resistor represents the physical resistance of the plate, but it was also used to sense and measure the current in SPICE. Half of the plates connect to the left capacitor terminal; the other half of the plates connect to the right terminal. The remaining sections of plates (which also connect to the terminals) do not have facing counterparts of opposite-polarity plates. These end sections are modeled by using one more horizontal segment of R_p resistance and L_p inductance; though, if geometry warrants it, these parameters can be easily adjusted to reflect the actual percentages of plates in the open dielectric region.

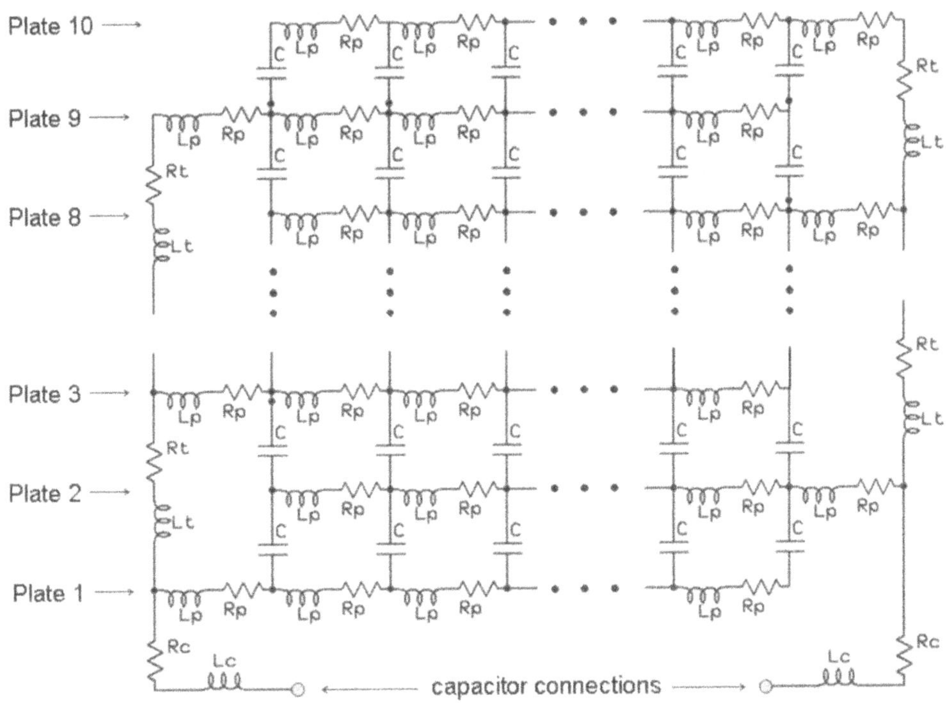

Figure 8.63 Partial schematics of the bedspring capacitor model. The model consists of 10 capacitor plates, 1 through 10. The lowest capacitor plate, connecting to the PCB, is plate 1. Plates 1, 3, 5, 7, and 9 are connected to the left terminal. Plates 2, 4, 6, 8, and 10 are connected to the right terminal. Each capacitor plate is divided into 10 equal segments (plus an end piece), represented by series *R-L* networks. At each internal plate node, a capacitor represents the dielectric material.

The capacitance of the dielectric material between the plates is represented by a uniform network of capacitors with values of C connecting each adjacent vertical node of the plates. There are 11 columns and 9 rows of the capacitance matrix; therefore, each C capacitance equals 1/99 of the total capacitance of the part. In the SPICE subcircuits these capacitors had two additional elements. Each capacitor had a parallel resistor to represent optional dielectric losses and a very small series resistance to sense and measure current. (Since not necessarily needed to represent series losses, which we usually neglect in the dielectrics, for this purpose independent voltage sources could also be used.) To increase clarity, these additional two elements are omitted from the partial schematics of Figure 8.63. The vertical end terminals are modeled by series R_t resistors and L_t inductors between every other plate connected to the terminal. The external connections, pads and vias are modeled by the series R_c and L_c elements. Note that the internal construction of MLCC parts is nominally symmetrical: therefore, the lowest capacitor plate (i.e., the plate that is close to the return-path plane in the PCB) is determined solely by the location of the connections of R_c and L_c in the model.

SPICE ac simulations were run on the model with various parameter combinations; the current was printed in each horizontal and vertical segment. The current magnitudes on the plates and in the dielectrics were plotted at various frequencies. Note that with the bedspring model we can achieve arbitrary geometrical resolution of the capacitor construction. One could construct a model in which each capacitor plate had its own entry; furthermore, each plate could be divided along its length into many more segments. To understand the current distribution-versus-frequency behavior of the part, this level of detail is not necessary (nor practical when there are several hundred plates). Even this simplified model of Figure 8.63 is too complex to use for PDN simulations, where we may need many of these models hooked up to the PCB. The bedspring model, however, provides a good insight into the variations of current distribution at different frequencies.

For sake of simplicity and clarity, the simulation results shown assume no dielectric losses; therefore, the capacitance of the insulating material is assumed to be frequency independent.

Figure 8.64 shows the simulated impedance magnitude and impedance real part of the bedspring model with the particular parameter values shown in Table 8.7.

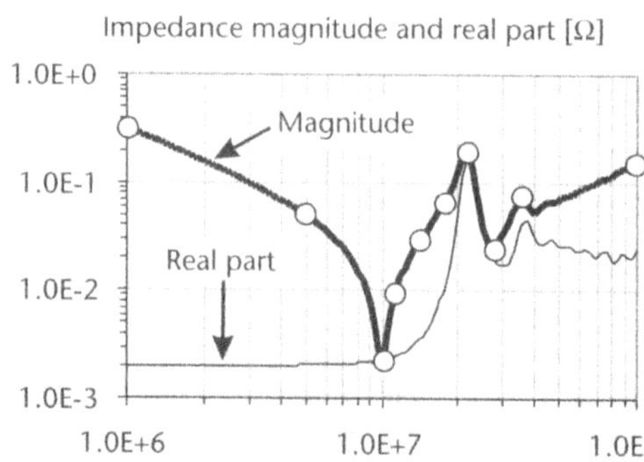

Figure 8.64 Simulated impedance magnitude and impedance real part at the capacitor connections terminals of the circuit shown in Figure 8.63. Circuit parameter values are shown in Table 8.7.

Table 8.7 Circuit Parameters for the Simulated Impedance Shown in Figure 8.64

Parameter	*Value*
C [nF]	5
L_p [nH]	10
R_p [mΩ]	1
L_t [pH]	100
R_t [mΩ]	0.1
L_c [pH]	100
R_c [mΩ]	1

Note that the purpose of this model and series of simulations was not to correlate to any particular measured results. Though this correlation could have been performed, it was clear that regardless of the correlation result this model is too complex to be used in actual simulations of PDNs. Since the sole purpose of the model was to study horizontal and vertical resonances in the structure, a series of 1D parameter sweeps were performed: the R_p, L_p, R_t, L_t, R_c, and L_c values were stepped through large ranges to see their effects on the impedance curve and current distribution (note that C only scales the frequency axis). Of the many simulated parameter combinations, the one shown in Figure 8.64 represents a case, where across one decade below and one decade above SRF, we can follow the gradual buildup of both vertical and horizontal resonances and can identify and follow key trends of current distribution variations.

With the selected parameters, the impedance magnitude curve shows the familiar V shape and strong secondary resonances characteristic of tall stacks of MLCCs mounted with low-inductance via geometry. The SRF is approximately 10 MHz and the first parallel resonance frequency (PRF) is 22 MHz. The impedance real part already confirms the measured trend shown in Figure 8.53: there is a visible (though small) increase of the resistance near SRF. Note that at lower frequencies, unlike in Figure 8.53, the impedance real part flattens out instead of increasing again; this is due to the fact that here we assumed lossless dielectrics.

Figure 8.64 has 10 markers placed along the magnitude curve. These markers show the frequency points, where Figures 8.66 through 8.75 illustrate the current distributions along the capacitor plates and inside the dielectrics. The marked frequencies are:

- 1 MHz, SRF/10;
- 5 MHz, one octave below SRF;
- 10 MHz, SRF;
- 12.3 MHz, slightly above SRF;
- 15.5 MHz, halfway between SRF and PRF;
- 17.4 MHz, slightly below PRF;
- 22 MHz, PRF;
- 27.5 MHz, frequency of the second minimum;

- 35 MHz, frequency of the second maximum;
- 100 MHz, 10 times SRF.

To help to interpret the current-distribution charts, Figure 8.65 shows the footprint of the capacitor-plate current charts and dielectric current charts with a sketch of a PCB, which defines the orientation and connections of the capacitor plates. The PCB is shown only to define the location of the lower and upper capacitor plates on the current-distribution charts ; otherwise, the board was not included in the simulation. Figures 8.66 through 8.75 show the current distributions with the same orientation as shown in Figure 8.65. In these figures, the lower-plate to right-terminal joint is at the front corner, the upper-plate to left-terminal joint is at the top corner. The current magnitudes along the conductive capacitor plates are shown in the (a) figures; the current magnitudes in the dielectric cells are shown in the (b) figures. Note the data-size difference between the two sets of graphs. The plate-current graphs plot 10 × 10 arrays, representing the 10 capacitor plates, with 10 segments on each plate. The dielectric-current graphs plot 11 × 9 arrays, representing the nine layers of eleven dielectric columns between the adjacent capacitor plates.

Figure 8.66 shows the current distribution at 1 MHz, 10 times below SRF. The current distribution in the dielectrics shows a uniform distribution. We have a total

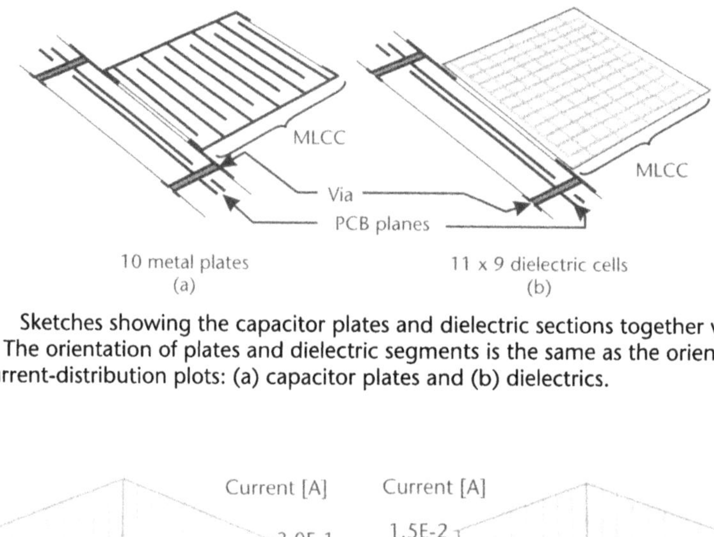

Figure 8.65 Sketches showing the capacitor plates and dielectric sections together with the PCB connection. The orientation of plates and dielectric segments is the same as the orientation on the following current-distribution plots: (a) capacitor plates and (b) dielectrics.

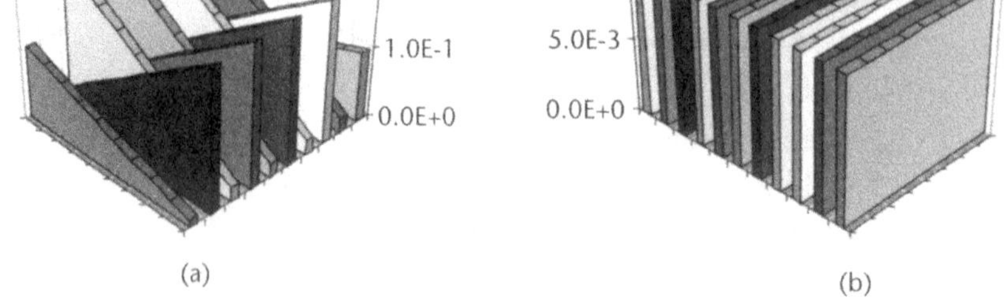

Figure 8.66 Current distribution (a) along the capacitor plates and (b) inside the dielectrics at 1 MHz.

of 99 dielectric cells in the model, and the 1A test current is evenly distributed; each dielectric cell carrying 1/99A at low frequencies. The current distribution in the capacitor plates is uniform, even though it manifests itself differently among them. We have five capacitor plates in the model connected to the left terminal and five plates connected to the right terminal. The current enters the plates from the terminals, then decreases linearly with distance as we move along the plate toward its open end (where the current drops to zero). In each horizontal segment the current is reduced due to displacement current leaking into the dielectric cells. This creates the interleaved triangle-shaped current profiles shown on this figure. Note also that in a symmetrical-plate stack, the outermost bottom and top plates carry only half the current compared to plates inside the stack, because the outside plates have a dielectric on only one of the sides where current could enter or exit the plates. If we sum the currents in the five plates at the left or at the right of the plot, the total sum is only 0.91A. This happens because the plot does not include the currents in the end pieces of the plates; the first segment where the current is measured along the plate occurs after the first of the 11 capacitors. Each capacitor carries 1/11 part or 9% of the current entering the plate.

Figure 8.67 shows the current distribution at 5 MHz, one octave below SRF. At this frequency, the distribution of the dielectric current is already not uniform. The sum of current magnitudes over the dielectric cells is still 1.0000A, but there is clearly more current in the dielectrics toward the upper plates and less current towards the lower plates. This phenomenon can also be seen when we closely compare the plate-current charts of Figures 8.66 and 8.67; current entering/exiting the lower plates is slightly less at 5 MHz. This change in current distribution at frequencies well below SRF tells us that the inductance corresponding to the particular current distribution already may start to change; in fact, the inductance increases, since more current flows further away from the return path. The current distribution variance is nearly linear with elevation within the stack; therefore, ESR does not yet increase noticeably: the fact that less current flows on the lower plates is balanced by the higher current density in the upper plates.

Figure 8.68 shows the current distribution at SRF. At this frequency the current magnitude around the bottom plates is already four times lower than the current

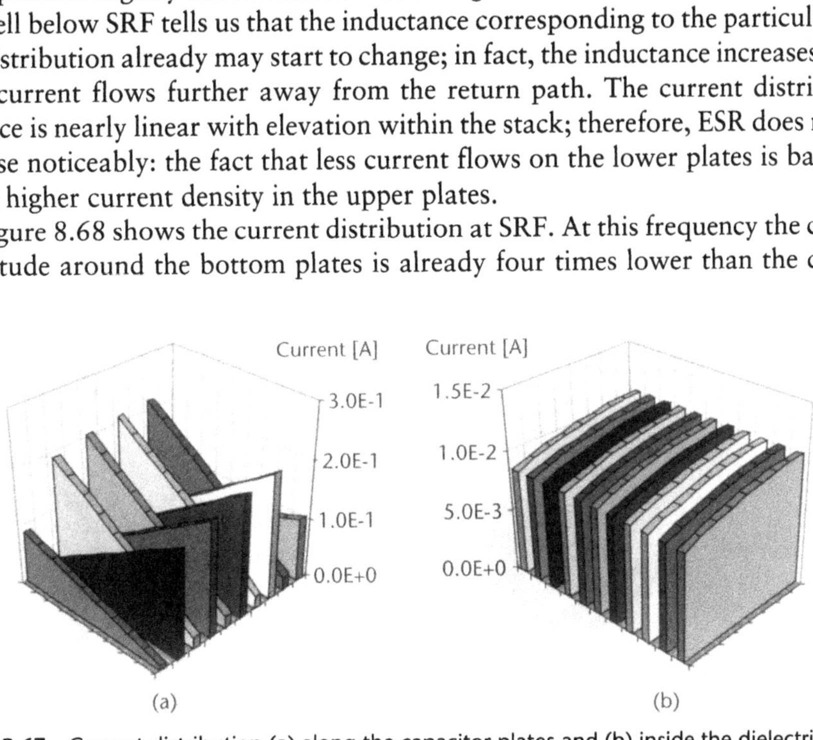

Figure 8.67 Current distribution (a) along the capacitor plates and (b) inside the dielectrics at 5-MHz frequency.

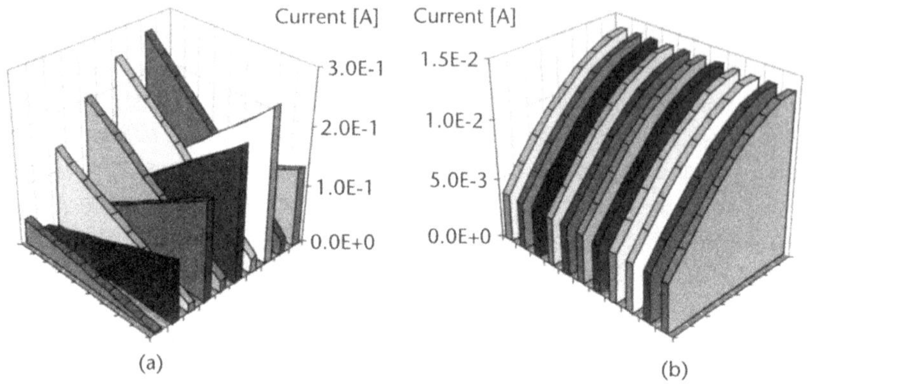

Figure 8.68 Current distribution (a) along the capacitor plates and (b) inside the dielectrics at the 10-MHz series resonance frequency.

around the upper plates. This result clearly defies conventional wisdom which suggests that the capacitive and inductive currents balance at SRF so that current fills the capacitor plates and dielectrics uniformly. The current variation with elevation within the plate stack starts to show traits of nonlinearity; the center of gravity of current flow ascends. This causes ESR to be higher (2.36 mΩ) at the series resonance frequency than the low-frequency value (2 mΩ). The sum of current magnitudes in the 99 dielectric cells starts to rise; the sum at this frequency is 1.0006A.

Figure 8.69 shows the current distribution at 12.3 MHz, one third on the way between the series resonance frequency and first parallel resonance frequency. At this frequency, we can observe the buildup of a full quarter-wave vertical resonance: current on the lower plates (close to the return path) is practically zero; so, the characteristic one-quarter sine wave current-distribution shape appears vertically inside the dielectrics. Also, the vertical distribution of current becomes strongly nonlinear; this explains the start of the sharp rise in ESR (see Figure 8.64). The sum of current magnitudes in the dielectric cells is 1.0147A.

Figure 8.70 shows the current distribution at 15.5 MHz, approximately halfway between SRF and PRF. The minimum line in the current distribution in the dielec-

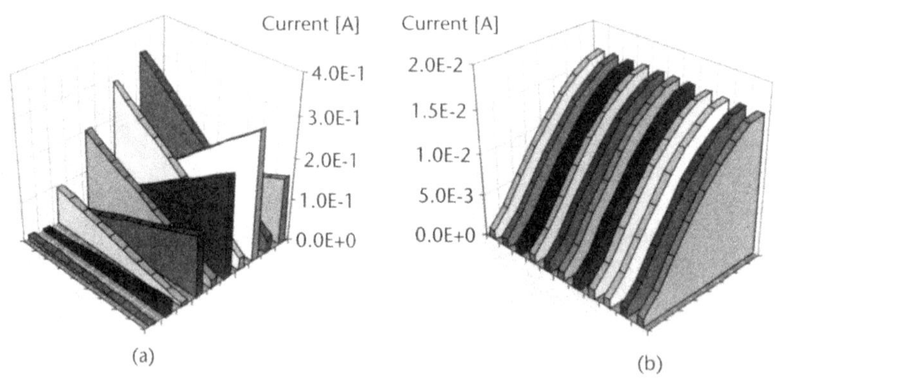

Figure 8.69 Current distribution (a) along the capacitor plates and (b) inside the dielectrics at 12.3-MHz frequency.

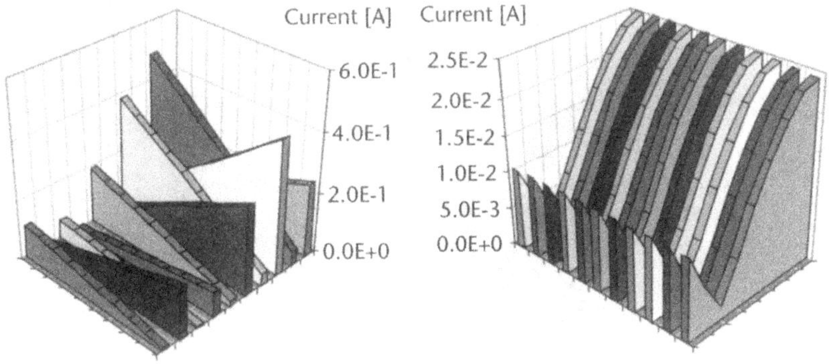

(a) (b)

Figure 8.70 Current distribution (a) along the capacitor plates and (b) inside the dielectrics at 15.5-MHz frequency.

trics has moved from the bottom plates further upwards in the stack. Note also that the resonance in the distributed LC circuit amplifies the current magnitude: the sum of current magnitudes over the 99 dielectric cells rose to 1.3754A; therefore, the vertical scales of both charts had to be increased (compared to earlier figures) to fit the data.

Figure 8.71 shows the current distribution at 17.4 MHz, slightly below the first parallel resonance frequency (PRF). The current minimum has moved further upwards in the stack of plates, and the current got further amplified. The sum of current magnitudes rose to 2.0291A, necessitating a further change of vertical scales. On the plate-current chart we clearly see the asymmetrical current distribution: the resonance pushes the current upwards in the stack; as a result, much more current flows on the upper plates. Therefore, not only does ESR rise, but, contrary to conventional assumptions, inductance increases (at first) compared to its low-frequency value.

Figure 8.72 captures the current distribution at 22 MHz, which is the first impedance maximum (PRF). The vertical resonance builds so that the current is minimal halfway along the stack and maximal both on the bottom and at the top of

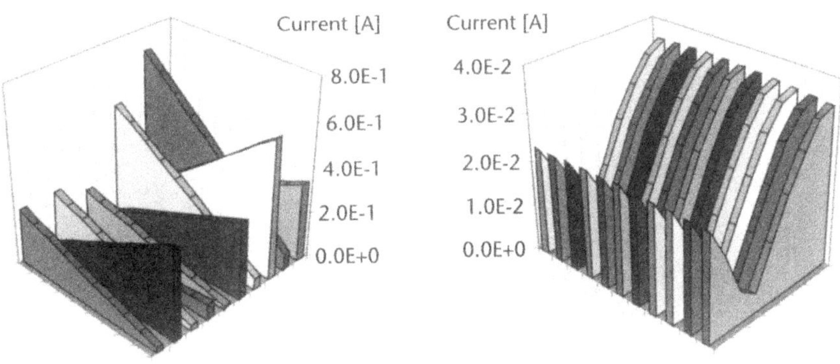

(a) (b)

Figure 8.71 Current distribution (a) along the capacitor plates and (b) inside the dielectrics at 17.4-MHz frequency.

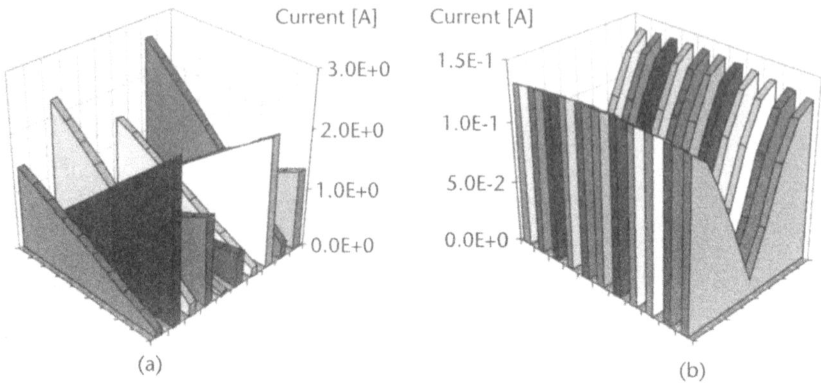

Figure 8.72 Current distribution (a) along the capacitor plates and (b) inside the dielectrics at 22-MHz frequency.

the capacitor stack. The resonance increased the currents further; so, the sum of current magnitudes is 9.3278A and the vertical scales had to be increased again. This is also the frequency where ESR in Figure 8.64 peaks.

Figure 8.73 refers to 27.5 MHz, which is the frequency of the second minimum. The vertical current distribution in the dielectrics develops a broader peak toward the bottom plates, so the resonance exhibits the broad maximum toward the return path. This is eventually responsible for the sudden drop of the extracted inductance. At the same time, the sum of current magnitudes also dropped to 3.1836A.

Figure 8.74 shows the current distribution at 34.7 MHz, at the second maximum frequency. Under this condition, more periods build up vertically, but more current stays towards the bottom plates also; the sudden fall of inductance has already finished. At this frequency the current distribution in the capacitor plates starts to show traits of nonuniform horizontal distribution; in the middle of the plate, the current is slightly above the linear estimate.

Figure 8.75 shows the current distribution at 100 MHz, 10 times above SRF. The current distribution in the dielectrics has multiple resonances, and horizontal spatial resonance becomes strong. Current crowding toward the bottom plates is pronounced.

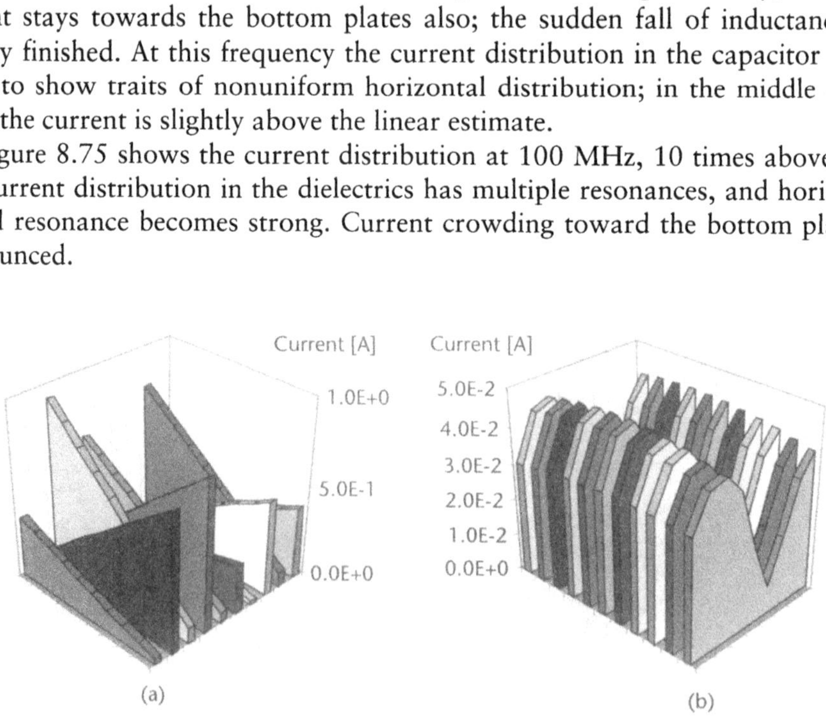

Figure 8.73 Current distribution (a) along the capacitor plates and (b) inside the dielectrics at 27.5-MHz frequency (second impedance minimum).

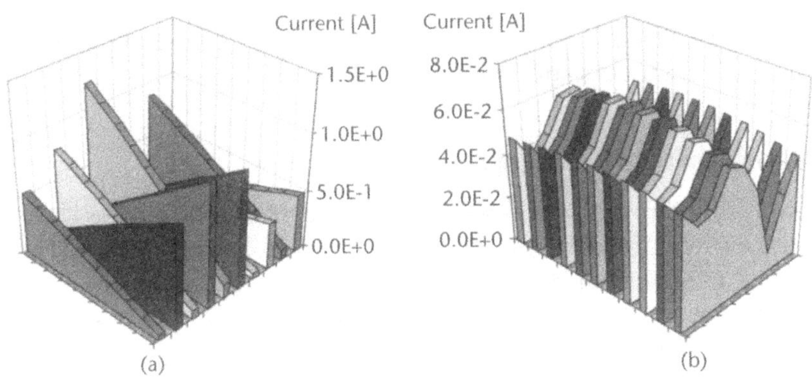

Figure 8.74 Current distribution (a) along the capacitor plates and (b) inside the dielectrics at 34.7-MHz frequency (second impedance maximum).

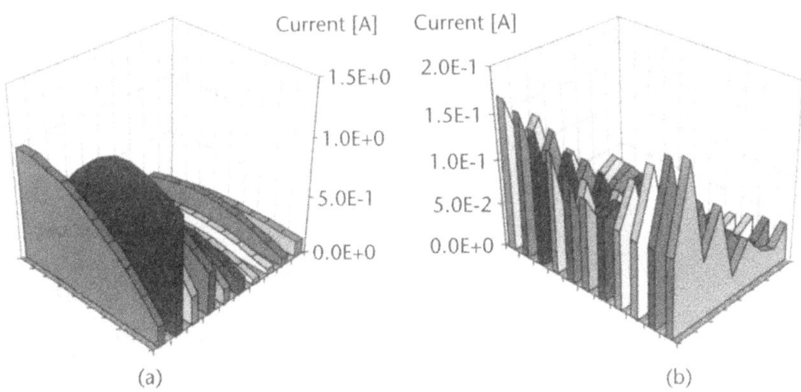

Figure 8.75 Current distribution (a) along the capacitor plates and (b) inside the dielectrics at 100-MHz frequency.

8.8 Causal Slow-Wave Model

The slow-wave causal model is built on the realization that a multilayer ceramic capacitor is a periodically loaded lossy transmission line [15]. The unloaded transmission line is formed by the two vertical terminals of the capacitor (assuming horizontal plate orientation) by removing the capacitor plates but leaving the dielectric material in place. This vertically oriented unloaded transmission line in itself is already lossy: the terminals have finite resistance, and the dielectric material has finite dielectric loss tangent. It is also known that causality dictates capacitance to change with frequency in proportion to the dielectric loss tangent. In an MLCC part, the large capacitance is achieved by interdigitated capacitor plates, attached in an alternating pattern to the opposite terminals. These capacitor plates form a set of periodically arranged lossy transmission lines, attached orthogonally to the capacitor terminals. As it will be shown, the multitude of capacitor plates will not only increase the total capacitance of the part, but it also behaves like a dielectric material with an increased loss tangent and additional frequency dependence of capacitance. The effective loss tangent is a mix of the loss tangent of the original dielectric material and the resistive loss of the capacitor plates.

8.8.1 The Unit-Cell Model

If we properly assign the dimensions and material constants, or if we perform a blind optimization of these parameters to match the measured behavior of a capacitor, we expect all of the major features will be captured simultaneously without the need to independently change and optimize capacitance, resistance, and inductance values. Also, the model is based on the physical properties of the structure; this guarantees that the model will be causal.

The sketch in Figure 8.76(a) defines the major dimensions of an MLCC relevant to our calculations. We assume that the rectangular capacitor body is W wide, L long, and H high. The illustration here shows a reverse-geometry capacitor, because for the correlation, we will use the measured data from the previous 10-μF 0508 MLCC example. The calculations and methodology, however, do not mandate a reverse-geometry capacitor. For other geometries, such as regular or interdigitated capacitors, L, W, and H can be changed appropriately and as needed.

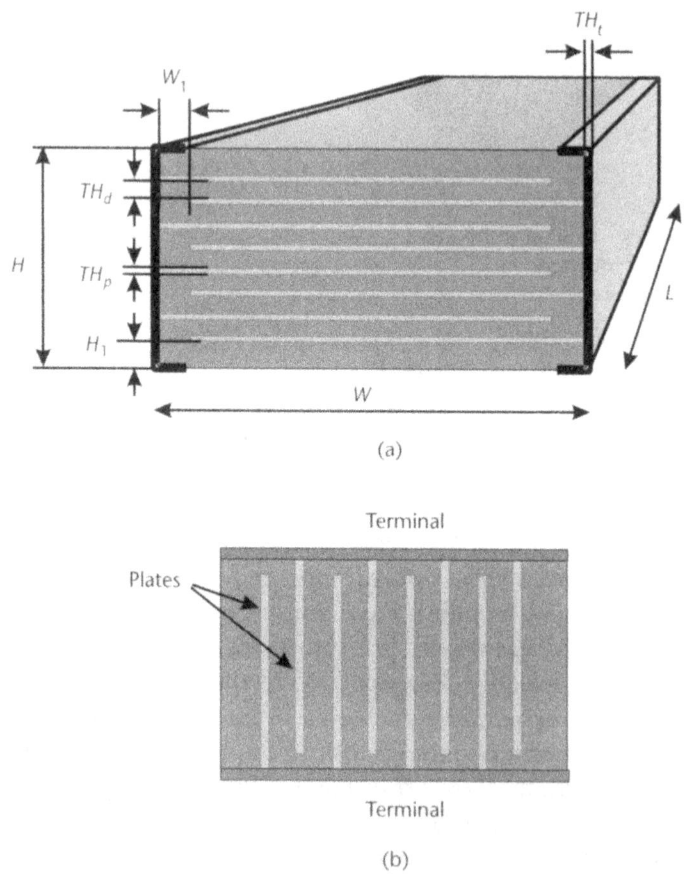

Figure 8.76 Side view of MLCC with (a) the important dimensions and (b) the loaded-line view of the same MLCC by turning it 90° and changing the aspect ratio.

To simplify the calculations, the capacitor cross section is assumed to be symmetrical, both horizontally and vertically. We assume the same H_1 cover thickness both on top and bottom. A similar horizontal symmetry assumes that capacitor plates are stopping at distance W_1 from the unconnected terminals at both sides. This symmetry is assumed only for sake of convenience; the procedure can be easily extended to different top and bottom cover thicknesses and/or for different end gaps at the left and right terminals.

The capacitor plate thickness is TH_p, each dielectric layer between the plates is TH_d thick. The vertical capacitor terminals are assumed to be TH_t in thickness. We further assume that the dielectric material has ε_d dielectric constant and tan_δ loss tangent at a given working frequency and that their frequency-dependent values are inter-related through the causality requirement and as it was shown in Section 4.5.5.2. The capacitor plates have a conductivity and permeability of σ_p, and μ_p, the terminal material has a conductivity and permeability of σ_t and μ_t.

The periodically loaded transmission-line model becomes apparent when we turn the capacitor sideways and distort the aspect ratio. As shown in Figure 8.76(b), the height (H) of the original capacitor body becomes the length of the transmission line, and the capacitor plates will represent a periodical loading along the transmission line, similar to capacitive irises in a waveguide.

The capacitance of the unloaded transmission line equals the capacitance of the capacitor body between the vertical terminals, without the capacitor plates:

$$C_0 = \varepsilon_0 \varepsilon_d L \frac{H}{W} \tag{8.14}$$

The propagation delay of the unloaded transmission line equals the propagation delay along the empty vertical capacitor body, with the capacitor plates removed but dielectric material in place:

$$t_{pd0} = \frac{H}{v} = H\sqrt{\varepsilon_0 \varepsilon_d \mu_0} \tag{8.15}$$

With C_0 and t_{pd0} known, we can calculate the characteristic impedance and the inductance of the unloaded vertical transmission line:

$$L_0 = \frac{t_{pd0}^2}{C_0} = \mu_0 H \frac{W}{L} \text{ and } Z_{00} = \sqrt{\frac{L_0}{C_0}} = \frac{120\pi}{\sqrt{\varepsilon_0}} \frac{W}{L} \tag{8.16}$$

The resistance of each terminal along its entire vertical length is:

$$R_t = \frac{1}{\sigma_t} \frac{1}{L} \frac{H}{TH_t} \tag{8.17}$$

From (8.14) through (8.17), we can calculate the parameters of the two end pieces, simply by scaling the unloaded transmission line length by the ratio of H_1/H for each end piece:

$$t_{pd\_end} = t_{pd0} \frac{H_1}{H} \tag{8.18}$$

The end-piece transmission lines are denoted by the suffix _end.

As shown in Figure 8.76, adjacent capacitor plates along the terminal will create the periodical loading. The impedance of the entire capacitor is observed at the left end of the transmission line; the right-hand side (originally the top side) of the structure is open.

The loading is created by the lossy, open-ended transmission lines formed by adjacent plates. Figure 8.77 shows the periodically loaded model expressed by a series of unit cells.

From (8.14) through (8.18), we can calculate the parameters of a transmission line formed by adjacent capacitor plates. This transmission line is denoted by the suffix _p.

$$C_p = \varepsilon_0 \varepsilon_d L \frac{W - 2W_1}{TH_d} \tag{8.19}$$

$$t_{pdp} = \frac{W - 2W_1}{v} = (W - 2W_1)\sqrt{\varepsilon_0 \varepsilon_d \mu_0} \tag{8.20}$$

$$L_p = \frac{t_{pdp}^2}{C_p} = \mu_0 \frac{TH_d (W - 2W_1)}{L} \text{ and } Z_{op} = \sqrt{\frac{L_p}{C_p}} = \frac{120\pi}{\sqrt{\varepsilon_0}} \frac{TH_d}{L} \tag{8.21}$$

$$R_p = \frac{1}{\sigma_p} \frac{W - W_1}{TH_p L} \tag{8.22}$$

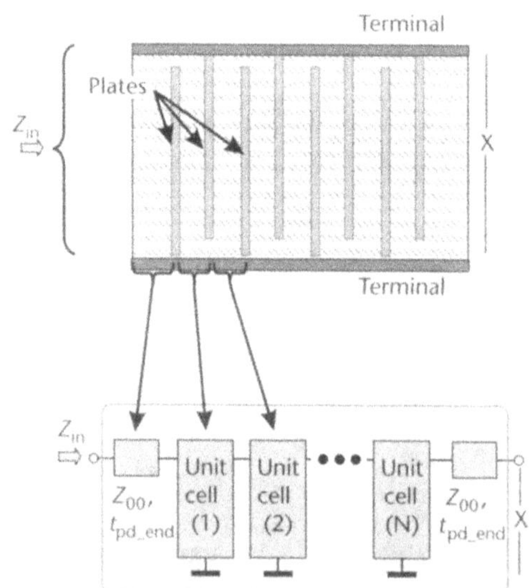

Figure 8.77 Slow-wave periodically loaded model of MLCC. The periodically loaded transmission line is broken down into symmetrical unit cells, each cell representing a length equal to the capacitor-plate pitch. The end pieces, corresponding to the bottom and top dielectric covers, are represented by unloaded transmission-line sections.

From N capacitor plates, we get N–1 pairs to create the periodical loading. For $N >> 1$, we can approximate the number of cells in the periodically loaded structure with N. The number of plates, the dielectric and plate thicknesses and the top/bottom cover thicknesses are inter-related through the following formula:

$$N = \frac{H - 2H_1}{TH_p + TH_d} \tag{8.23}$$

The conductive and dielectric losses of the transmission lines, in theory, are frequency dependent. The δ skin depth in conductors is defined as:

$$\delta = \sqrt{\frac{1}{\pi f \sigma \mu}} \tag{8.24}$$

The skin depth in copper reaches 2 μm at 1-GHz frequency. To meet the requirements of the high ceramic firing temperature, capacitor plates often use materials with conductivity lower than that of copper. The lower conductivity increases the skin depth. This means the resistance of individual nonferromagnetic capacitor plates is skin-depth limited and is relatively frequency independent up to hundreds of megahertz frequencies. This is the case for thin silver-palladium capacitor plates. With the change to nickel plates over the past decade, the permeability of the ferromagnetic material has to be taken into account. Nickel has five times higher resistivity and 100 times higher permeability, with respect to copper. With nickel plates, the skin depth at 1 GHz is 0.22 μm, which is thinner than the capacitor plates in many of today's MLCCs. Assuming nickel electrodes with 1-μm thickness, skin depth equals the capacitor-plate thickness at 50 MHz. Below 50 MHz, the ac resistance of the capacitor plates will gradually bottom out at their dc resistance.

The terminals are usually much thicker than the capacitor plates, therefore their resistance and inductance may show a somewhat stronger frequency dependence. In addition to a thin composite layer to promote adhesion, the terminal material of MLCCs is usually tin-covered copper.

The dielectric losses are represented by a parallel conductance G in the transmission-line model:

$$G = 2\pi f C \tan\_\delta \tag{8.25}$$

In (8.25), we can substitute the appropriate capacitance and loss-tangent values for the unloaded transmission line of terminals or the lossy transmission line of capacitor plates.

This generic model links the geometry and material properties of an MLCC to a causal electrical model, which can be used to calculate the impedance of the capacitor directly. Though this model is still too complex to include in an actual circuit simulator in multiple copies, it is very suitable for correlation purposes. We can use any of the computer math packages to obtain the input impedance of the periodically loaded transmission-line circuit (which represents the impedance of the capacitor).

8.8.2 The Lossy Transmission-Line Model

The model derived in Figures 8.76 through 8.78 is generic, and as such, it is valid over a wide range of parameters. When we look at the actual geometry and resulting model numbers for a typical MLCC, we can achieve substantial simplifications without a major loss of accuracy.

Since we are interested in the input impedance of the structure that is open at its end (on the top), the second end piece of unloaded transmission line can be totally neglected. The top end piece is open terminated; therefore, only the static capacitance of the end piece matters. Not having capacitor plates in the end piece, for large N, its static capacitance is orders of magnitudes lower than the total capacitance, and therefore it can be rightfully ignored.

The end piece on the bottom (left) is in series to the external connections; therefore, it cannot be completely ignored. We can still simplify the left end piece in Figure 8.77 by using the earlier arguments and then neglect its parallel capacitance and parallel conductance. This leaves us with its series inductance and series resistance. These values can be obtained from (8.16) and (8.17), by substituting H_1 for H.

$$L_{0\_end} = \mu_0 \frac{H_1 W}{L} \tag{8.26}$$

$$R_{t\_end} = \frac{1}{\sigma_t} \frac{H}{L\,TH_t} \tag{8.27}$$

The unit cells can be simplified in a similar way. Capacitances and conductance of the series unloaded transmission-line pieces can be ignored, leaving only a series L-R term. For one unit cell, using the suffix _uc, and substituting $TH_p + TH_d$ for H_1, we get:

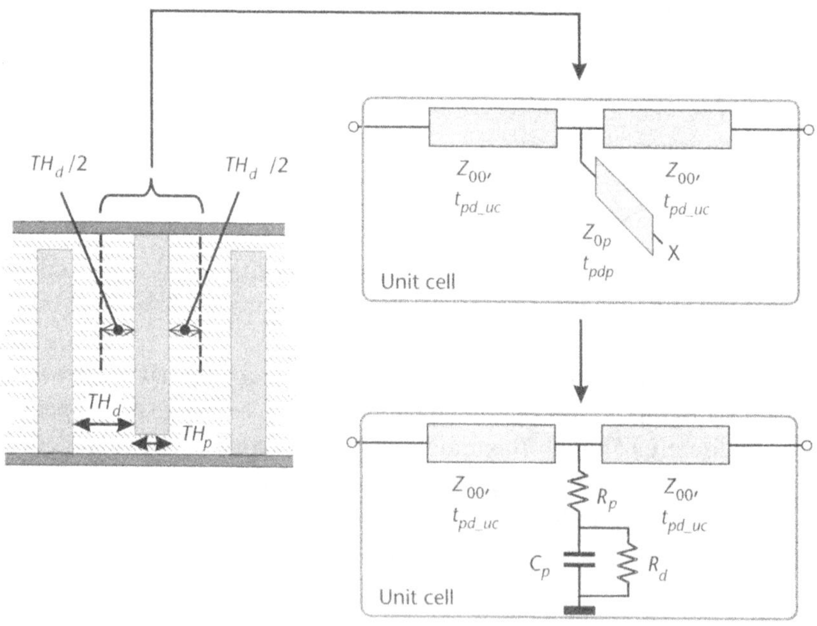

Figure 8.78 Generating unit-cell parameters from the geometry.

$$L_{0\_uc} = \mu_0 \frac{(TH_p + TH_d)W}{L} \tag{8.28}$$

$$R_{t\_uc} = \frac{1}{\sigma_t} \frac{TH_p + TH_d}{L\,TH_t} \tag{8.29}$$

The open-ended loading transmission line formed by adjacent capacitor plates can be simplified by neglecting its inductance. The C_p capacitance and R_p series plate resistance of the transmission line were already given in (8.19) and (8.22). There is one remaining element, though, that we should not ignore: the parallel conductance of the loading open-ended transmission line. It can be calculated from (8.25), by substituting the values for one plate pair:

$$G_p = 2\pi\, f\, C_p \tan\_\delta \tag{8.30}$$

These simplifications lead to the equivalent circuit of the unit cell shown in Figure 8.79. The shunt capacitance has its own G_p conductive loss term, originated from the dielectric loss tangent, and an R_p series resistive loss term, originated from the resistance of the adjacent capacitor plates. At any given frequency, the series and parallel loss terms can be combined into a single term. The schematics in Figure 8.79(b) show a parallel equivalent, where G_p' represents a combination of R_p and G_p.

Note that during the transformation, in a general case, both the capacitance and conductance will change. Assuming that tan_δ is small, we get:

$$C_p' = \frac{C_p}{1+\left(\frac{\omega}{\omega_p}\right)^2} \quad \text{and} \quad G_p' = G_p + \omega C_p' \frac{\omega}{\omega_p} \tag{8.31}$$

where $\omega_p = 1/(R_p C_p)$.

For frequencies where $\omega_p >> \omega$, the formulas can be further simplified to:

$$C_p' \approx C_p \quad \text{and} \quad G_p' \approx G_p + \omega C_p \frac{\omega}{\omega_p} \tag{8.32}$$

By combining (8.30) and (8.32), we get:

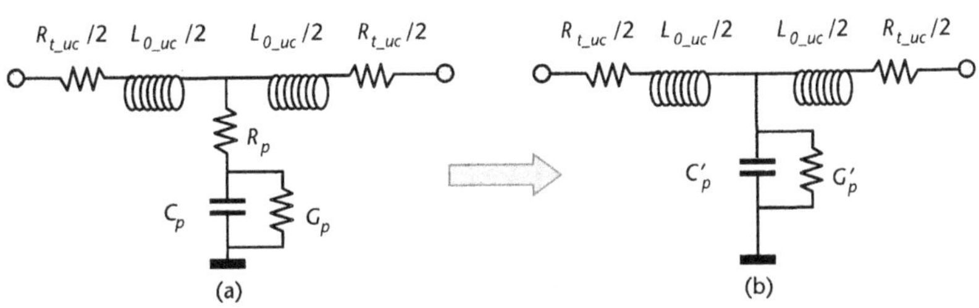

Figure 8.79 Equivalent schematics of the simplified unit cell. Part (a) neglects the capacitance and conductance of the series transmission line and neglects the inductance of the parallel transmission line. Part (b) combines the series and parallel loss terms around the shunt capacitance.

$$G'_p = \omega C'_p \tan\_\delta + \omega C'_p \frac{\omega}{\omega_p} \tag{8.33}$$

From (8.33) and (8.25) we can calculate an effective loss tangent for the lossy transmission line:

$$\tan\_\delta' = \frac{\tan\_\delta + \dfrac{\omega}{\omega_p}}{1 + \left(\dfrac{\omega}{\omega_p}\right)^2} \tag{8.34}$$

In the final step, we can realize that the cascaded unit cells represent the ladder equivalent of a uniform lossy transmission line. To obtain the per-unit-length transmission-line equivalent parameters, we multiply the unit-cell parameters by N. The inductance and the resistance are then simply the inductance and resistance of the (H–2H_1) section of the vertical terminals. The capacitance will become approximately the full capacitance of the part itself; though, as indicated by (8.31), it drops sharply above the ω_p corner frequency. Moreover, to obey causality, the capacitance also drops slightly with frequency due to the finite loss tangent. The parallel conductance can be calculated from the full capacitance and the effective loss tangent.

This eventually leads us to the simplified equivalent circuit of Figure 8.80. Note that this simple circuit is causal, so it works on many time-domain and frequency-domain simulators.

Note that the important aspect of this simplified model is not the lossy transmission line itself, but rather the unique way that it captures the convoluted effect of conductive and dielectric losses in the frequency-dependent capacitance and conductance per unit length.

8.8.3 Correlations

We return to the example shown in Figure 8.53. The 10-μF 0508 MLCC part was measured in a fixture, shown in Figure 8.81. The small test board had 22 layers. Capacitor sites are on both top and bottom, connecting to 15 × 10-mm (600 × 400-mil) rectangular plane shapes. The horizontal layout of the site is shown in Figure 8.78(b). The capacitor pads for the 0508 site are on the bottom side (layer 22), connecting to power planes on layers 20 and 21 with a set of blind vias.

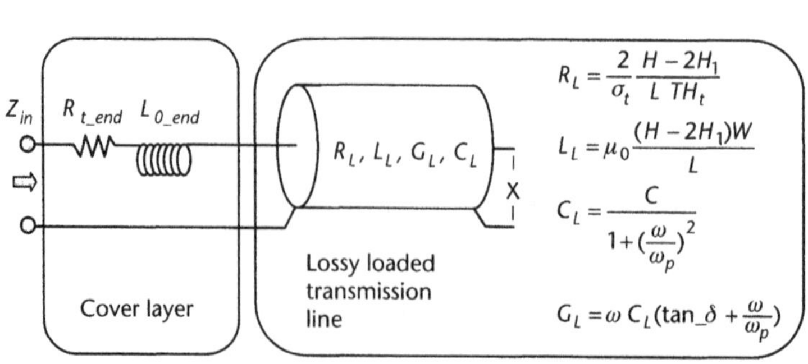

Figure 8.80 Simplified causal equivalent circuit of a multilayer ceramic capacitor.

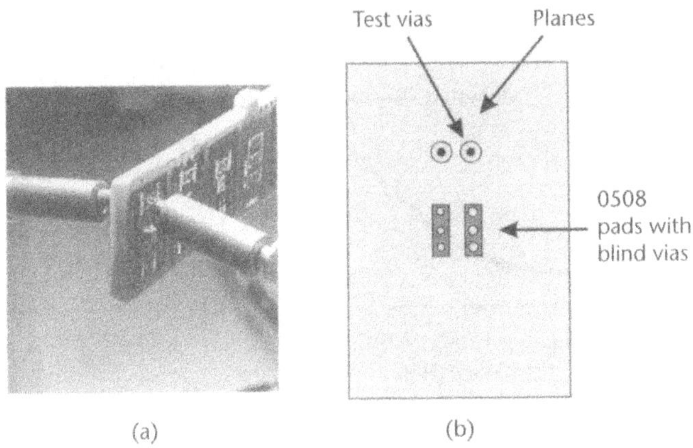

Figure 8.81 Geometry of test fixture for measuring the 10-μF 0508 MLCC sample: (a) photo of the fixture and (b) layout of the test fixture.

First, the test site was characterized with the capacitor pads open and shorted with two vector network analyzers: 4395A in the 100-Hz–10-MHz frequency range and 4396A in the 100-kHz–1,800-MHz frequency range. Figure 8.82 shows the measured impedance magnitude and phase of the open test site and the extracted capacitance versus frequency. Note the straight slope of the capacitance curve on the linear-logarithmic scale; this indicates an approximately 2% of loss tangent of the FR4 material. The average capacitance is around 120 pF at 10 MHz.

Figure 8.83 shows the measured data and extracted parameters with the capacitor pads shorted. The graphs show composite data from the two VNAs. The expressions approximating the real part of impedance and inductance:

$$R(f) = R_{dc}\left(1+\left(\frac{f}{f_R}\right)^N\right) \quad \text{and} \quad L(f) = L_{inf} + \frac{\Delta L}{1+\left(\frac{f}{f_R}\right)^N} \tag{8.35}$$

where R_{dc} = 1.7 [mΩ], f_R = 12 [MHz], L_{inf} = 140 [pH], ΔL = 80 [pH], N = 0.8.

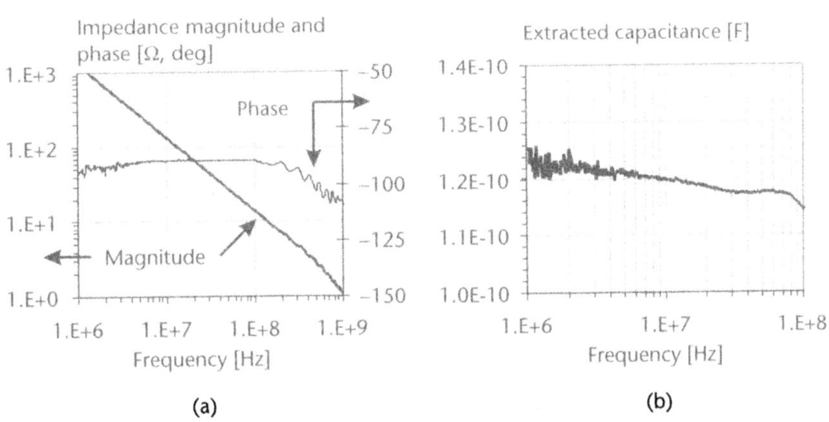

Figure 8.82 Characterization results of open fixture: (a) impedance magnitude and phase and (b) extracted capacitance versus frequency.

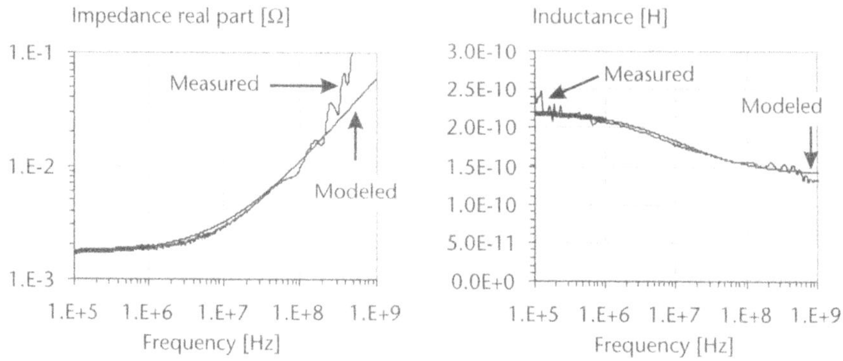

(a) (b)

Figure 8.83 Characterization results of shorted fixture: (a) impedance real part and (b) extracted inductance. The smooth lines are approximating curves based on (8.32).

The parallel capacitance of the open test site and the series resistance and inductance of the shorted test site have to be included when we are looking for correlation of the lossy transmission-line capacitor model to measured data. Note, however, that the shorted test-site data also reflects resistance and inductance of the short itself, not only the test fixture. Since the short is later replaced with the capacitor, this small part of resistance and inductance will be double counted.

The unit-cell model captures the physical properties of the capacitor and converts them into an electrical model. Strictly speaking, we would need to know how many capacitor plates the part has and concatenate the same number of unit cells. However, often we do not know the number of capacitor plates in the part; clearly this information is not necessary. The per-unit electrical parameters of concatenated unit cells saturate beyond a number of cells, resulting in diminishing change as we add more cells. The saturation curve does depend (weakly) on the characteristics of the unit cell, but usually 50–100 cells will result in a sufficiently accurate approximation. Figure 8.84 illustrates the convergence with the unit cells used to describe the capacitor sample in Figure 8.53. For this exercise, the unit cells have been

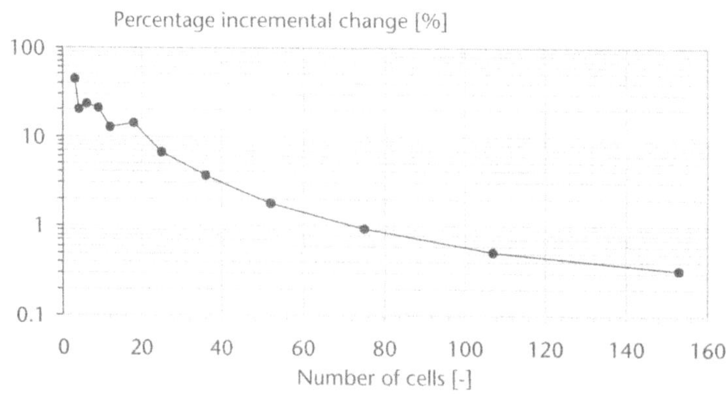

Figure 8.84 Saturation curve of concatenated unit cells. Vertical axis: percentage change between consecutive iteration with different number of unit cells. Horizontal scale: number of unit cells.

adjusted so that the given number of unit cells always sum to the same total characteristics of the sample capacitor.

As opposed to the capacitance curve of the FR4 material of the test fixture, the extracted capacitance versus frequency in Figure 8.53 exhibits multiple sections of frequency ranges with an approximately constant slope. This behavior suggests that, for each of these frequency ranges a different dielectric loss tangent should be applied. This is illustrated in Figure 8.85, where we compare the correlations with a single dielectric loss tangent value (0.015) versus three different loss tangent values (0.025, 0.0135, and 0.01). With three loss tangent values applied separately for each frequency range and combined linearly to continue to enforce causality, we can properly capture the shape of the impedance real part (and also the frequency dependency of the capacitance) over three decades of frequencies.

Below the series resonance frequency (SRF), the capacitance and impedance real part are primarily coming from the dielectric material; therefore, the extracted capacitance versus the frequency curve provides a useful guide to the number of segments we may need to use to properly describe the frequency dependency of the dielectric material. Near to and above SRF, on the other hand, we have no direct indication about the possible change of the loss tangent. We can, for instance, assume that the loss tangent just below SRF continues unchanged above SRF as well. This approach was followed in the correlation results shown here. As an alternative solution, we can assume additional frequency segments above SRF with their respective and unknown loss tangent values; using this method, we obtain the values by optimized curve fitting to the measured data.

After we have obtained the capacitor parameters' frequencies below SRF, we can put together the full model: cascaded unit cells, end piece, and fixture. These models are based on the physical parameters of the capacitor. In this case, however, the measured part was not cross sectioned, and the internal geometry and material constant data was not available from other sources. Instead, the correlation was performed by the manual and automated optimization of the model with seed values based on reasonable assumptions. The number of parameters to be optimized is large enough that a reasonable correlation should be possible to achieve. However, not having the true numbers for the capacitors, a very accurate correlation was not

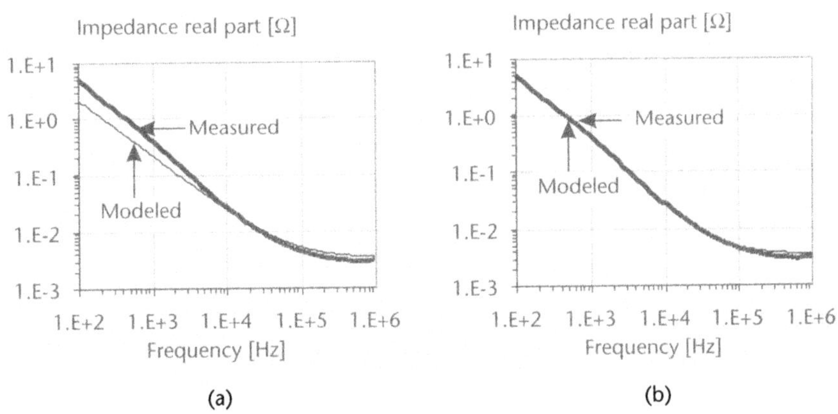

Figure 8.85 Correlation of impedance real part below the series resonance frequency: (a) approximation using one loss tangent value and (b) approximation using three different dielectric loss tangent values.

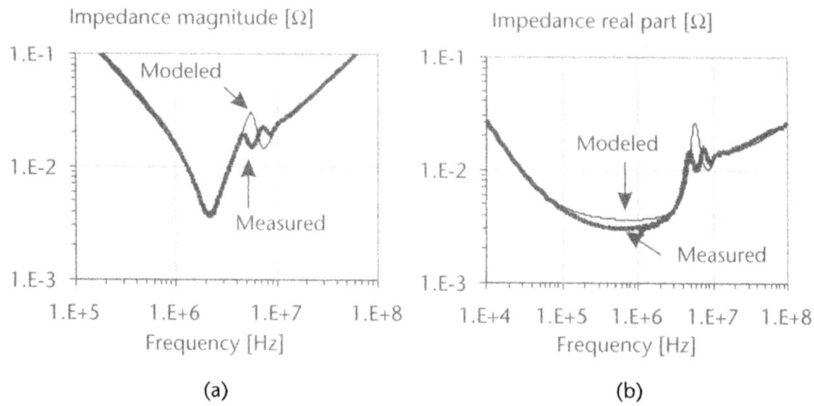

Figure 8.86 Correlation with unit-cell model: (a) impedance magnitude and (b) impedance real part.

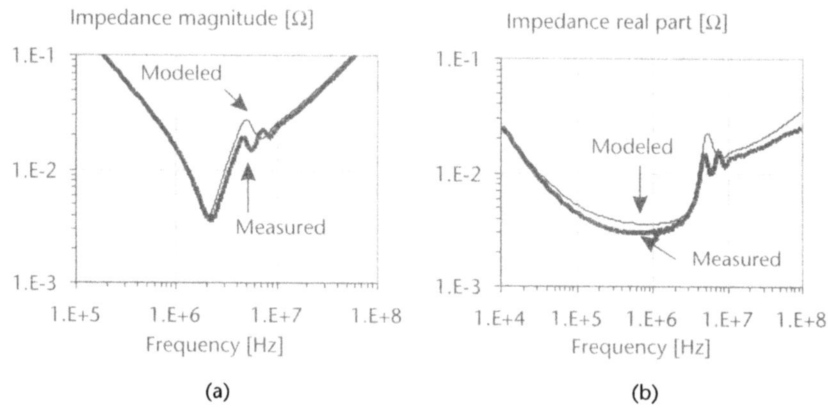

Figure 8.87 Correlation with lossy transmission-line model: (a) impedance magnitude and (b) impedance real part.

the goal. Instead, different parameter settings have been assumed to determine if the model can properly describe and capture the important signatures of the measured data plots: increased ESR at SRF, sudden increase of ESR above SRF, and slow rise of ESR above the secondary resonances.

As an illustration, Figure 8.86 shows the correlation after a brief optimization.

The same set of measured data was also correlated to the lossy transmission-line model of Figure 8.80. Figure 8.87 shows the correlation after a brief manual optimization of the parameters.

References

[1] Novak, I., et al, "Distributed Matched Bypassing for Board-Level Power Distribution Networks," *IEEE Trans. on Advanced Packaging*, Vol. 25, No. 2, May 2002, pp. 230–243.

[2] Novak, I., "On the Uniqueness of the Inductance of Bypass Capacitors," *Proceedings of DesignCon East 2005*, Worcester, MA, September 19–21, 2005.

[3] Smith, L. D., et al., "Power Distribution System Design Methodology and Capacitor Selection for Modern CMOS Technology," *IEEE Trans. on Advanced Packaging*, Vol. 22, No. 3, August 1999, p. 284.

[4] Smith, L. D., and D. Hockanson, "Distributed SPICE Circuit Model for Ceramic Capacitors," *Proceedings of the 51st Electronic Components and Technology Conference*, Orlando, FL, May 29–June 1, 2001.

[5] Novak, I., S. Pannala, and J. R. Miller, "Overview of Some Options to Create Low-Q Controlled-ESR Bypass Capacitors," *Proceedings of the 13th Topical Meeting on Electrical Performance of Electronic Packaging*, Portland, OR, October 2004.

[6] American Technical Ceramics, "Vertical Orientation of ATC Chip Capacitors on Stripline," Application Note #001-828, Rev. B, 6/98.

[7] AVX, "Low Inductance Capacitor Array," Data Sheet, June 2000.

[8] Novak, I., and J. R. Miller, "Frequency-Dependent Characterization of Bulk and Ceramic Bypass Capacitors," *Proceedings of EPEP2003*, Princeton, NJ, October 27–29, 2003.

[9] Herbert, J. M., *Ceramic Dielectrics and Capacitors,* New York: Gordon and Breach Science Publishers, 1985.

[10] Uchino, K., *Ferroelectric Devices*, New York: Marcel Dekker, 2000.

[11] Ishida, H., "Measurement Method of ESL in JEITA and Equivalent Circuit of Polymer Tantalum Capacitors," *Proceedings of DesignCon 2005*, Santa Clara, CA, January 31–February 3, 2005.

[12] Hock, J., and A. Ritter, "A Measurement Technique for High Frequency Low Inductance Decoupling Capacitors," *Proceedings of DesignCon 2005 East*, Worcester, MA, September 19–22, 2005.

[13] Novak, I., "A Black-Box Frequency Dependent Model of Capacitors for Frequency Domain Simulations," *Proceedings of DesignCon East 2005*, Worcester, MA, September 19–21, 2005.

[14] Motulsky, H., and A. Christopoulos, *Fitting Models to Biological Data Using Linear and Non-Linear Regression,* Cambridge, U.K.: Oxford University Press, 2004.

[15] Novak, I., G. Blando, and J. R. Miller, "Slow Wave Causal Model for Multi Layer Ceramic Capacitors," *Proceedings of DesignCon2006*, Santa Clara, CA, February 6–9, 2006.

CHAPTER 9

Characterization and Modeling of Inductors, DC-DC Converters, and Systems

Chapter 8 covered the modeling of bypass capacitors, which are the most common components in PDNs. Other components in a PDN are dc sources (batteries, ac-dc converters, dc-dc converters), inductors, and resistors. In PDNs, all of these components may be optional.

By definition, the PDN assumes a dc source. However, if we focus only on the characterization of a filter network for a low-power analog pin that takes the input power from a high-power rail, then the power comes from the high-power rail; therefore, it may be outside the scope of the filter circuit. Sometimes the dc source is not an integral part of the PDN; for instance, when the circuit uses a battery or a third-party detachable dc source (such as a separate ac-dc converter). In these cases, we may have a specification for the source and the associated interconnect, and the PDN characterization can focus on the remaining passive network. When the dc source is part of the PDN, we can have one or more ac-dc or dc-dc converters feeding the network. Most often, a single dc-dc converter provides input power. For high-power or redundant solutions, multiple sources can be connected to the input.

Inductors are also optional in PDNs. In high-power and wideband PDNs, we usually need to minimize inductance in the circuit; therefore, we do not use discrete inductor components. Nonetheless, inductance will remain present in the PDN because it is associated with the wires, PCB planes and vias, connector and socket pins, and planes, traces, and vias in packages. These inductances are inevitable side effects of the physical construction; their characterization and modeling was covered in the previous chapters. Discrete inductors are introduced intentionally to provide additional isolation between PDN domains. An inductor in series to a parallel bypass capacitor provides a second-order lowpass function; therefore, it is very efficient to filter out high-frequency noise. To avoid peaking with low-loss capacitors, ferrite beads with intentionally raised losses are popular. For magnetic storage purposes in dc-dc converters, low-loss inductors are required. We will cover these two options separately.

Discrete resistors are not typical in PDNs. Similar to inductance, resistance is also an inevitable side effect of conductors; so, our goal is to minimize them in order to minimize losses. There are situations, however, when the combination of low-loss inductors and capacitors creates unwanted peaking and ringing. If we cannot find components with the required medium or high ESR to suppress the peaking, we can use series discrete resistors. Series discrete *R-C* components can be used for instance, to suppress plane modal resonances [1], or a discrete resistor with a few ohms value can be used in bulk filtering. As controlled-ESR capacitors with user-selectable ESR values become more widely available [2], the usage of discrete resistors in PDNs is expected to continue to decrease. Sometimes a resistor is also used in series to decoupling inductors to enhance low-frequency filtering.

When we need discrete resistors in PDNs, we use resistors with values in the low ohms range. Given low resistance values, the primary parasitics is the series inductance of the part; so, their equivalent circuit is a series *R-L* network. Therefore, the characterization and modeling can be accomplished as it was for power/ground vias and shorted planes.

9.1 Characterization and Modeling of Inductors

Figure 9.1 shows how we use inductors in PDNs. In contrast to bypass capacitors, inductors are in series to the power path; therefore, normally none of their terminals is grounded. The pad and body capacitances, therefore, will show up as high-frequency limitations, creating parallel resonance with the inductor. The box around the inductor symbol is a reminder that the ideal inductor can be replaced with any of the more detailed models.

Figure 9.2 shows three of the simplest inductor models. For now we assume that the elements in these equivalent circuits are linear and frequency independent *L*, *R* and *C* values. The simplest of the three, a single inductor can be used in a limited frequency range (below the parallel resonance) or when the series losses are of low importance. The series *L-R* model adds a loss term and extends the validity of this approximate model. The *L-R-C* model adds the parasitic body (or winding) capacitance to the model, so that we can capture the parallel resonance at higher frequencies. The body and winding capacitance has little nonlinearity and frequency

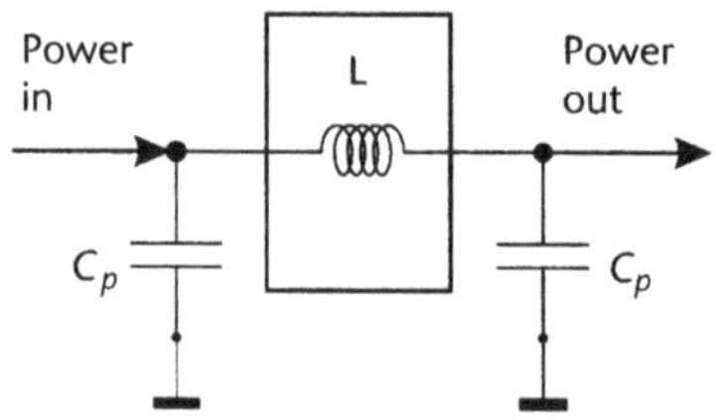

Figure 9.1 Equivalent circuit of an inductor in PDN.

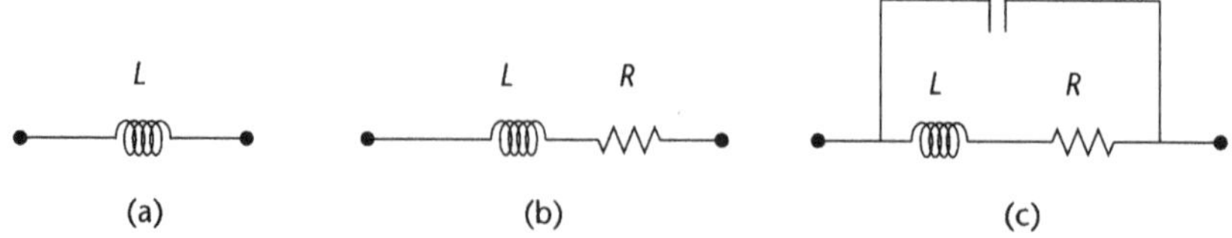

Figure 9.2 Simple *L-R-C* inductor models: (a) single *L* model, (b) series *L-R* model, and (c) series *L-R* with parallel *C*.

dependence. With frequency-independent resistance and inductance, these models are suitable to approximate inductors without ferromagnetic cores.

To achieve low volume for a given inductance, ferromagnetic cores are common in inductors used in PDN applications. Ferrites will create frequency dependent inductance and series loss and will require more complex equivalent circuits.

Before we look at more detailed models, we present measured data on different inductor samples to show how the equivalent-circuit elements behave as a function of frequency and bias.

9.1.1 Lossy Ferrite Inductors

As capacitors with high-dielectric constant ceramics and thin layers have a strong sensitivity to dc and ac bias voltages, inductors with high-permeability core materials have a similar sensitivity to dc and ac currents. The first example uses a small-size low-current ferrite bead. Figure 9.3 shows the inductance extracted from the impedance imaginary part measured on a 0805-size ferrite bead. The measurement was done with an Agilent 4294A impedance analyzer. The sample was measured in the 0A–0.1-A dc bias range in 0.01-A increments. Note that the maximum-rated dc current of the part is 0.4A, but the built-in bias generator of the impedance analyzer limited the dc bias range to 0.1A.

The same data that is plotted in 3D in Figure 9.3 is replotted in different ways in Figure 9.4. Figure 9.4(a) shows a few selected traces from the 3D plot. Up to about 1 MHz, the traces for different dc bias current run in parallel; there is little change of inductance with frequency. There is a sharp drop of inductance above 20 MHz. Figure 9.4(b) plots the inductance at 100 kHz versus dc bias current.

The impedance real part of the same ferrite bead is plotted in Figure 9.5. In the 0–0.1-A dc bias range, there was only a small variation of impedance real part with bias. The measured dc resistance at low frequencies is 0.385Ω, which matches well with the 0.5Ω maximum dc resistance specification. The impedance real part stays at its dc value up to about 1 MHz. Above a few megahertz, the resistance increases sharply. The relationship of the real and imaginary parts is shown in Figure 9.6. At medium and high frequencies, where the inductive reactance drops, the series loss increases, thus maintaining a minimum amount of impedance for series isolation. The series losses are also useful in decreasing the Q of the ferrite bead with any low-ESR capacitor we may have connected to it at either end. The parallel resonance of the part occurs at 80 MHz; it is marked by the sharp drop of reactance.

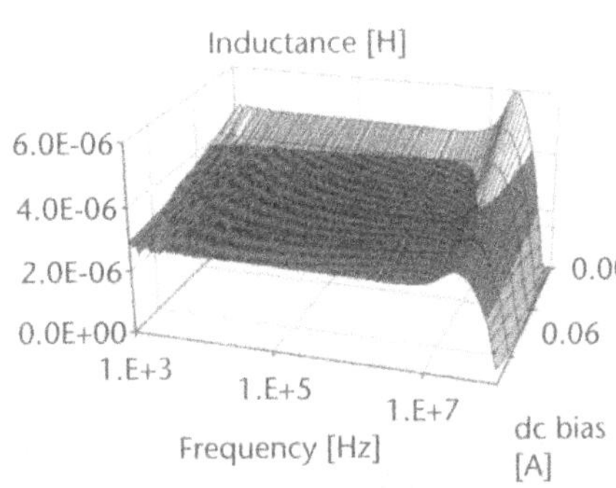

Figure 9.3 Inductance of a lossy ferrite bead as a function of dc load current.

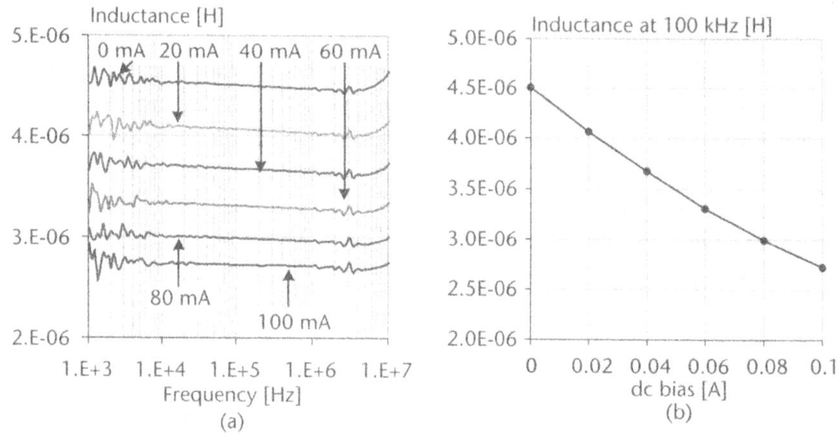

Figure 9.4 Inductance of a lossy ferrite bead as a function of dc load current: (a) inductance versus frequency with various dc bias values and (b) induction at 100 kHz as a function of dc bias.

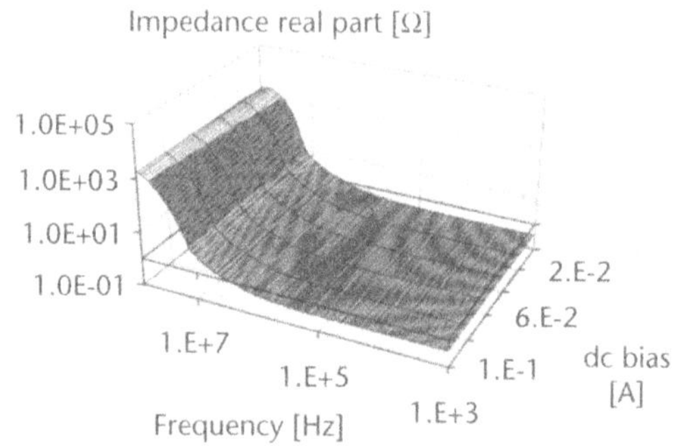

Figure 9.5 Impedance real part of a lossy ferrite bead as a function of frequency.

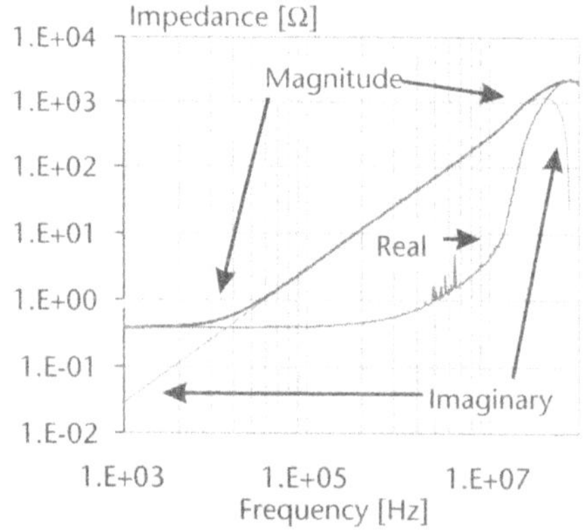

Figure 9.6 Comparison of impedance real, imaginary, and magnitude as a function of frequency. The dc bias is zero.

The previous plots were collected with a low ac source voltage of 10 mVrms. This source voltage was low enough that lowering it further did not change the impedance plots, but it was still large enough to suppress excess noise on the data. With higher ac source voltage, the impedance plot changes. The extracted inductance as a function of frequency and ac source voltage is shown in Figure 9.7. The DUT was the same that we used for the previous plots, and the dc bias current was zero.

Similar to Figure 9.4, Figure 9.8 replots the data to show the trends differently. We can notice that while traces at low frequencies in Figure 9.4(a) ran in parallel traces (with only an offset due to dc bias current) in Figure 9.8 exhibit slightly different slopes with different ac source voltages. Also, a higher source voltage results in a higher measured inductance. However, we need to keep in mind that a constant source voltage (versus frequency) does not necessarily mean constant voltage across the DUT. Just like the constant RF power of a VNA, when we measure a capacitor, a given source voltage will produce frequency dependent ac voltage across the DUT. With capacitors below the series resonance, a constant source voltage results in decreasing DUT voltage as frequency goes up. With inductors at frequencies below their parallel resonance frequencies, the trend is reversed: from a constant source voltage, increasing frequency results in increasing voltage across the DUT. At frequencies where the DUT impedance is much less than the source impedance of the measuring instrument, the frequency dependency is close to linear. As inductive reactance increases with frequency, there will be a cutoff frequency between the inductance and source impedance above which the voltage across the DUT saturates close to the source voltage.

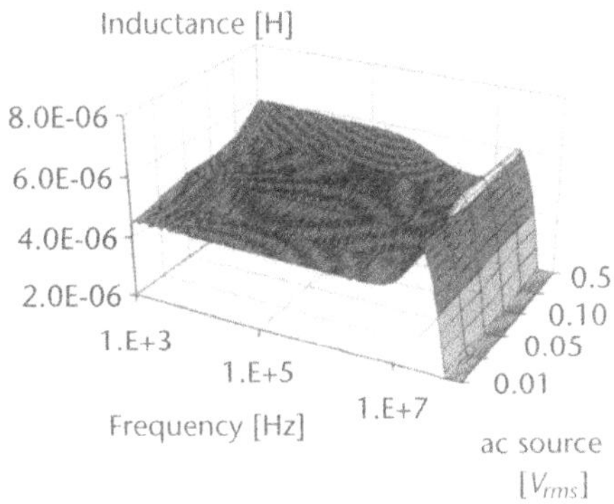

Figure 9.7 Inductance of a lossy ferrite bead as a function of ac source voltage.

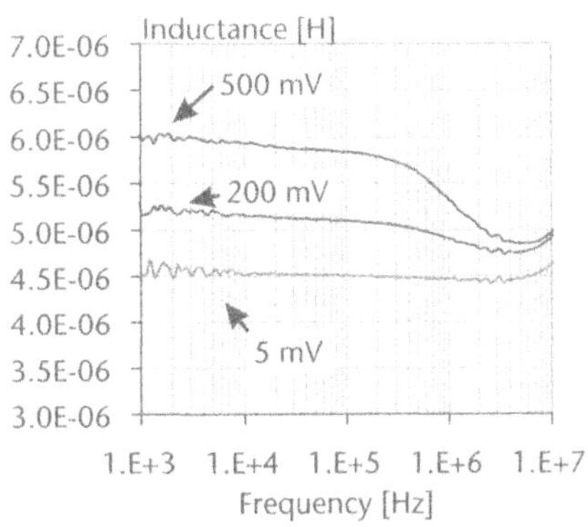

Figure 9.8 Inductance of a lossy ferrite bead as a function of ac source voltage.

While dc bias current did not change the impedance real part appreciably, with ac source voltage this is not the case. Figure 9.9 shows the impedance real part versus frequency of the same DUT with zero dc bias and with different ac source voltages. The surface plot shows a visible distortion in the transition frequency range. This is shown also in Figure 9.10, which plots a few lines from the 3D plot.

9.1.2 Low-Loss Ferrite Inductors

The second example uses a small-size power inductor. The part comes in a 7.6-mm^2 plastic house, and it is rated for 2.5-A maximum dc current.

In Figure 9.11 the extracted inductance versus frequency and dc bias current is plotted. The inductance is very stable both against frequency and bias. The impedance real part shows similar low sensitivity to dc bias, as shown in Figure 9.12. The peak at 80 MHz occurs at the parallel resonance between the inductance and body capacitance.

The plots as a function of ac source voltage are not shown, as they are identical to Figures 9.11 and 9.12; the data shows no appreciable variation with bias in the measured range.

9.1.3 Linear Inductor Models

We take the measured data of the lossy ferrite bead from Section 9.1.1 to show how various models composed of linear elements correlate. Figure 9.13 shows the magni-

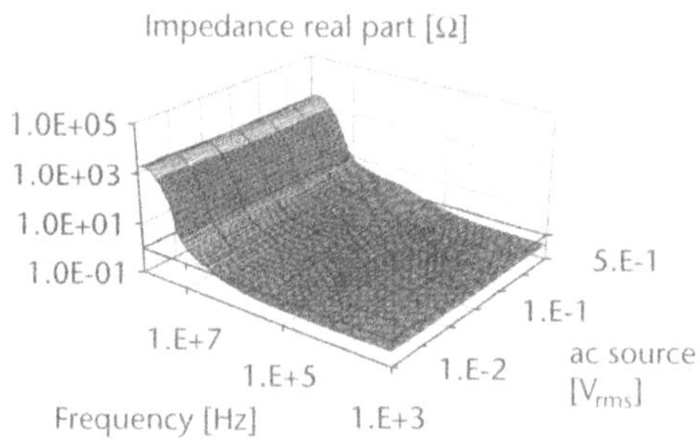

Figure 9.9 Impedance real part of a lossy ferrite bead as a function of ac source voltage.

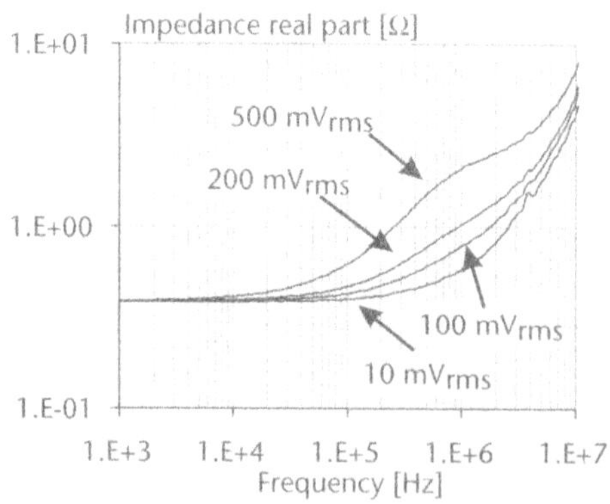

Figure 9.10 Impedance real part of a lossy ferrite bead as a function of ac source voltage.

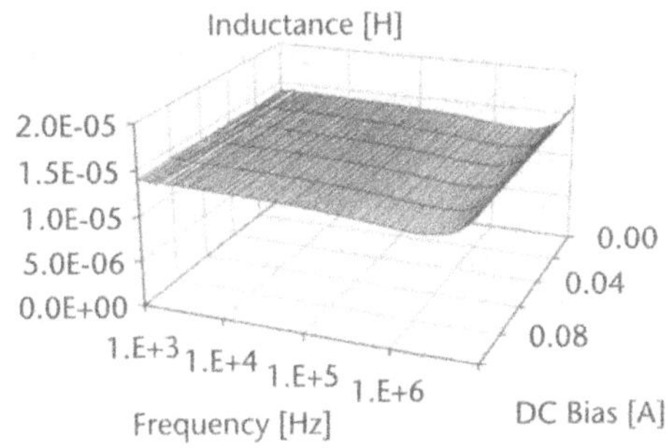

Figure 9.11 Inductance of a small power inductor as a function of dc load current.

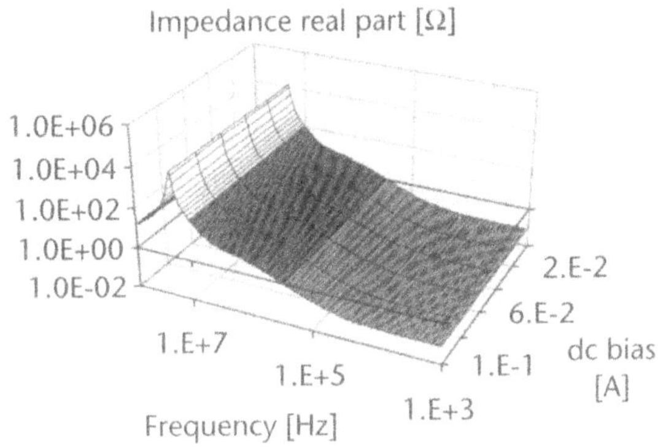

Figure 9.12 Impedance real part of a power inductor as a function of dc load current.

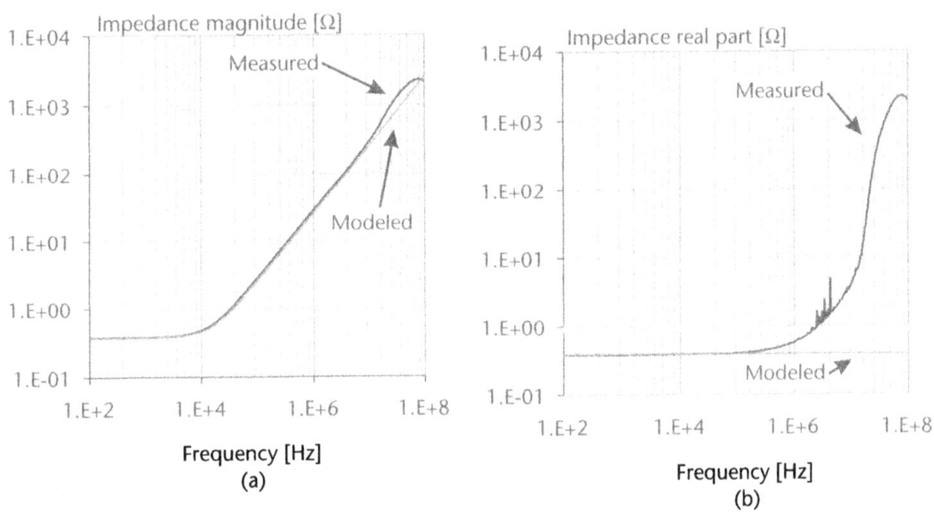

Figure 9.13 Correlation of (a) lossy ferrite bead magnitude and (b) impedance real part with models from Figure 9.2(b).

tude and real part correlation, using the equivalent circuits from Figure 9.2(b). The impedance magnitude shows reasonable correlation, except at the high end of the frequency range, where the body capacitance creates a parallel resonance. The modeled real part, however, is constant with frequency, as opposed to the sharp increase in the measured real part above 1 MHz. This error becomes important when we combine the inductor model with low-ESR capacitors.

Both the low-frequency and high-frequency losses can be better approximated with two resistors in the equivalent circuit. Figure 9.14 shows a four-element linear equivalent circuit after [3]. Figure 9.15 plots the correlation of the magnitude and real part with $C = 0.6$ pF, $L = 3.5\ \mu$H, $R_1 = 2100\Omega$, $R_2 = 0.29\Omega$.

The rise of the impedance real part is now captured, except the change in the inflection point around 10 MHz. The four-element model can also capture the parallel resonance with its Q. This is illustrated in Figure 9.16, where we show the magnitude and phase correlation of the small-size power inductor from Section 9.1.2 with $C = 3$ pF, $L = 15\ \mu$H, $R_1 = 10{,}000\Omega$, $R_2 = 0.065\Omega$.

If we parameterize the inductance, the linear model can also approximate the effect of dc bias. Figure 9.17 reproduces the inductance versus dc bias data from Figure 9.4(b), together with two possible approximations. Figure 9.17(a) shows a second-order polynomial approximation, while Figure 9.17(b) uses an exponential approximation. The large dots indicate the measured data; the thin lines show the approximation curves. The approximation functions were obtained with the trendline function of the spreadsheet over the measured bias range of 0 to 0.1A. The second-order polynomial is:

$$L = ax^2 + bx + c \tag{9.1}$$

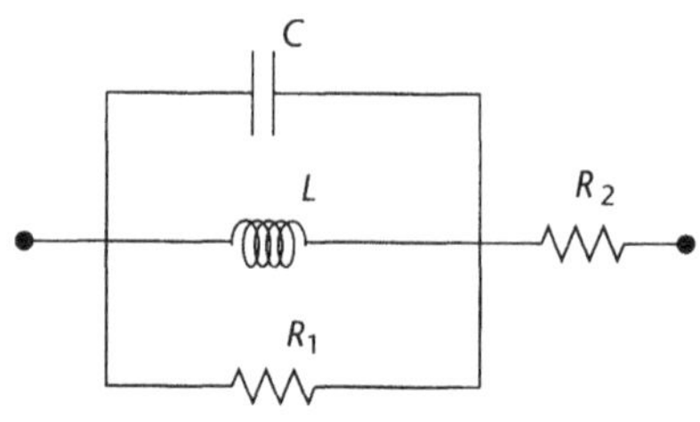

Figure 9.14 Four-element linear inductor model.

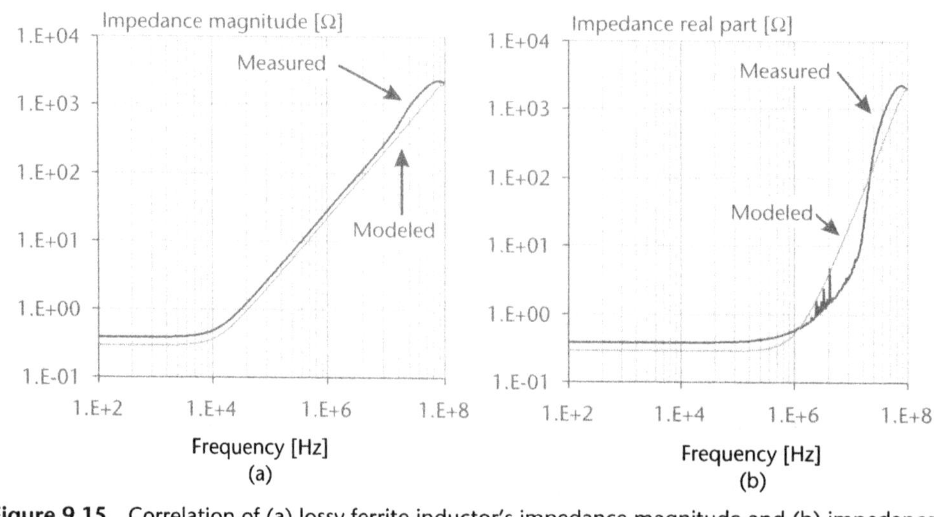

Figure 9.15 Correlation of (a) lossy ferrite inductor's impedance magnitude and (b) impedance real part with the four-element linear model from Figure 9.14.

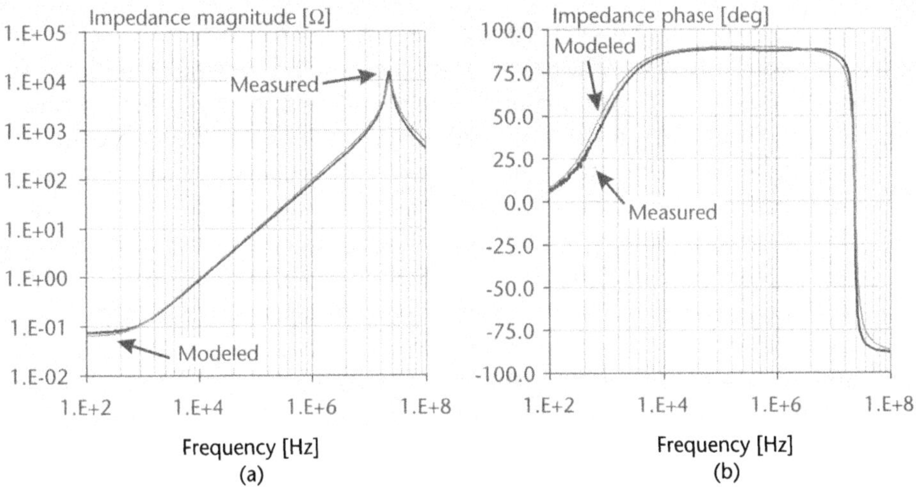

Figure 9.16 Correlation of (a) the small-size power inductor's impedance magnitude and (b) phase from Section 9.1.2 with the four-element linear model from Figure 9.14.

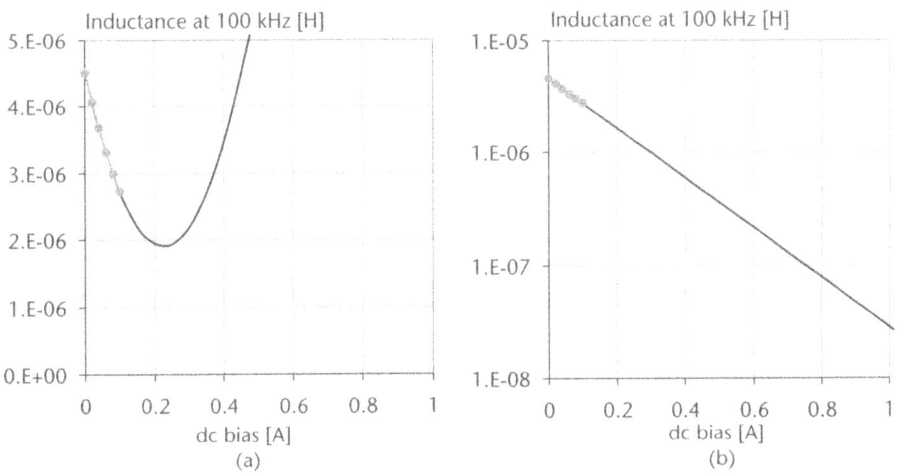

Figure 9.17 (a) Second-order and (b) exponential approximation of the inductance versus dc bias from Figure 9.4(b).

where L is the inductance in H, x is the dc bias current in A, a = 5.0619E-05, b = −2.2904E-05, and c = 4.5052E-06.

The exponential approximation is:

$$L = ae^{bx} \tag{9.2}$$

where L is the inductance in H, x is the dc bias current in A, a = 4.4980E-06, b = −5.0566.

The two approximations illustrate a potential pitfall and the proper approximation approach. Even though both approximations follow the measured data well, beyond the measured range the two approximations behave very differently. To show this difference, we plot the approximation functions up to a 1-A bias. The second-order function reaches a minimum at a 0.25-A bias, beyond which its value monotonically rises. Unless we make absolutely sure that the approximation formula does not get used outside the measured range, using this approximation may result in unrealistic inductance values at a higher bias. The exponential approximation does not change its trend: within and beyond the measured range, it has the same rate of change. This is emphasized by using a logarithmic vertical scale, since the exponential function becomes a straight line. The measured data on this scale falls on the same straight line, suggesting that inductance drops exponentially with increasing dc bias. Without further known data points we cannot guarantee that the trend will stay the same at higher bias values; but, at least, it becomes more reasonable to assume that the exponential approximation will follow the component's behavior properly outside the measured range.

9.1.4 Frequency-Dependent Inductor Models

Some component vendors offer frequency-dependent simulation models for their inductors. Reference [4] suggests an equivalent circuit shown in Figure 9.18.

In the equivalent circuit C, R_1, and R_2 are assumed to be constants, whereas L_{var}, R_{var1}, and R_{var2} are frequency dependent through five constants (k_1 ... k_5):

$$R_{var1} = k_1\sqrt{f},\ R_{var2} = k_2\sqrt{f} \tag{9.3}$$

where f is frequency in hertz and R_{var1} and R_{var2} are in ohms.

$$L_{var} = k_3 - k_4 \log(k_5 f) \tag{9.4}$$

where L_{var} is inductance in henries and f is frequency in hertz.

We apply this model to the data from Figures 9.7 through 9.9, with a brief manual optimization of the eight parameters. The result is shown in Figure 9.19. The parameter values for the correlation: C = 2.7 pF, R_1 = 1,600Ω, R_2 = 0.4Ω, k_1 = 1.3E-8, k_2 = 1,500, k_3 = 5.5, k_4 = −0.1, k_5 = 1.0E-8.

Both the magnitude and phase show excellent correlation in the entire measured frequency range.

Bias dependencies, nonlinearities, and frequency dependencies can be taken into account by using a network of SPICE building blocks.

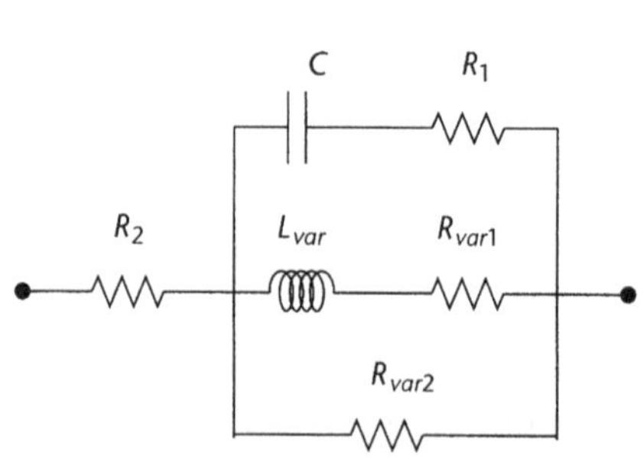

Figure 9.18 Six-element frequency-dependent linear model from [4].

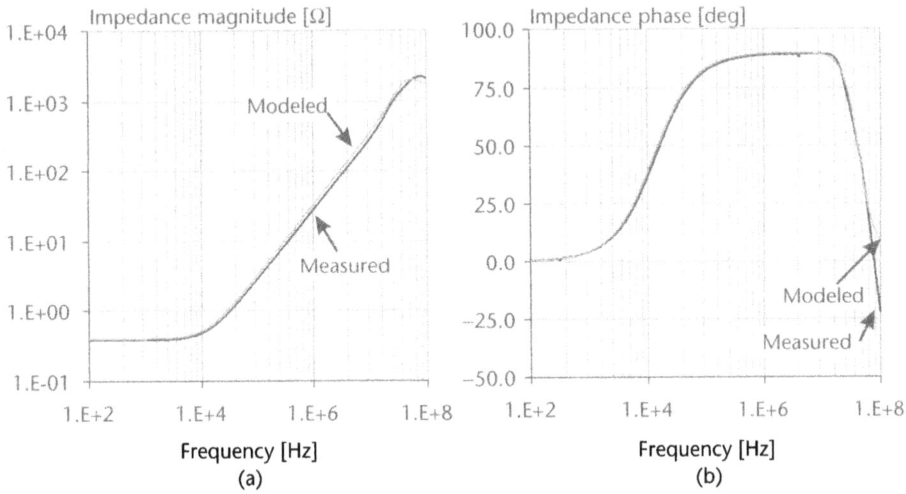

Figure 9.19 Correlation of the six-component frequency-dependent model from Figure 9.18 on data from Figures 9.7 through 9.9: (a) impedance magnitude and (b) phase of impedance.

9.2 Characterization and Modeling of Power Converters

As it was shown in Chapter 7, power-source modules, such as dc-dc converters or analog voltage regulators, can be characterized with various input, output, and transfer functions. The primary frequency domain characterization item is the small-signal output impedance curve. This parameter provides the best opportunity to have simulated or measured data for the characterization. As module sizes and the number of on-module filter components continue to shrink, we can expect that the input impedance and transfer functions of power modules will gain importance, and their characterization and simulation processes will be elaborated.

There are many different voltage-regulator topologies. One classification splits them into analog and switching modes of operation. The analog voltage regulators come in two subclasses: series and shunt regulators. The series analog regulators have a controlled resistor in series to the load, between the input and output. Their power efficiency depends on the input-output voltage differential. The closed feedback loop keeps the output voltage constant against changes in input voltage and load current. The small-signal equivalent block schematic is shown in Figure 9.20(a). The output voltage of switching regulators is controlled by adjusting the duty cycle of the switching. The control mechanism can be based on monitoring either the output voltage or the inductor current, leading to different compensating schemes. A typical buck regulator can be characterized with the small-signal equivalent block schematic shown in Figure 9.20(b).

The error amplifier and the three feedback impedances (Z_1, Z_2, and Z_3) create the frequency-dependent feedback loop around the reference voltage and output voltage. In Figure 9.20(b) the diamond-shaped block, labeled *M*, is the duty-cycle modulator, which also contains the switches between the input voltage and return. We have dashed boxes around the *L* output inductor and *C* output capacitor to indicate that they may be composite elements (for instance, the output capacitor symbol may represent multiple capacitors in parallel). Note that their proper mod-

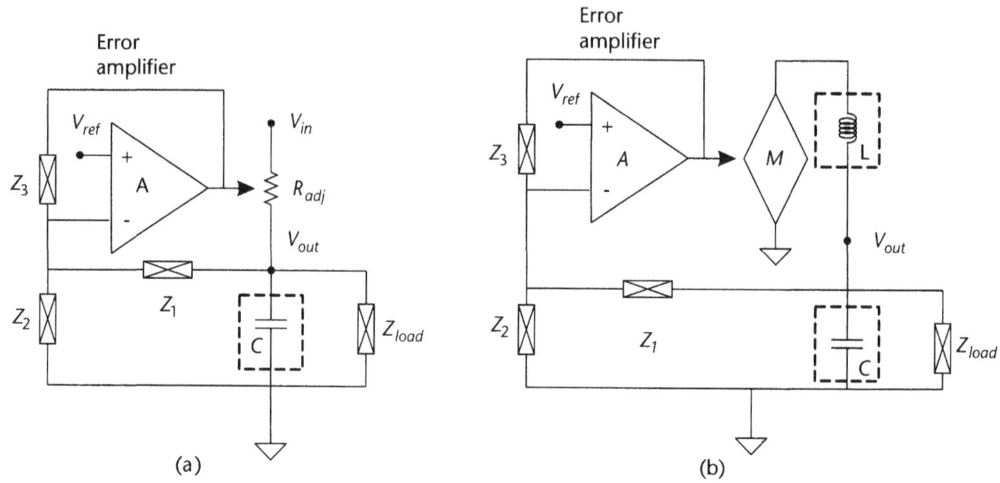

Figure 9.20 Small-signal equivalent circuits of (a) analog and (b) switching-mode voltage regulators.

eling may require more complex circuits, as was shown earlier. The Z_{load} component in the equivalent circuit represents the entire load network, various bypass capacitors with their interconnects, as well as the actual load itself. It is important to realize that the load impedance is in the feedback loop; therefore, it may affect loop stability and frequency characteristics.

Analog and sampled feedback loops and control circuits have well-established literature describing small-signal loop parameters. These equivalent circuits assume that we know the circuitry and its parameters, which is many times not the case for the end user. References [5, 6], for instance, describe analytical solutions.

9.2.1 Small-Signal Output Impedance of DC-DC Converters

The small-signal output impedance of dc-dc converters does depend on input voltage, output-voltage setting, and load current. In well-designed converters with analog linear control loops, the output impedance should not vary greatly, as the load current varies over the entire allowed range. As an illustration, Figure 9.21 shows the small-signal output impedance surface of a 5-A nonisolated point-of-load (POL) converter. The impedance was measured with constant 3.3-V input and 1.0-V output voltages, over the frequency range of 100 Hz–10 MHz. The converter was fed from a bench supply, and one 100-μF low-ESR POSCAP was connected across the converter inputs. The output was connected to an electronic load and the VNA ports, and no external capacitors were added. The load current was varied from 0A to 5A in 0.2-A increments. The tight variations can also be seen on the 2D charts of Figure 9.22. Figure 9.22(a) shows the full measured frequency range; Figure 9.22(b) is enlarged between 100 Hz and 10 kHz. Note that above 20 kHz all traces run on top of each other. The largest spread occurs around 2 kHz, where the impedance magnitude varies with load current in the 8.6–14.4-mΩ range.

This converter showed similar performance with other input and output voltages within the allowed limits.

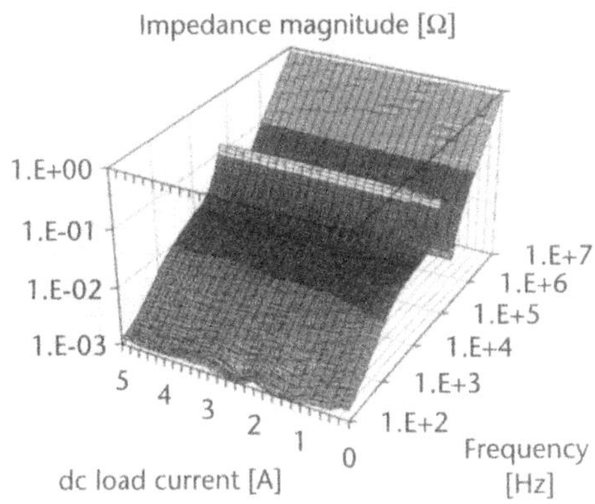

Figure 9.21 Small-signal output impedance surface of a stable point-of-load converter; dc load current was stepped from 0 to 5A in 0.2-A increments.

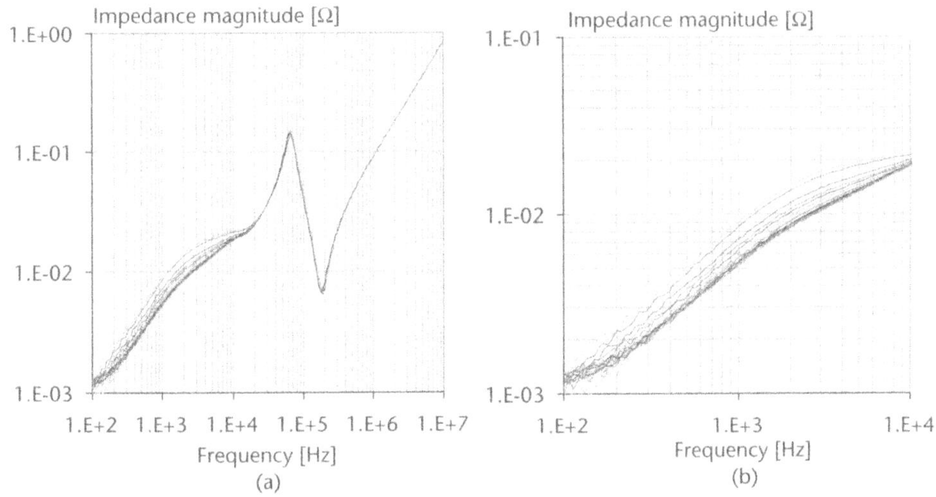

Figure 9.22 Small-signal output impedance of a stable point-of-load converter; dc load current was stepped from 0 to 5A in 0.2-A increments: (a) full frequency range of 100 Hz–10 MHz and (b) enlarged frequency scale.

If the analog converter loop is not designed properly, the small-signal output impedance profile may show significant variations with changes in load current, input voltage, or output voltage. It is useful to test multiple units, and if possible, to conduct tests over the allowed temperature range. Figure 9.23 shows the impedance plot of a nonisolated 20-A converter with 12-V input and 2.5-V output voltages. Several units were tested at room temperature, over the specified input voltage range. The impedance profile shown in the figure was from an unstable unit; this unit performed acceptably at nominal input voltage, but exhibited peaking with 13-V input voltage. The surface plot was taken with dc load currents in 0.5-A increments in the 0-A to 20-A range. Note that the impedance profile is smooth below 2A and above 16A. The unstability shows up in two ways in the impedance profile: a sharp peak builds up at 40 kHz, and the low-frequency impedance plateau fluctuates between 0.1 and 4 mΩ. Note that as the load current changes, the 40-kHz

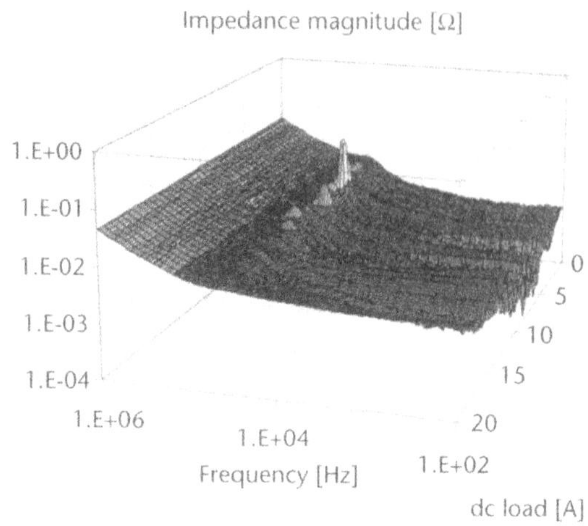

Figure 9.23 Small-signal output impedance surface of an unstable converter.

peaks and the low-frequency plateau fluctuate periodically. Moreover, the 40-kHz peaks appear when the low-frequency plateau is lower.

Some dc-dc converters have wide input and output voltage ranges, and in certain portions of these ranges, the output impedance profile may exhibit irregularities. Figure 9.24 shows two impedance profiles measured on the same nonisolated 20-A dc-dc converter, with two different output-voltage settings. The input voltage was 12V.

The output voltage was adjustable over the 1.0- to 3.5-V range. Figure 9.24(a) shows the output impedance with 1.2-V dc output voltage. There is a steplike bay in the surface below 10 kHz in the 1–5-A load range, and a 20-mΩ peak at 2 kHz with 0–1-A and 5–20-A currents, but otherwise the impedance profile appears smooth. Figure 9.24(b) shows the impedance profile of the same converter with 1.5-V output voltage. There are sharp peaks in the 1–5-A load-current range at 8 kHz, with peak magnitudes up to 60 mΩ.

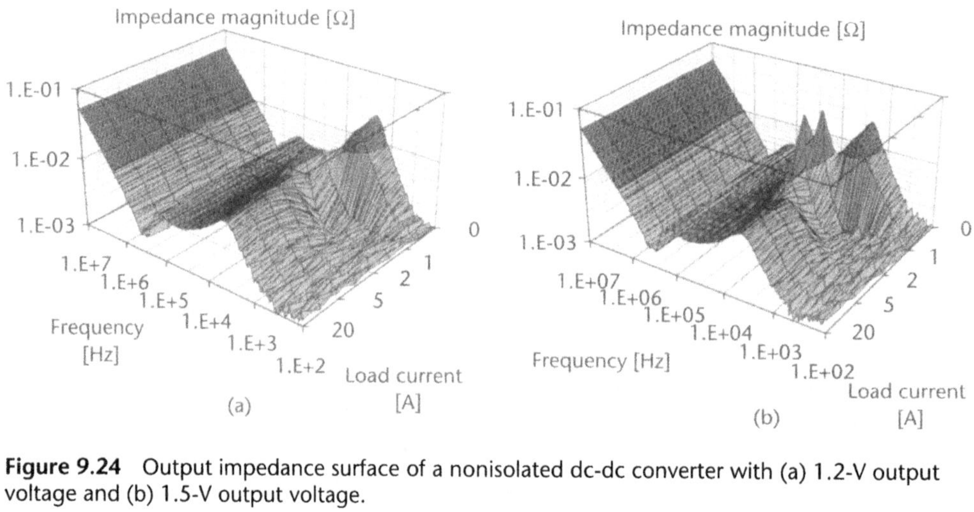

Figure 9.24 Output impedance surface of a nonisolated dc-dc converter with (a) 1.2-V output voltage and (b) 1.5-V output voltage.

A change in the input voltage may equally result in distortions of the small-signal output-impedance profile. Figure 9.25 shows an illustration from an isolated converter, with nominally 48-V input and 1.5-V output voltages. Figure 9.25(a) shows the output impedance at a nominal input voltage, while the load current was varied from 0A to the maximum 20A. Note the ripples and sharp peaks that show up between 30 kHz and 100 kHz at different load currents. The graph in Figure 9.25(b) shows the same converter with a high input voltage. Note the shifts in the peaks and ripples.

Digital power is a promising new trend in power conversion. It enables not only a versatile on-the-fly control of the usual parameters, such as output voltage and turn-on and turn-off delays and ramps, but ultimately enables the implementation of nonlinear control loops. Nonlinear control loops make it possible to improve transient behavior. Though this seems to be all and only good, there is a hidden catch: the validation of a system design then becomes more difficult. The validation of a high-performance system design should be never skipped, especially when off-the-shelf converters are used. Since the dynamic performance of the converter depends on the external capacitors we connect to the converter (it is likely that during manufacturing, the converter was tested with a different set of capacitors than in our design), we have to validate the converter together with our design. As it was illustrated above, a well-behaving analog control loop with a properly designed set of external capacitors will show a more-or-less constant impedance profile, regardless of the static load current and input voltage. This means that the validation is reduced to checking the small-signal output impedance profile over the entire input-parameter range, such as input voltage, output voltage, and dc load current. The small-signal output impedance can be obtained in a straightforward manner; furthermore, since it does not require injecting a test signal in series to the control loop, in general it is easier to perform than measuring the Bode plots for the classical loop stability.

When the control loop is made purposely nonlinear, the validation cannot be done any longer with small-signal output impedance measurements. A nonlinear loop control may tighten the loop for larger output deviations, and loosen it for small deviations. This means that any measured small-signal output impedance may

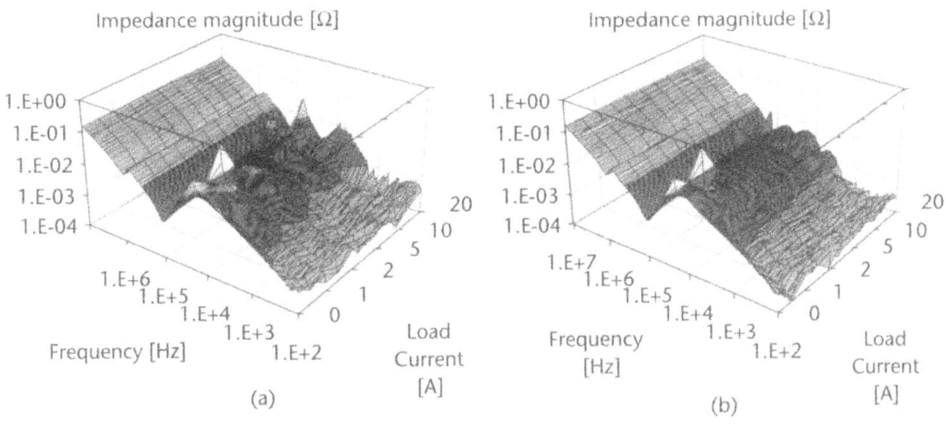

Figure 9.25 Output impedance surface of an isolated 48-V–1.5-V dc-dc converter with (a) nominal input voltage and (b) high input voltage.

show unrealistic impedance values [7]. An illustration of this is shown in Figure 9.26. Note the low-frequency impedance magnitude of about 18 mΩ. If this were linear impedance, a 1-A load-current step would produce an 18-mV output transient, whereas a 10-A load step would produce 180-mV transients. A quick time-domain scan showed that these transients did not occur: larger load-step currents did not produce proportionally larger transients (that is the purpose of the nonlinear loop). While the nonlinear control loop helps to improve the transient behavior, it also requires us to scan against dc load current and load-step magnitude. The frequency parameter of the impedance plots is replaced by the slewrate or rise time of the excitation current step. For a full validation, the testing must be performed in the time domain which, due to its wideband nature, is more prone to external noise contaminating the data. Digital control loops, on the other hand, can perform loop diagnostics, including stability and response function tests, in the background.

9.2.2 Black-Box Modeling of Output Impedance

A large class of stable small-signal output impedance profiles can be modeled without much knowledge about the internal structure of the converter and regulator loop. The simple models shown here are also applicable to analog voltage regulators and switching converters alike.

Frequency-domain characterizations may simply omit the dc source voltage. If the regulator loop has high dc gain, the dc and low-frequency output impedances will be very low. This suggests that the simplest linear model might be a short or a low-value resistance in series to an inductance, representing the connection inductance. The equivalent circuit from Figure 9.2(b) can be used for this simplest case.

In Figure 9.27 we show how the *R-L* linear model matches the measured small-signal output impedance of the dc-dc converter data shown in Figures 9.21 and 9.22. The magnitude curves show a reasonable agreement up to about 1 kHz.

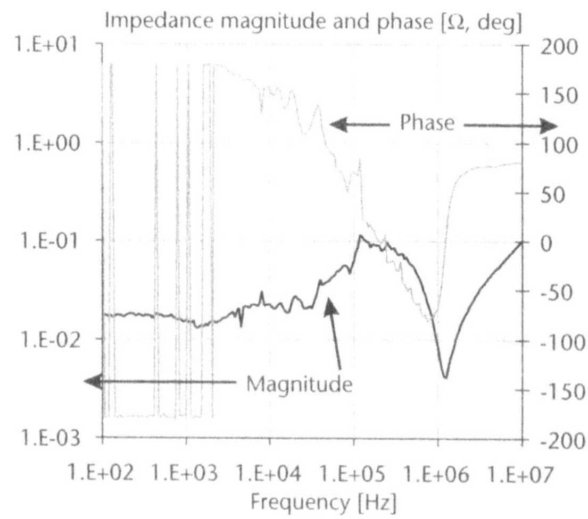

Figure 9.26 Measured small-signal output impedance of a dc-dc converter with nonlinear digital control loop.

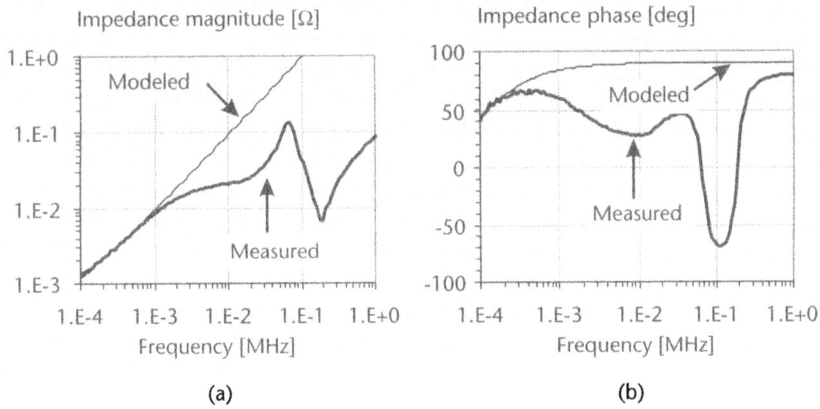

Figure 9.27 Correlation of an *R-L* model and measured output impedance from Figure 9.28. R = 1 mΩ, L = 1.5 μH: (a) impedance magnitude and (b) phase of impedance.

The phase curves agree well below 300 Hz. The higher-frequency signatures cannot be modeled with just the two free parameters of the *R-L* equivalent circuit.

Reference [8] suggests a combination of a black-box and a physical-based model for switching regulators. It is based on the realization that the output inductance can be modeled by a series *R-L* circuit, and it feeds the output capacitors, which are modeled by the usual three-element *C-R-L* network. Eventually the two networks are connected in parallel. If we neglect the feedback loop, we get the equivalent circuit shown in Figure 9.28.

The measured impedance data from Figure 9.21, fitted on the model of Figure 9.28 produces the correlation shown in Figure 9.29. With five free parameters in the model, we can capture the series resonance of the output capacitors, as long as there is just one type of capacitor on the output. The model still misses the antiresonance peak both in frequency and magnitude. This peak is missed because the equivalent circuit does not take into account the frequency-dependent loop gain in the feedback control circuit, which manifests itself as further damping at mid-frequencies, below the switching frequency.

We can synthesize and approximate impedance profiles more accurately with multiple capacitors, inductors, and resistors. The impedance profiles of many converters with stable analog loops can be matched by a few parallel-connected *C-R-L* networks. To illustrate this possibility, we use the same impedance profile from Figure 9.21. The upper bound of the impedance profile was matched to the cumulative impedance of three parallel-connected *C-R-L* components in an Excel spreadsheet. The matching can be done either by manual iteration, or by using the built-in solver

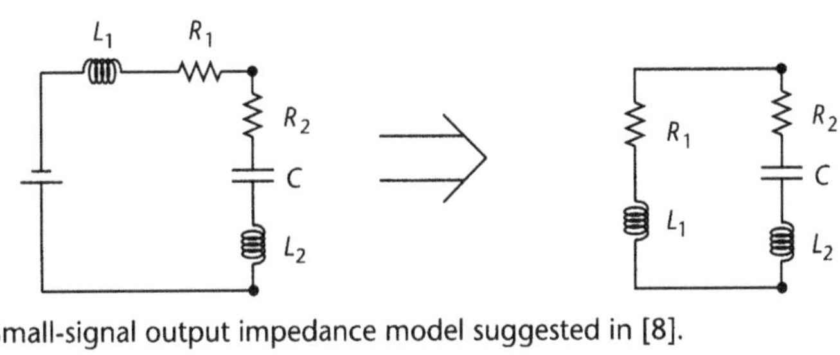

Figure 9.28 Small-signal output impedance model suggested in [8].

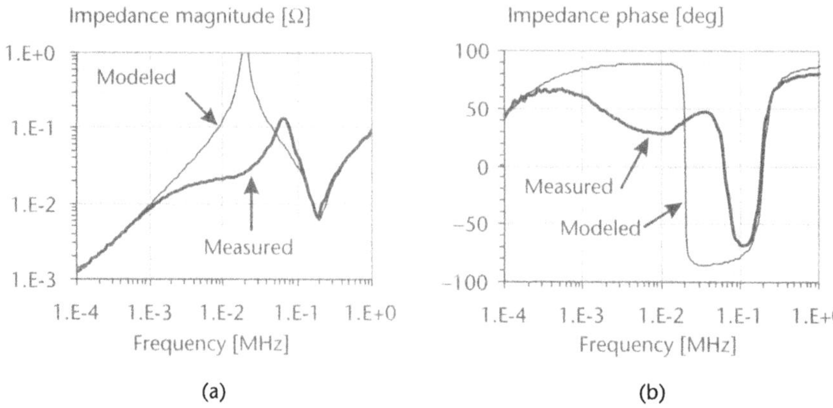

Figure 9.29 Correlation of measured data from Figure 9.21 and model of Figure 9.27 with the following parameters: R_1 = 1 mΩ, L_1 = 1.5 μH, C = 40 μF, R_2 = 6 mΩ, and L_1 = 16 nH: (a) impedance magnitude and (b) phase of impedance.

function to minimize the relative difference between the measured and modeled values. Table 9.1 shows the component values, and Figure 9.30 shows the correlation.

This model is an extension of the earlier five-element model. We use the same R-L circuit to represent the lowest frequencies (C_1) and the same *C-R-L* network to model the series resonance of output capacitors (C_3), and just add one parallel capacitor (C_2) to account for the damping of the control loop. The capacitance in this case has to be very high, 0.1F or higher, to provide a good match. Alternately, as for C_1, we could use an *R-L* network in place of C_2, by making its capacitance infinite. Note that the measured and modeled magnitude traces coincide almost completely. Note also that while C_3 in this model represents physical capacitors, C_1 and C_2 are equivalent circuits of the output network and control loop, and therefore have no direct physical meaning.

9.3 Modeling and Characterizing Systems

In the previous sections and in Chapter 8 we looked at the characterization and modeling of the various components and building blocks of PDNs. When we apply those principles to system characterization, we may use the same measuring methods, simulation tools and characterization methods, and concatenate the result to get to the system description. Along the way, we may need to decide about the accuracy options of measurements and the complexity of simulations and modeling. For

Table 9.1 Component Values for Three-Capacitor Model of Converter from Figure 9.21

	C_1	C_2	C_3
C [F]	inf.	0.1	4E-5
R [mΩ]	1	20	6
L [H]	1.5E-6	0.15E-6	1.6E-8

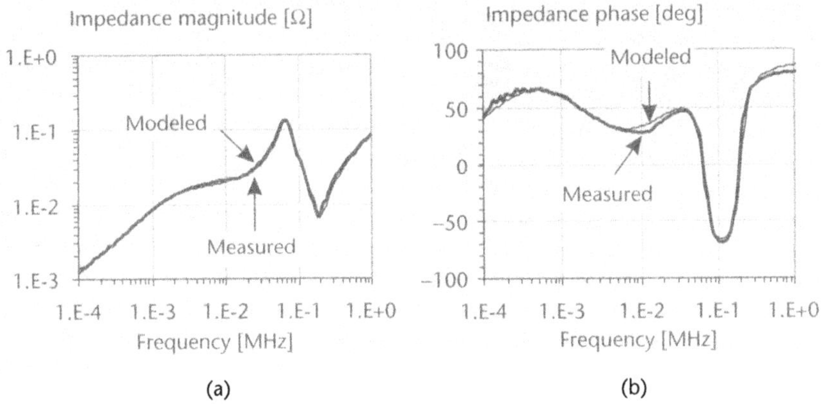

Figure 9.30 Correlation of three-capacitor converter model: (a) impedance magnitude and (b) phase of impedance.

a full source-to-silicon PDN path the applicable bandwidth is very wide, spanning from dc to many gigahertz. The application of measuring methods and models that provide the full bandwidth for all of the building blocks becomes a practical burden on measurement and modeling resources; often, it is not necessary. As shown through several examples throughout the text, apart from possible resonances in predictable frequency ranges, PDN blocks behave like lowpass filters with cutoff frequencies, dependent on the size and application. When we perform full-system characterization and modeling, we can make use of this fact by setting the complexity and bandwidth of each block to fit the location of the result. The bedspring capacitor model of Section 8.7 helps us to understand how and why the inductance and resistance of a ceramic capacitor mounted on a PCB changes with frequency. However, this level of detail is not necessary or practical when we look at the PDN impedance presented to the silicon, because we have PCB connections and a package along the way. Similarly, as we will show in the following sections, the high-frequency details of the silicon may generate un-noticeably small differences if we stay on the PCB.

The characterization of full-system PDNs can be a major undertaking that requires a lot of details. The following sections will provide a few selected illustrations in which we will point out trade-offs and potential pitfalls.

9.3.1 Return Path and Rail Coupling in Flip-Chip BGA Package

In this section, we show the quality check of high-speed signal returns from a complex multilayer organic package. The flip-chip package had multiple power rails: one relatively noisy rail for TTL-level signaling and a quieter rail for a high-speed parallel bus. There were unsplit ground planes within a layer with split plane shapes for the two supply rails. Traces for each signal class were routed between ground and their corresponding power-plane shapes. Self- and transfer impedances on the bare package were measured on the bump side with wafer probes. Figure 9.31 shows the self impedances of the two supply rails. The impedance of the bare package looks similar to that of a bare PCB, and it can be modeled and simulated by the same techniques. The larger sizes of plane shapes result in a pronounced parallel res-

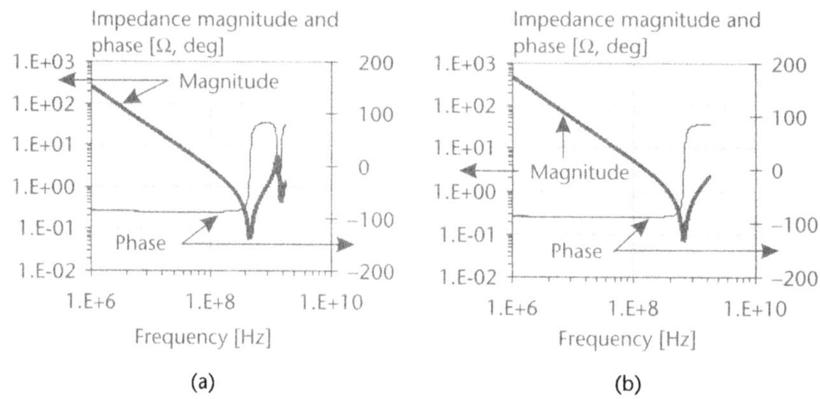

Figure 9.31 Self-impedances of two supply rails on a bare multilayer flip-chip package: (a) TTL rail and (b) high-speed rail.

onance on the TTL supply rail at 1 GHz. Due to its smaller size, up to 2 GHz there was no modal resonance on the high-speed rail.

The plane shapes of the two supply rails were in the same layer. Through the split, there is coupling between the two supply rails. Figure 9.32 shows the transfer impedance between the two supply rails at two different locations, measured on a bare package. The two impedance curves are very similar at low frequencies. The transfer impedance between two bumps close across the split shows strong high-frequency coupling, whereas between remote bump pairs, the high-frequency coupling is weak.

The coupling of several traces running in the vicinity of the plane split and the two supply plane cavities were also measured. Figure 9.33 shows the transfer impedance from some of the high-speed traces, all routed over the high-speed plane, but at varying distances from the plane split. All traces were terminated on the ball side with surface-mount resistors, and the coupling was measured with wafer probes on the bump side. Figures 9.33(a) and 9.33(b) show the low-frequency and high-frequency coupling, respectively. The line numbering follows the approximate distance of the trace from the plane split. Line 1 exhibits the highest coupling; note that it was

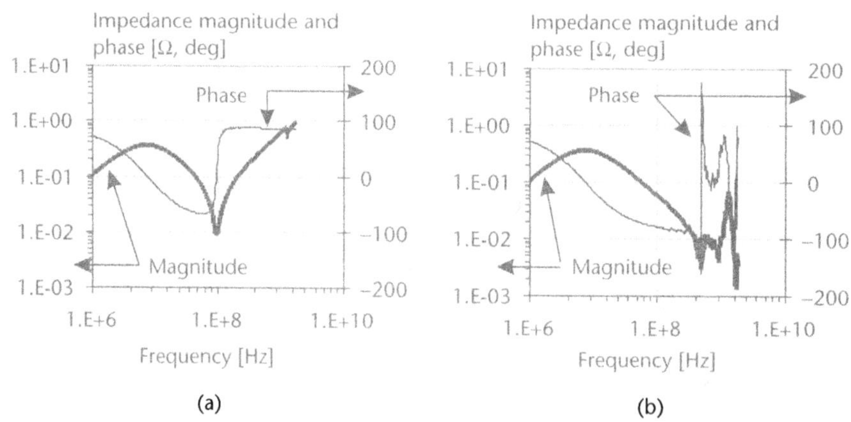

Figure 9.32 Transfer impedance between the two supply rails in the bare package: (a) test points close across the split and (b) test points far away from the split.

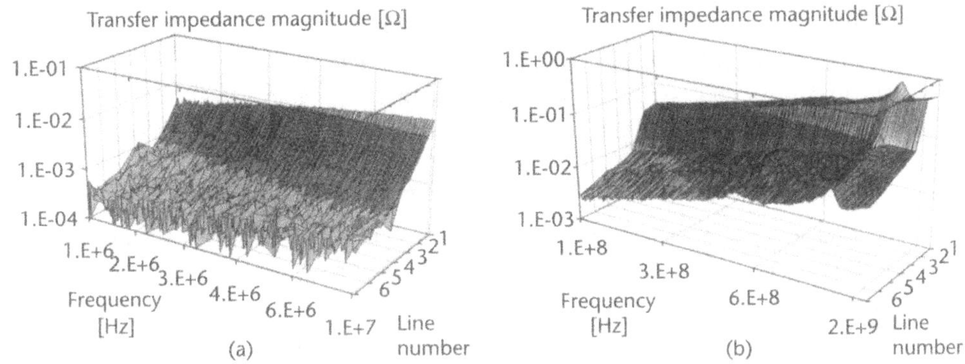

Figure 9.33 Transfer impedance from terminated high-speed traces to the TTL plane cavity: (a) 1–10-MHz frequency range and (b) 100-MHz–2-GHz frequency range.

close (but not the closest) to the split along its full length. Also note the resonance ridge on the right plot occurs above 1 GHz: this indicates increased coupling at the modal resonance frequency of the TTL planes.

To get a comparison baseline, the coupling was also measured from the same traces to the high-speed plane shape cavity. The transfer impedances with the same frequency scales are shown in Figure 9.34. For the traces being close to the plane split, the transfer impedance is almost the same to either supply-rail cavity. This indicates that noise from the TTL supply rail can easily find its way to the high-speed trace running close to the split.

If this were a pair of split planes with a straight gap separating them and straight traces running in parallel to the gap, it would be easy to model and to simulate the effect of the return-path coupling. In that case, a 2D field solver may be sufficient to give the proper result. The actual package, however, had a jagged plane split, with several inevitable antipad cutouts further complicating the current distribution along the edges and near the traces. Moreover, the different traces had different angles and length near the split before departing to another layer through via transition. Attempting to simulate and model the coupling through power domains of this

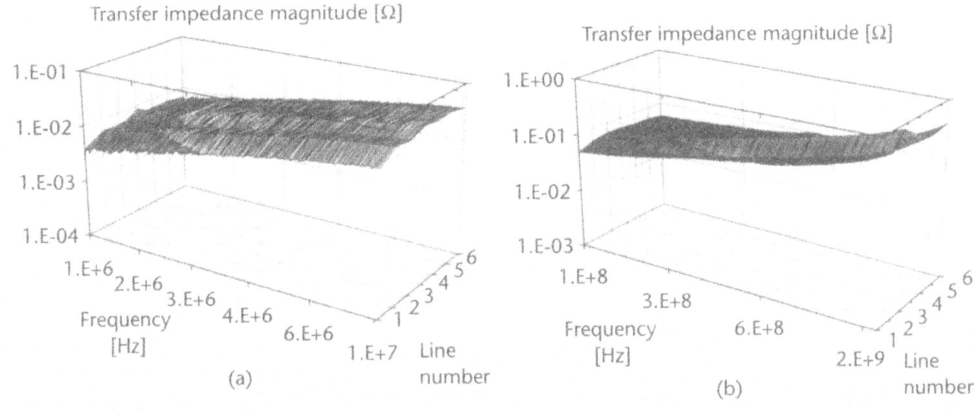

Figure 9.34 Transfer impedances between the high-speed traces and their corresponding plane cavities: (a) low-frequency range and (b) high-frequency range.

structure would have required the inclusion of a large portion of the package with many small details of the geometry. In this particular case it, was faster and simpler to obtain the necessary data by measurements.

9.3.2 Core and DIMM Memory Rails with Various Populations

In this section we show illustrations from core and memory supply rails measured with various component populations. Supply rails for cores in CPUs, FPGAs, and large ASICs are often point-of-load connections: they feed only one or a few loads of the same kind. The core logic may draw substantial current, sometimes in excess of 100A at voltages of 1.5V or less. The core logic may exhibit large current fluctuations, but package capacitors and the package and connection inductances tend to limit the bandwidth of current transients entering the board. This means we usually need low impedance on the board from the core supply rail only at low frequencies.

For the purposes of PDN characterization, supply rails of dual-in-line memory modules (DIMM) behave similarly to core supply rails: the popular DDR-I and DDR-II memory modules require substantial current on the 2.5-V or 1.8-V supply rail, with significant current fluctuations. The on-module bypass capacitors and the DIMM socket inductance form a low-pass filter with a cutoff frequency in the order of 1 MHz, and therefore, for the purposes of feeding the memory with clean power, the required impedance on the main board has to be maintained only up to this corner frequency (with a suitable overhead margin).

Since many of the CPUs and memory modules are socketed, we may be interested in assessing the impact that these large components have on the impedance of the populated main board.

Figure 9.35 compares self- and transfer impedances of a 50-A CPU core rail with and without a CPU installed. We show the data only in the 1-MHz–1,800-MHz range, because at low frequencies there was no measurable difference. Even at high frequencies, the measured self-impedance across the core planes near the CPU

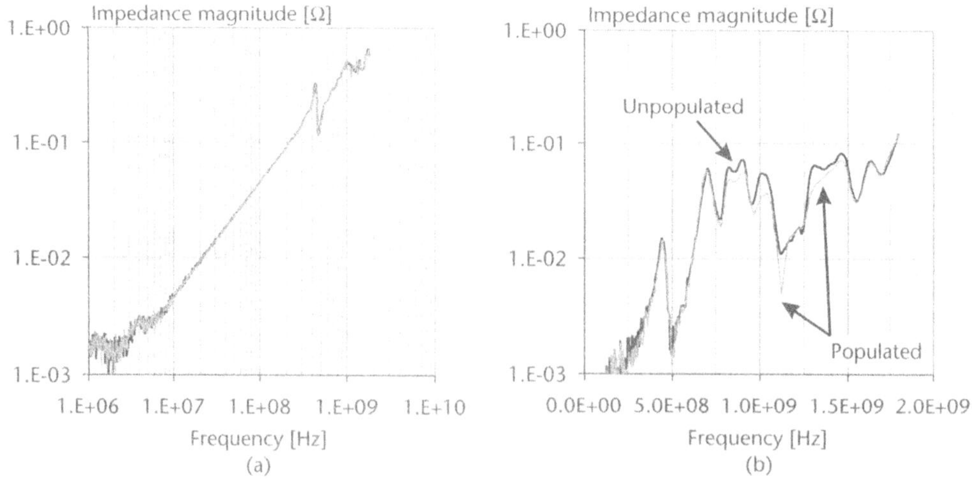

Figure 9.35 Self- and transfer impedance magnitudes of a CPU core rail with and without CPU in place. (a) Self-impedances and (b) transfer impedance across the socket. The two traces on the self-impedance plot completely overlap and therefore are not labeled separately.

socket showed no measurable difference with (an unpowered) versus without CPU. We can see a small difference only in the transfer impedance, which is measured between two test points on the opposite sides of the socket. Note the horizontal scale; to show the small difference, the linear frequency scale is used for a transfer-impedance plot. This tells us that for the purposes of modeling board PDN, the presence of the CPU can be ignored. However, on an unpopulated board or boards with insufficient bypassing, we will see more variation of impedance with CPU versus without CPU.

The following two figures show self-impedances of the DDR-I memory rail on the same CPU module, measured in the 1–1,800-MHz frequency range in various population configurations. The baseline data was taken with empty DIMM sockets. In one test, shown in Figure 9.36, one, two, and four DIMM modules were plugged in, and the self-impedance on the main board was remeasured at the same test point, close to the sockets. It was found that the capacity of the DIMM and the module vendor made no measurable difference, as long as the on-module bypass capacitor components were similar. There was a small and predictable change as the number of inserted modules increased. To show the small differences clearly, the measured dataset is presented in two graphs: a 1–100-MHz range with logarithmic horizontal and vertical scales and a 1–500-MHz range with linear scales.

At lower frequencies, the on-module bypass capacitors somewhat reduce the impedance on the main board. As expected, there is more reduction with more memory modules plugged in, and the difference gradually diminishes below 1 MHz and above 100 MHz. At higher frequencies there is one notable change: the parallel resonance frequency formed by the plane capacitance and the cumulative inductance of DIMMs shifts upwards as more memory is added. This upward shift is expected, because the memory modules at high frequencies behave like additional parallel inductance on the supply rail, reducing the overall inductance and therefore shifting the resonance frequency upwards.

To see the upper bounds of possible self-impedance changes, the main board was remeasured with memory modules with shorted rails. Bare boards of the

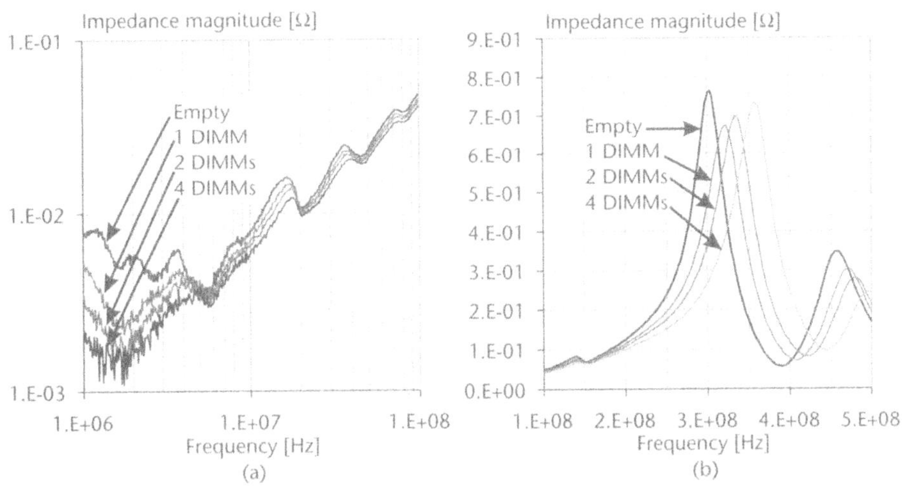

Figure 9.36 Self-impedance of a DDR-I memory rail with zero (labeled empty), one, two, and four DIMMs plugged in: (a) 1–100 MHz with logarithmic scales and (b) 100–500 MHz with linear scales.

DIMM modules were used with solid copper-strip shorts across all bypass components and chips. The self-impedance at the same test point, with the same number of shorted modules plugged in, is shown in Figure 9.37. The data is presented in the same way as in Figure 9.36.

There are only two minor differences between Figures 9.36 and 9.37. The first difference is that the shorted modules create proportionally lower impedances at low frequencies, so that the differences do not diminish below 1 MHz. The second difference is that with shorted modules, the *Q* of the parallel resonance peak is somewhat higher; otherwise there is no noticeable difference in the resonance frequency values. This comparison suggests that in the particular system the high-frequency self-impedance of the main board's memory rail can be approximated with shorted memory modules; the bypass capacitor values and memory chips make hardly any difference. At low frequencies the modules behave as a single capacitor, where the capacitance is the sum of all the on-module capacitances. For the purposes of board-impedance simulations, the DIMM modules can be modeled by a single *C-R-L* model, evenly split, and distributed along the supply pins.

Figure 9.38 shows the measured impedance of an 80-A core rail with multiple plane pairs. In addition to low-Q bulk capacitors, only one value of an MLCC part was used, which is called Big-V design. The high-current rail had multiple power planes in the stackup. A combination of inevitable horizontal plane resistance, component placement, and plane allocation made it possible to almost completely eliminate the impedance minimum at the SRF of the MLCC parts, effectively creating a flat *R-L*-type response. With zero horizontal plane resistance between capacitors, the simulated impedance minimum at the SRF of MLCCs would be 0.06 mΩ. This supply rail was also a signal reference, and therefore the modal plane resonances were suppressed by *RC* termination along the plane edge. The graph in Figure 9.38(a) shows the self-impedance of a fully populated board. The graph in Figure 9.38(b) compares the measured impedance profiles with and without dissipative edge termination (DET) components on the rail.

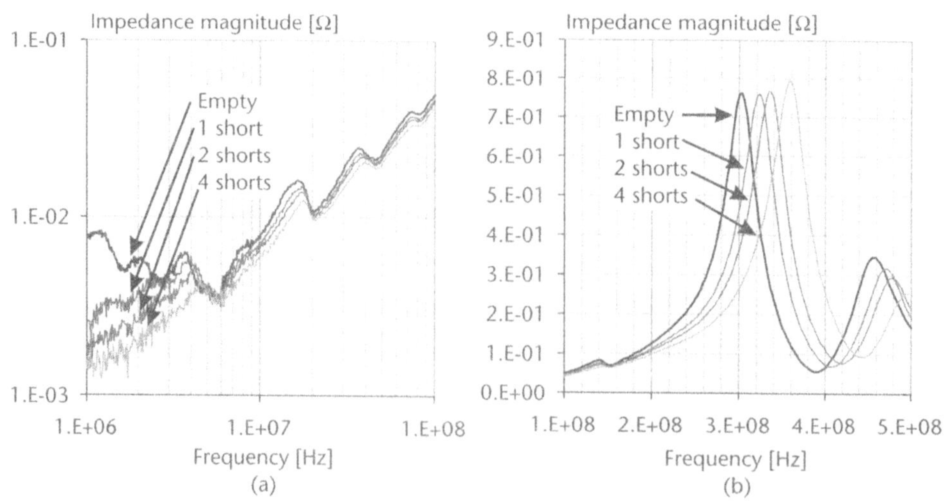

Figure 9.37 Self-impedance of a DDR-I memory rail with zero (labeled Empty), one, two, and four shorted DIMM boards plugged in: (a) 1–100 MHz with logarithmic scales and (b) 100–500 MHz with linear scales.

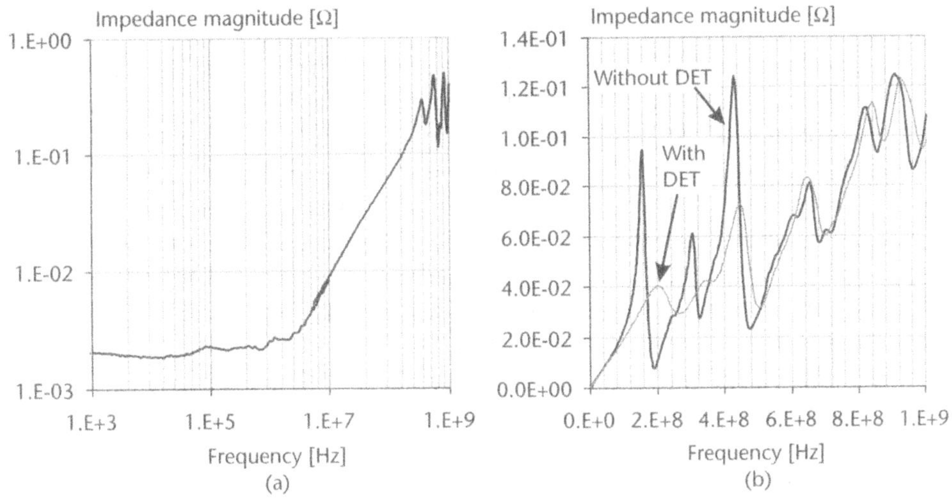

Figure 9.38 (a) Measured self-impedance profile on an 80-A CPU core rail, with Big-V design. (b) High-frequency impedance profile of a similar core rail, with and without dissipative edge termination (DET).

Figure 9.39 shows the combination impedance profile of the CPU core rail from Figure 9.38. The PDN impedance was measured in the 100-Hz–1,800-MHz frequency range with two VNAs. Bare board and fully populated boards with and without input power applied are compared.

9.3.3 Detailed Characterization on High-Speed Supply Rail

In this section, we show characterization results on a supply rail feeding a set of 10 switch ASICs, running at 2.5- and 5.0-Gbps speeds. The supply rail was rated for a maximum of 16A. The board had two thin-laminate plane pairs in the stackup close to the surfaces. The 35-μm (1-oz) ground plane was unsplit, the 35-μm (1-oz) power

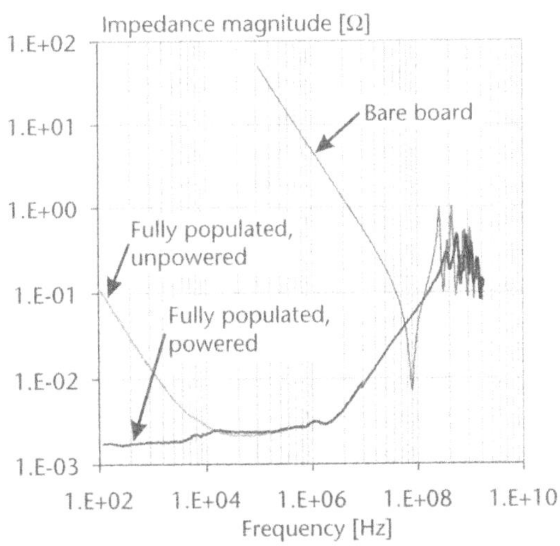

Figure 9.39 Self-impedance magnitude comparison of bare board, fully populated unpowered, and fully populated powered boards.

plane had a T shape to feed the ASICs: the horizontal arm covered the 10 uniformly spaced switch ASICs, the bottom end of the short vertical flap connects to the dc-dc converter. There were only two kinds of bypass capacitors on the supply rail: low-inductance POSCAPs with medium ESR and a large number of low-inductance ceramic capacitors. The capacitors were uniformly distributed around the ASICs. The approximate top view of the plane shape with the major component locations is shown in Figure 9.40. The 10 ASICs are labeled A1 through A10, the dc-dc converter is labeled D. There were multiple test points for PDN measurements on this rail, three of them are identified on the sketch as T1 through T3. The full horizontal length of the shape was 14.8 inches.

The supply rail was measured in several different configurations. First the bare board was measured to check the static capacitance and modal resonances of the planes. A board with just the PDN components (no ASIC) and a fully built board were also measured with no input power and with input power applied. Figure 9.41 shows the composite result at T2, measured by two different VNAs. On the figure only the fully built board is labeled, because the results from the board with only PDN parts were identical. Note that the bare board modal resonances were suppressed by the low-inductance capacitors. The presence of chips made no difference at any frequency; the applied power changed the self-impedance only below 2 kHz.

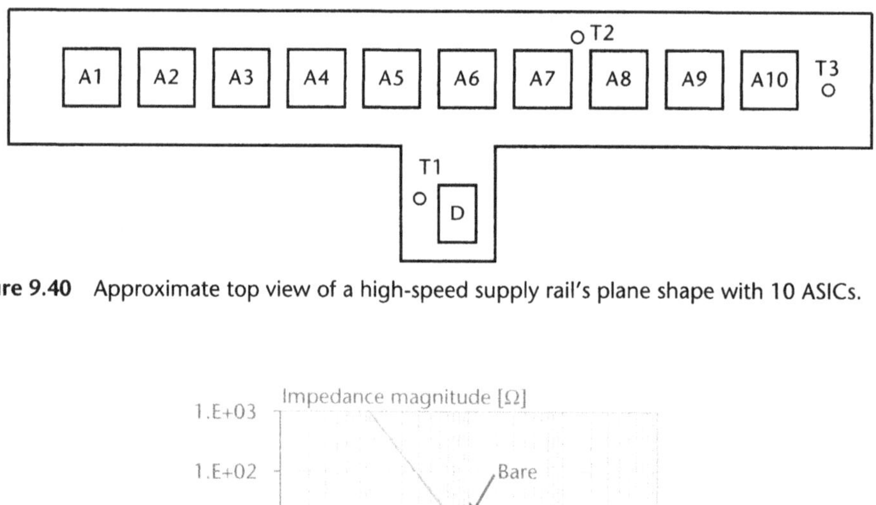

Figure 9.40 Approximate top view of a high-speed supply rail's plane shape with 10 ASICs.

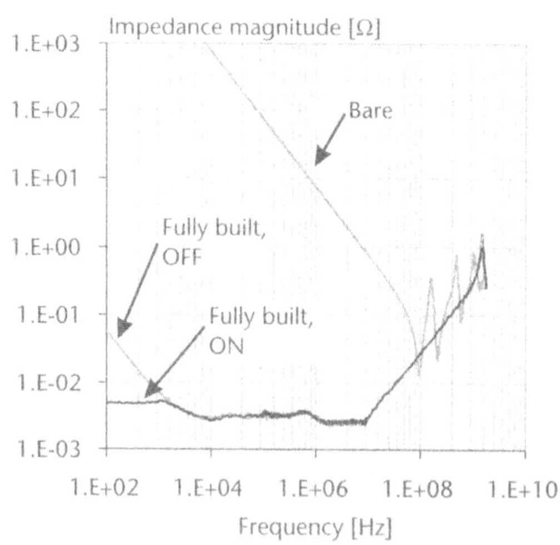

Figure 9.41 Impedance magnitude of a SerDes supply rail measured on bare and fully populated boards with and without input power.

The bare-board impedance was correlated to two different simulation methods: SPICE grid (Section 4.3.2) and analytical formulas (Section 4.2.1.2). The correlation is shown in Figure 9.42 in two different ways. The logarithmic frequency scale of Figure 9.42(a) exposes both the low-frequency and high-frequency differences. At low frequencies, the SPICE grid simulation runs in parallel to the measured data; this indicates that by adjusting the dielectric constant, we could improve the match. The lossy analytical formula, however, as shown in Section 4.2, overestimates capacitance at low frequencies; therefore, the slope changes. At high frequencies, both simulations account for the modal resonances, though the resonance frequencies are slightly misaligned due to the noncausal model with frequency independent dielectric constant and loss. The populated board, however, successfully suppresses the modal resonances by the low-inductance bypass capacitors around the ASICs; so, in this case, the modeling of modal resonances is less important.

The impedance was measured at various test points across the T shape. The most noticeable difference appeared at low frequencies, powered due to the resistance in the planes. Figure 9.43 shows the low-frequency impedance at the T1, T2, and T3 test points: at the dc-dc converter, at the end of the T shape, and in the middle of the T shape.

As expected, the lowest impedance was measured near the dc-dc converter and the highest impedance was measured at the end of the T shape. Above 50 kHz, however, the trend changes; since there were no low-inductance capacitors near the dc-dc converter, the impedance becomes the highest due to plane inductance.

To see the impact of active devices on the PDN impedance, transfer impedance measurements were made between various points. Figure 9.44 shows the transfer impedance between the end and the middle of the T-shaped plane. A bare board, a populated board with bypass capacitors only, and a fully populated board were measured with and without input power. There was no noticeable difference between the boards with and without switch chips; therefore, only the fully populated boards are labeled. The only visible difference occurred below 100 kHz, where

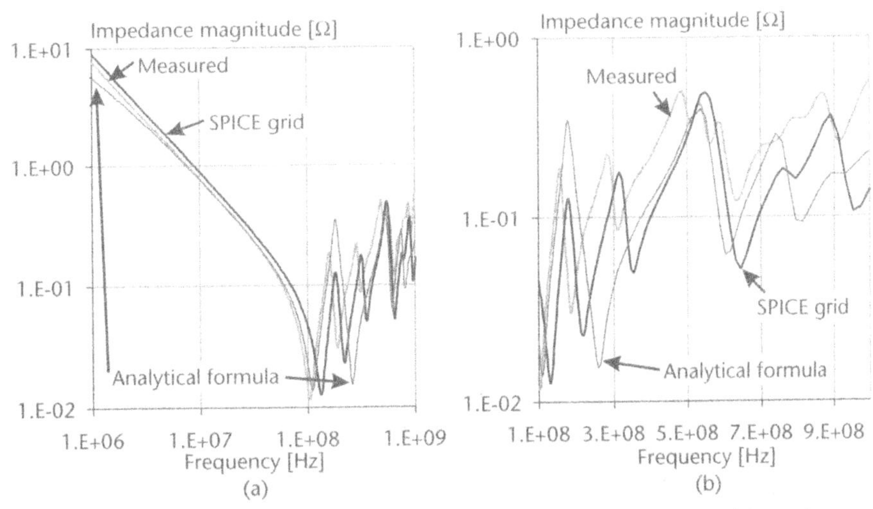

Figure 9.42 Correlation of bare-board impedance to SPICE-grid and analytical-formula simulations: (a) full frequency range with logarithmic scale and (b) high-frequency plot with linear scale.

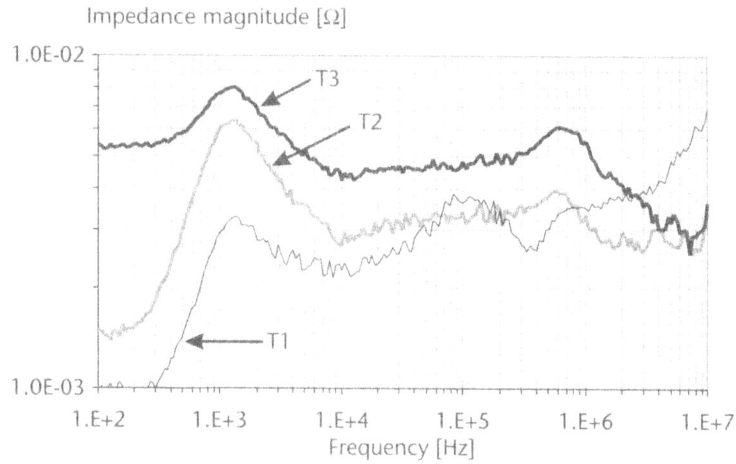

Figure 9.43 Variation of low-frequency impedance with location.

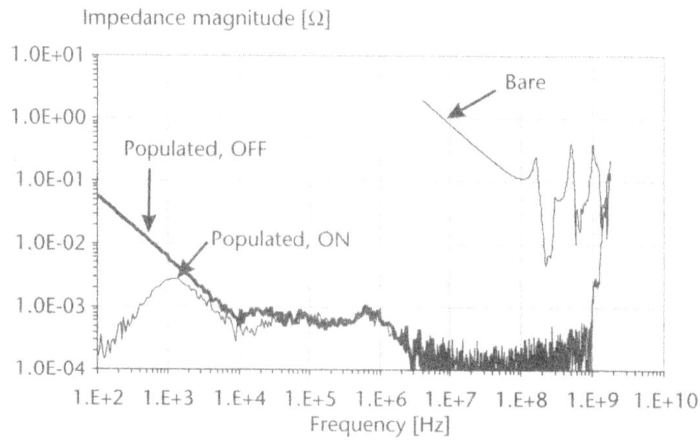

Figure 9.44 Transfer impedance between the end and middle of T shape.

the active loop of the dc-dc converter lowers the transfer impedance with input power applied.

To see the presence of the chips on the board PDN impedance, we need to measure the transfer impedance under the BGA chips. Figure 9.44 showed the transfer impedance between two chips measured at test points next to the BGA packages. Figure 9.45 shows the transfer impedance between the same two chips, measured with wafer probes in the middle of the ASIC pin fields on the back side of the board. The graph shows the transfer impedance on a bare board and a populated board, without chips, and a populated board with chips, with and without input power. In the 100–1,000-MHz frequency range there is a clear difference between the board impedance with and without mounted chips, though applying input power does not have a significant impact. Applying input power creates a different impedance below 100 MHz; the resonance peak between the chip capacitance and board inductance appears at 90 MHz with no input power. The peak shifts to 50 MHz when input dc power is applied.

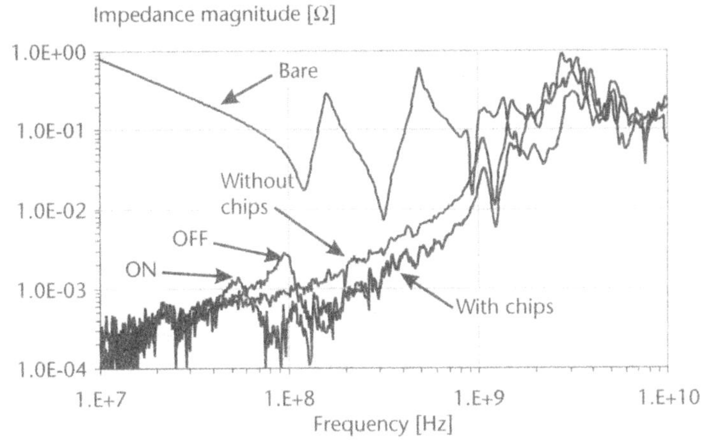

Figure 9.45 Measured transfer impedance between two adjacent switches.

The graph in Figure 9.46 shows the cause of this resonance shift. The three traces show the impedance magnitudes of the bare package and a package with silicon, with and without input power applied. The measurements were made with wafer probes at the same balls where the board measurements were made for the graph in Figure 9.45. Note that when power is applied, the impedance at low frequencies drops, indicating an increase of capacitance. This increasing capacitance shifts the resonance frequency in the impedance of the assembled board from 90 MHz to 50 MHz.

The fully populated board was also simulated. Figure 9.47 shows the correlation results. The lossy SPICE grid captures the plane inductance and the cumulative effect of bypass capacitors and shows good agreement. At medium frequencies, the lossy plane model allows us to capture the proper value of the impedance plateau: without losses, the cumulative ESR of the bulk capacitors would push the impedance below 1 mΩ.

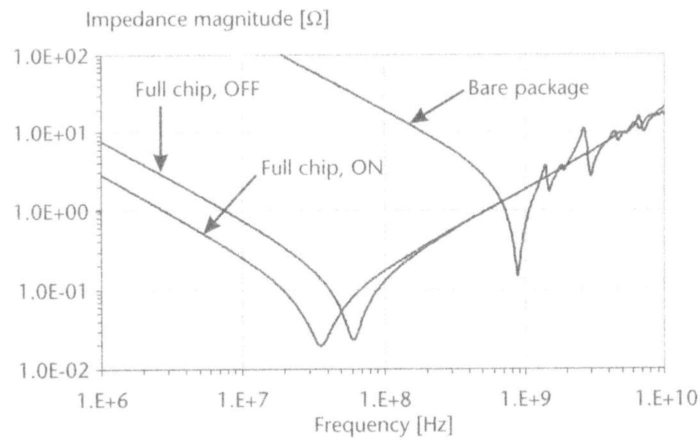

Figure 9.46 Supply-rail impedance of a packaged switch chip.

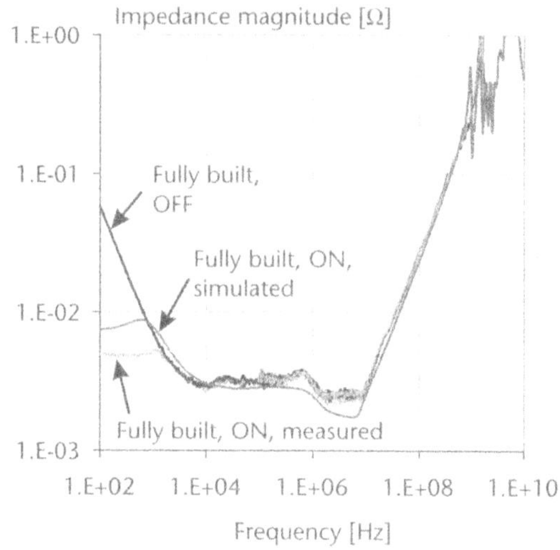

Figure 9.47 Correlation of populated board with and without power.

References

[1] Novak, I., "Reducing Simultaneous Switching Noise and EMI on Ground/Power Planes by Dissipative Edge Termination," *IEEE Trans. on Advance Packaging,* Vol. 22, No. 3, August 1999, pp. 274–283.

[2] Ishida, H., et al., "Controlled-ESR Capacitors Have Arrived," *Proceedings of DesignCon 2007*, Santa Clara, CA, January 29–31, 2007.

[3] TDK Corp., "TDK Chip Beads Equivalent Circuits for SPICE," http://www.components.tdk.com, 2005.

[4] Coilcraft, "SPICE Model-0805PS," Document 267, http://www.coilcraft.com, 2003.

[5] Ng, S. W., and Y. S. Lee, "A Unified Small-Signal Simulation Algorithm for SMPS," *IEEE Trans. on Circuits and Systems—I: Regular Papers*, Vol. 51, No. 3, March 2004.

[6] Mihajlovic, Z., B. Lehman, and C. Sun, "Output Ripple Analysis of Switching DC–DC Converters," *IEEE Trans. on Circuits and Systems—I: Regular Papers*, Vol. 51, No. 8, August 2004.

[7] Novak, I., "DC-DC Converters: What Is Wrong with Them?" *Proceedings of DesignCon 2007*, Santa Clara, CA, January 29–31, 2007.

[8] Johnson, H., "Voltage Regulator Droop," *EDN*, September 14, 2006.

About the Authors

Istvan Novak is a distinguished engineer at Sun Microsystems, Inc. Besides signal integrity design of high-speed serial and parallel buses, he is engaged in the design and characterization of power-distribution networks and packages for Sun servers. He creates simulation models and develops measurement techniques for power distribution. Istvan has more than 20 years of experience with high-speed digital, RF, and analog circuits and system design. He is a fellow of the IEEE for his contributions to the fields of signal-integrity, RF measurements, and simulation methodologies.

Jason R. Miller is currently a senior staff engineer at Sun Microsystems, Inc., where he works on ASIC packaging, interconnect characterization, and system simulation. He received a Ph.D. and an M.S. in electrical engineering from Columbia University. He has authored or coauthored over 30 technical articles on topics such as high-speed modeling and simulation.

Index

S

W

Z

Recent Titles in the Artech House Microwave Library

Active Filters for Integrated-Circuit Applications, Fred H. Irons

Advanced Techniques in RF Power Amplifier Design, Steve C. Cripps

Automated Smith Chart, Version 4.0: Software and User's Manual, Leonard M. Schwab

Behavioral Modeling of Nonlinear RF and Microwave Devices, Thomas R. Turlington

Broadband Microwave Amplifiers, Bal S. Virdee, Avtar S. Virdee, and Ben Y. Banyamin

CMOS RFIC Design Principles, Robert Caverly

Computer-Aided Analysis of Nonlinear Microwave Circuits, Paulo J. C. Rodrigues

Design of FET Frequency Multipliers and Harmonic Oscillators, Edmar Camargo

Design of Linear RF Outphasing Power Amplifiers, Xuejun Zhang, Lawrence E. Larson, and Peter M. Asbeck

Design of RF and Microwave Amplifiers and Oscillators, Pieter L. D. Abrie

Digital Filter Design Solutions, Jolyon M. De Freitas

Distortion in RF Power Amplifiers, Joel Vuolevi and Timo Rahkonen

EMPLAN: Electromagnetic Analysis of Printed Structures in Planarly Layered Media, Software and User's Manual, Noyan Kinayman and M. I. Aksun

Essentials of RF and Microwave Grounding, Eric Holzman

FAST: Fast Amplifier Synthesis Tool—Software and User's Guide, Dale D. Henkes

Feedforward Linear Power Amplifiers, Nick Pothecary

Foundations of Oscillator Circuit Design, Guillermo Gonzalez

Frequency-Domain Characterization of Power Distribution Networks, Istvan Novak and Jason R. Miller

Fundamentals of Nonlinear Behavioral Modeling for RF and Microwave Design, John Wood and David E. Root, editors

Generalized Filter Design by Computer Optimization, Djuradj Budimir

High-Linearity RF Amplifier Design, Peter B. Kenington

High-Speed Circuit Board Signal Integrity, Stephen C. Thierauf

Intermodulation Distortion in Microwave and Wireless Circuits, José Carlos Pedro and Nuno Borges Carvalho

Introduction to Modeling HBTs, Matthias Rudolph

Lumped Elements for RF and Microwave Circuits, Inder Bahl

Lumped Element Quadrature Hybrids, David Andrews

Microwave Circuit Modeling Using Electromagnetic Field Simulation, Daniel G. Swanson, Jr., and Wolfgang J. R. Hoefer

Microwave Component Mechanics, Harri Eskelinen and Pekka Eskelinen

Microwave Differential Circuit Design Using Mixed-Mode S-Parameters, William R. Eisenstadt, Robert Stengel, and Bruce M. Thompson

Microwave Engineers' Handbook, Two Volumes, Theodore Saad, editor

Microwave Filters, Impedance-Matching Networks, and Coupling Structures, George L. Matthaei, Leo Young, and E.M.T. Jones

Microwave Materials and Fabrication Techniques, Second Edition, Thomas S. Laverghetta

Microwave Mixers, Second Edition, Stephen A. Maas

Microwave Radio Transmission Design Guide, Trevor Manning

Microwaves and Wireless Simplified, Third Edition, Thomas S. Laverghetta

Modern Microwave Circuits, Noyan Kinayman and M. I. Aksun

Modern Microwave Measurements and Techniques, Second Edition, Thomas S. Laverghetta

Neural Networks for RF and Microwave Design, Q. J. Zhang and K. C. Gupta

Noise in Linear and Nonlinear Circuits, Stephen A. Maas

Nonlinear Microwave and RF Circuits, Second Edition, Stephen A. Maas

QMATCH: Lumped-Element Impedance Matching, Software and User's Guide, Pieter L. D. Abrie

Phase-Locked Loop Engineering Handbook for Integrated Circuits, Stanley Goldman

Practical Analog and Digital Filter Design, Les Thede

Practical Microstrip Design and Applications, Günter Kompa

Practical RF Circuit Design for Modern Wireless Systems, Volume I: Passive Circuits and Systems, Les Besser and Rowan Gilmore

Practical RF Circuit Design for Modern Wireless Systems, Volume II: Active Circuits and Systems, Rowan Gilmore and Les Besser

Production Testing of RF and System-on-a-Chip Devices for Wireless Communications, Keith B. Schaub and Joe Kelly

Radio Frequency Integrated Circuit Design, John Rogers and Calvin Plett

RF Design Guide: Systems, Circuits, and Equations, Peter Vizmuller

For further information on these and other Artech House titles, including previously considered out-of-print books now available through our In-Print-Forever® (IPF®) program, contact:

Artech House
685 Canton Street
Norwood, MA 02062
Phone: 781-769-9750
Fax: 781-769-6334
e-mail: artech@artechhouse.com

Artech House
46 Gillingham Street
London SW1V 1AH UK
Phone: +44 (0)20 7596-8750
Fax: +44 (0)20 7630 0166
e-mail: artech-uk@artechhouse.com

Find us on the World Wide Web at: www.artechhouse.com

www.ingramcontent.com/pod-product-compliance
Lightning Source LLC
LaVergne TN
LVHW062113200926
848484LV00007B/1658